METHODS AND PROTOCOLS IN FOOD SCIENCE

Series Editor
Anderson S. Sant'Ana
University of Campinas
Campinas, Brazil

For further volumes:
http://www.springer.com/series/16556

Methods and Protocols in Food Science series is devoted to the publication of research protocols and methodologies in all fields of food science.

Volumes and chapters will be organized by field and presented in such way that the readers will be able to reproduce the experiments in a step-by-step style. Each protocol will be characterized by a brief introductory section, followed by a short aims section, in which the precise purpose of the protocol will be clarified.

Plant Chemical Compositions and Bioactivities

Edited by

Mostafa Gouda

College of Biosystems Engineering and Food Science, Zhejiang University, Hangzhou, China; Department of Nutrition and Food Science, National Research Centre, Cairo, Egypt

Xiaoli Li and Yong He

College of Biosystems Engineering and Food Science, Zhejiang University, Hangzhou, China

Humana Press

Editors
Mostafa Gouda
College of Biosystems Engineering
and Food Science
Zhejiang University
Hangzhou, China

Department of Nutrition and Food Science
National Research Centre
Cairo, Egypt

Yong He
College of Biosystems Engineering
and Food Science
Zhejiang University
Hangzhou, China

Xiaoli Li
College of Biosystems Engineering and Food
Science
Zhejiang University
Hangzhou, China

ISSN 2662-950X ISSN 2662-9518 (electronic)
Methods and Protocols in Food Science
ISBN 978-1-0716-3937-5 ISBN 978-1-0716-3938-2 (eBook)
https://doi.org/10.1007/978-1-0716-3938-2

This Humana imprint is published by the registered company Springer Science+Business Media, LLC, part of Springer Nature.
The registered company address is: 1 New York Plaza, New York, NY 10004, U.S.A.

Paper in this product is recyclable.

Preface to the Series

Methods and Protocols in Food Science series is devoted to the publication of research protocols and methodologies in all fields of food science. The series is unique as it includes protocols developed, validated, and used by food and related scientists, as well as theoretical basis are provided for each protocol. Aspects related to improvements in the protocols, adaptations, and further developments in the protocols may also be approached.

Methods and Protocols in Food Science series aims to bring the most recent developments in research protocols in the field as well as very well established methods. As such the series targets undergraduate, graduate, and researchers in the field of food science and correlated areas. The protocols documented in the series will be highly useful for scientific inquiries in the field of food sciences, presented in such way that the readers will be able to reproduce the experiments in a step-by-step style.

Each protocol will be characterized by a brief introductory section, followed by a short aims section, in which the precise purpose of the protocol is clarified. Then, an in-depth list of materials and reagents required for employing the protocol is presented, followed by a comprehensive and step-by-step procedures on how to perform that experiment. The next section brings the do's and don'ts when carrying out the protocol, followed by the main pitfalls faced and how to troubleshoot them. Finally, template results will be presented and their meaning/conclusions addressed.

The *Methods and Protocols in Food Science* series will fill an important gap, addressing a common complain of food scientists, regarding the difficulties in repeating experiments detailed in scientific papers. With this, the series has a potential to become a reference material in food science laboratories of research centers and universities throughout the world.

Campinas, Brazil ***Anderson S. Sant'Ana***

Preface to the Series

[illegible]

Preface

The international community's attention has lately focused on the emerging technologies for fast and accurate analysis of food products and their quality, which are considered to be ecofriendly and have significant reproductivity impacts. The potential chemical-free and accurate methods that can track the accumulation of these compounds during the industrial process could open the way through the smart systems for producing high-quality and precise final products. Furthermore, scientists are trying to find instant and non-invasive techniques to explore the scientific principles of phytochemicals' functional principles, especially bioactive macromolecules and micromolecules. As an important example, the quality of tea product during its fermentation process is based on the accumulation of some compounds like theaflavins (TF), thearubicin (TR), and theabrownin (TB), which have significant impacts on its food industrial systems. In addition, many studies have reported that these kinds of phytochemicals have high sensitivity to the different analytical protocols because of their isomers' activity and their unique chemical structure. Besides, the plant and animal health-related key components like carbohydrates, lipids, and proteins are precursors of essential pathways and are novel sources of bioactive molecules that are affected by oxidation and other physical and chemical processes. These physicochemical changes could affect the quality of the final food product. Therefore, tracking these mechanisms could positively affect the field of the food chemistry and its industrial scale research, especially in the functional prebiotic carbohydrate, protein, and lipid production. As an example, microalgae lipids provide several aspects in the food science and nutrition area as a viable alternative source in food technology. They could have up to 40% eicosapentaenoic acid (EPA) and docosahexaenoic acid (DHA) of its fatty acids as functional bioactive molecules. And the real-time study of their combination with the other natural safe molecules (e.g., polyphenols and flavonoids) could raise their bioactivities, while caring about their storage, processing, and quality efficiency by using the novel analytical technologies.

The vision for this handbook is to provide food researchers the current state of advanced destructive and non-destructive techniques for measuring the chemical composition, functional bioactivities, and physicochemical changes and explaining the functional mechanisms for facilitating the tracking of the functional molecules during the different industrial processes. Besides, the development as well as providing practical methodologies that demonstrate the integration of the analytical methods and protocols are discussed in details. The chapters deal with a critical discussion of both laboratory and food industrial topics, with each chapter containing both a discursive section along with a detailed methods section. The handbook's topics include sample preparation, emerging extraction technologies protocols, analytical methodologies for phytochemicals profiling; multi-way models in food data analysis; and the integration of omics for biochemical analyses understanding.

Along with the various chemical-free and chemical-dependent methods covered in the individual chapters, the handbook also includes an extensive bibliography that details current existing literature in the fields like:

- Food materials science
- Natural bioactive products physicochemical analyses
- Nano- and microsensors and sensation
- Green chemistry
- Analytical methods
- Electrochemical sensors

Hangzhou, China *Mostafa Gouda*
Cairo, Egypt
Hangzhou, China *Xiaoli Li*
Hangzhou, China *Yong He*

Contents

Contributors

HASSAN IDRIS ABDU • *Shaanxi Province Key Laboratory of Bioresources, College of Bioscience and Bioengineering, Shaanxi University of Technology, Shaanxi, China*

ASHFAQ AHMAD • *Department of Biology, College of Arts and Sciences, Khalifa University, Abu Dhabi, UAE*

FARID E. AHMED • *GEM Tox Labs, Institute for Research in Biotechnology, Greenville, NC, USA*

NANCY C. AHMED • *GEM Tox Labs, Institute for Research in Biotechnology, Greenville, NC, USA*

SYED SALMAN ASHRAF • *Advanced Materials Chemistry Center (AMCC), Khalifa University, Abu Dhabi, UAE*

MARJAN ASSEFI • *Marie Curie Science Research Center, Greensboro, NC, USA*

DUYGU AYDEMIR • *Department of Medical Biochemistry, School of Medicine, Koç University, Istanbul, Turkey*

MUNIR ABBA DANDAGO • *Department of Food Science and Technology, Kano University of Science and Technology, Wudil, Nigeria*

UMAR ABDULBAKI DANHASSAN • *Department of Agricultural and Environmental Engineering, Faculty of Agriculture, Bayero University Kano, Kano, Nigeria*

YULUN DONG • *Zhejiang University, Hang Zhou, China*

KAMEL MOHAMED ELTOHAMY • *Key Laboratory of Environment Remediation and Ecological Health, Ministry of Education, College of Environmental and Resources Sciences, Zhejiang University, Hangzhou, China*

SEVCAN ERŞAN • *University of California, Los Angeles, CA, USA*

SHIYAN FANG • *College of Mechanical and Electronic Engineering, Northwest A&F University, Yangling, Shaanxi, China*

MOSTAFA GOUDA • *College of Biosystems Engineering and Food Science, Zhejiang University, Hangzhou, China; Department of Nutrition and Food Science, National Research Centre, Cairo, Egypt*

YONG HE • *College of Biosystems Engineering and Food Science, Zhejiang University, Hangzhou, China*

ZHENXIONG HUANG • *Zhejiang university, Hangzhou, China*

LAILA A. HUSSEIN • *Department of Nutrition & Food Science, National Research Centre, Giza, Egypt*

XUMING JIANG • *College of Mechanical Engineering, Zhejiang University of Technology, Hangzhou, China*

PEI JINJIN • *College of Biosystems Engineering and Food Science, Zhejiang University, Hangzhou, China*

XUSHENG JI • *College of Biosystems Engineering and Food Science, Zhejiang University, Hangzhou, China*

KAI-UWE LEWANDROWSKI • *Center For Advanced Spine Care of Southern Arizona, Tucson, AZ, USA; Division Personalized Pain Research and Education, PME, Tucson, USA*

JUNMENG LI • *College of Mechanical and Electronic Engineering, Northwest A&F University, Yangling, Shaanxi, China*
LEI LIN • *College of Biosystems Engineering and Food Science, Zhejiang University, Hangzhou, China*
MING LIN • *College of Mechanical Engineering, Zhejiang University of Technology, Hangzhou, China*
XIAOLI LI • *College of Biosystems Engineering and Food Science, Zhejiang University, Hangzhou, China*
XUELUN LUO • *College of Biosystems Engineering and Food Science, Zhejiang University, Hangzhou, China*
ALI MOKHTAR MAHMOUD • *National Research Council of Italy, Institute of Biomolecular Chemistry, Pozzuoli, Italy*
MAHMOUD A. MOHAMMAD • *USDA/ARS Children's Nutrition Research, Houston, TX, USA*
MUSTAPHA MUHAMMAD NASIRU • *Department of Food Science and Technology, Faculty of Renewal Natural Resources, Federal University Dutsin-Ma, Katsina, Nigeria*
SANUSI SHAMSUDEEN NASSARAWA • *Department of Food Science and Technology, Faculty of Agriculture, Bayero University Kano, Kano, Nigeria*
JIYU PENG • *College of Mechanical Engineering, Zhejiang University of Technology, Hangzhou, China*
GOWHAR RASHID • *Amity Medical School, Amity University Haryana, Gurugram, Haryana, India*
NASSARAWA ISMA'IL SANUSI • *Department of Agricultural Extension and Community Deployment, Faculty of Agriculture and Agricultural Technology, Federal University Dutsin-Ma, Katsina, Nigeria*
ALIREZA SHARAFSHAH • *Marie Curie Science Research Center, Biochemistry Department, Greensboro, USA*
ABUBAKAR SHITU • *Department of Agricultural and Environmental Engineering, Faculty of Agriculture, Bayero University Kano, Kano, Nigeria*
MUSA ABUBAKAR TADDA • *Department of Agricultural and Environmental Engineering, Faculty of Agriculture, Bayero University Kano, Kano, Nigeria*
MOHAMED FARAG TAHA • *School of Agricultural Engineering, Jiangsu University, Zhenjiang, China; College of Biosystems Engineering and Food Science, Zhejiang University, Hangzhou, China; Department of Soil and Water Sciences, Faculty of Environmental Agricultural Sciences, Arish University, North Sinai, Egypt*
NURIYE NURAY ULUSU • *Department of Medical Biochemistry, School of Medicine, Koç University, Istanbul, Turkey*
YAN WANG • *College of Mechanical and Electronic Engineering, Northwest A&F University, Yangling, Shaanxi, China*
KEQIANG YU • *College of Mechanical and Electronic Engineering, Northwest A&F University, Yangling, Shaanxi, China*
HINA ZAMAN • *Department of Chemical and Life Sciences, Qurtuba University of Science and Information Technology, Khyber Pakhtunkhwa, Pakistan*
SHUAI ZHANG • *College of Biosystems Engineering and Food Science, Zhejiang University, Hangzhou, China*

Wenkai Zhang • *College of Biosystems Engineering and Food Science, Zhejiang University, Hangzhou, China*
Zhihao Zhan • *College of Biosystems Engineering and Food Science, Zhejiang University, Hangzhou, China*
Yanru Zhao • *College of Mechanical and Electronic Engineering, Northwest A&F University, Yangling, Shaanxi, China*
Yinglei Zhao • *College of Biosystems Engineering and Food Science, Zhejiang University, Hangzhou, China*
Fengle Zhu • *Zhejiang University of Technology, Hangzhou, China*
Songming Zhu • *College of Biosystems Engineering and Food Science, Zhejiang University, Hangzhou, China*

[illegible]

Chapter 1

Application of FT-IR, and Hyperspectral Analytical Protocols for Tea Leaves Natural Catechins and Caffeine Detection and Evaluation

Zhenxiong Huang, Xiaoli Li, and Mostafa Gouda

Abstract

The contents and distribution of catechins and caffeine, strongly reflect the physiology and nutrition of tea. In recent years, spectroscopy technology has been widely applied to the detection of tea content. Based on spectroscopy technology (Fourier transform infrared (FT-IR) spectroscopy and hyperspectral imaging technique) and chemometrics, rapid and nondestructive determination model of catechins and caffeine content can be developed. In this chapter, we present a protocol for determining catechin and caffeine content in tea leaves by spectroscopy technology coupled with chemometrics. This analytical method could provide quick and efficient detection of catechin and caffeine in fresh tea leaves, which have great impacts on the cultivation of tea trees and the selection of raw materials for tea processing.

Key words Tea leaves, Caffeine, Catechin, FT-IR spectroscopy, Hyperspectral imaging, HPLC

1 Introduction

Tea has a long history in the world and being used in about 160 countries every day for drinking [1]. By reason of the pleasant flavor and various health promoting functions contributed, tea has become one of the most consumed nonalcoholic beverages [2]. Catechins and caffeine are two significant bioactive chemicals in tea leaves, and also are the basis of high-quality tea leaves. Catechins have been shown to have a great impact on the color, fragrance, oral astringency, and bitterness of tea. Moreover, catechins can inhibit bacteria and resist insects by reacting with protein. Caffeine, which can stimulate the central nervous system and enhance the excitability of the cerebral cortex, is one of the most important biological ingredients of tea [3]. Therefore, it is necessary to determine the catechins and caffeine contents of leaves because it can offer suggestions for optimizing high-quality tea processing by improving the raw material.

Mostafa Gouda et al. (eds.), *Plant Chemical Compositions and Bioactivities*, Methods and Protocols in Food Science,
https://doi.org/10.1007/978-1-0716-3938-2_1,

The traditionally detected in contents of tea catechins are by chromatography and capillary electrophoresis methods [4, 5], and caffeine are detected by the iodometric method and ultraviolet spectrophotometry [6, 7]. Among them, high-performance liquid chromatography (HPLC) is a widely accepted method. However, these methods are time-consuming, complicated, and reagent-consuming in large-sample detection [8]. In recent years, spectroscopy technology has proven to be a powerful analytical tool in the determination of tea components. Hyperspectral imaging technique is an emerging technology that integrates both digital image and spectroscopy technology in one system to acquire spatial and spectral information of tea leaves [9]. A spectrum for each pixel and an image for each narrow band can be acquired which enable this system to reflect componential and constructional characteristics of tea leaves and their spatial distributions [10]. FT-IR spectroscopy is used to obtain high spectral resolution data of absorption over a wide infrared (IR) spectral range where the specific frequency of molecular absorption that matches the transition energy of the vibrational frequency of a bond or group in catechins and caffeine of tea leaves [11]. The peaks at around 2852 and 2924 cm^{-1} could be assigned to C–H bonds of CH_3 groups, which are related to caffeine [12]. The main absorption peaks of samples in the fingerprint region were at 825, 1036, 1145, 1317, and 1519 cm^{-1} consistent with the spectral peaks of catechins [13]. The absorption peak at 1369 cm^{-1} consistent with the spectral peaks of caffeine [14]. Based on FT-IR spectroscopy and hyperspectral imaging technique, quantitative detection and distribution visualization of catechins and caffeine in tea leaves can be achieved.

Here, we explain the potential of FT-IR spectroscopy for determination of catechins and caffeine with no organic reagent extraction. Meanwhile, hyperspectral imaging technique was employed to determinate the catechins and caffeine content and map the spatial distribution of catechins and caffeine in tea leaves.

2 Materials

The chemicals and solvents used must be of HPLC analytical grade. Unless otherwise specified, all processes and solutions can be carried out at room temperature.

2.1 Chemicals

1. Acetic acid, from Thermo-Fisher (Waltham, MA, US).
2. Acetonitrile, from Thermo-Fisher (Waltham, MA, US).
3. High-purity water, from Watsons (Hongkong, China).
4. Potassium bromide, from Labio (Shanghai, China).

2.2 Mobile Phases

1. Mobile phase A was acetic acid, acetonitrile, and high-purity water in proportion of 0.5:3:96.5 (v/v/v).
2. Mobile phase B was acetic acid, acetonitrile, and high-purity water in proportion of 0.5:30:69.5 (v/v/v) (*see* **Notes 1–3**).

2.3 HPLC Materials

An HPLC analysis system (LC-20AD, Shimadzu, Kyoto, Japan) included an SIL-20A automatic injection valve and an SPD-20A UV-visible detector. The extracted solution of samples were injected into a Shim-packed column (e.g., 5 μm, 4.6 mm × 250 mm, or similar) (*see* **Note 4**).

2.4 FT-IR Spectrometer

An FT-IR spectrometer (Nicolet Is10, Thermo Fisher Scientific, Waltham, Mass.) equipped with a TGS detector. The scanning range was 7800–349 cm^{-1} (*see* **Note 5**).

2.5 Hyperspectral Image System

Hyperspectral image system included ImSpectorV10E (Specim, Finland) (*see* **Note 6**), a 12-bit CCD camera (Hamamatsu, Japan), two 150W tungsten halogen lamps (Fiber-Lite DC950 Illuminator, Dolan Jenner Industries Inc., USA), and a conveyor system (*see* **Note 7**). The scanning range was 380–1030 nm. Noise of spectra are usually removed during the analysis process.

3 Methods

3.1 Extracting of Samples for HPLC

1. Tea leaves are freeze-dried (FreeZone, Labconco, Kansas City, Mo.) and milled into a powder with a grinder (FW100, TY Instrument Co. Ltd., Shanghai, China).
2. The ground of tea leaves were sieved through a 40-mesh sifter.
3. 0.1 g of sieved tea powder and 25 mL of pure water was added into test tube and stirred.
4. The tube was placed in an 85 °C water bath for 20 min.
5. The supernatant was filtered with a 0.22 μm microporous hydrophilic membrane.

3.2 Operating Procedures for HPLC

1. Inject the filtered and degassed mobile phase into the storage tank.
2. Rinse the metal filter with the mobile phase, and then immerse the filter in the mobile phase of the storage tank.
3. Place the liquid storage tank at a certain height to avoid unnecessary falls caused by handling and other operator movements.
4. Start the HPLC in the following order: pump → detector → HPLC software → set the software parameters according to the preplanned method, such as analysis time, detection wavelength, flow rate, etc.

5. Start the pump and run for 5 min, mainly to eliminate bubbles in the system. After completion, close all exhaust valves.
6. Running the mobile phase at a fixed rate according to the predetermined speed, and follow the baseline until it stabilizes. Then, monitoring can be carried out on computer software.
7. After the baseline is stable, set the running parameters of tea sample in the software, such as flow rate and analysis time. The analysis time may vary due to factors such as sample, flow rate, and column length.
8. After setting the parameters, inject a fixed volume of sample solution into the injection valve, and then start the injection command in the software. The sample solution in the injector enters the chromatographic column with the mobile phase.
9. Monitor various readings through software, and when the detector detects all peak values of the sample, stop running and set a new injection. Before analyzing the next sample, it is recommended to allow 5–10 min for the mobile phase to pass through the chromatographic column in order to clean the residue of the previous sample. After confirming baseline stability, proceed to the next injection.
10. After the analysis is completed, first turn off the detector, then turn off the pump, and finally turn off the software.

3.3 HPLC Data Analysis

1. Linear standard calibration curves of catechins and caffeine were established according to the injection volume of the standards and HPLC peak areas.
2. Catechins and caffeine contents in the tea extract were initially identified by comparing the retention times with those of the standards.
3. Catechins and caffeine contents were then calculated from the corresponding peak areas and the external standard calibration curves.

3.4 Constructing Determination Models Based on FT-IR Spectra

1. Ball milling were utilized to milled freeze-dried leaves into powder (*see* **Note 8**).
2. 0.01 g of sample was uniformly mixed with 0.49 g of dried KBr powder (*see* **Note 9**). For this step, we recommend grinding under an infrared heat lamp to minimize the change in moisture content.
3. 0.1 g of evenly mixed powder was pressed at 15 MPa for 15 s using a tableting machine (*see* **Note 10**).
4. Turn on FT-IR spectrometer power supply, display, computer, and other power switches in order.

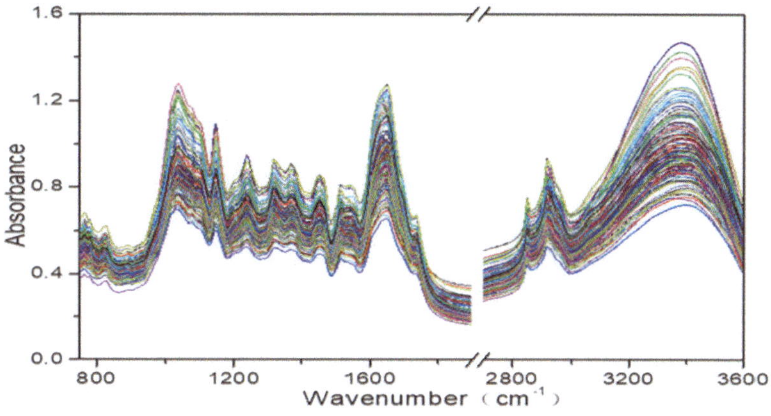

Fig. 1 FT-IR spectra of tea leaves

5. Double click on the desktop shortcut key "OMINIC" to enter the OMINIC workstation.
6. After the instrument is automatically initialized, when the "optical table status" appears in the upper right corner, the instrument can be preheated for about 10 min before measurement.
7. Click on "Experimental Settings" under the "Collection" menu. Scanning time and spectral resolution were set before collecting spectra. For this step, we recommend 32 times and 4 cm^{-1} for scanning time and spectral resolution separately. Final format: select "transmittance".
8. Setting the spectral range of 4000–400 cm^{-1} in the "Optical Table".
9. Opening the sample room cover, place the blank control on the sample rack in the sample room, and cover the sample room cover. Click "Collect Background" under the "Collect" menu to pop up a dialog box. Click "OK" to perform background scanning (*see* **Note 11**).
10. Opening the sample chamber cover, remove the blank control, place the sample prepared by the appropriate method on the sample rack in the sample chamber, and cover the sample chamber cover. Click on "Collect Samples" under the "Collect" menu bar to scan the samples. After the data collection is completed, the "Data Collection Complete" window will pop up and click "Yes". The spectra of tea leaves are shown in Fig. 1 (*see* **Note 12**).
11. Selecting "Save" under the "File" menu to open the "Save As" window. Selecting "Save Type" as "CSV Text (*. CSV)" from the drop-down menu, and entering the name of the saved file, and click "Save".

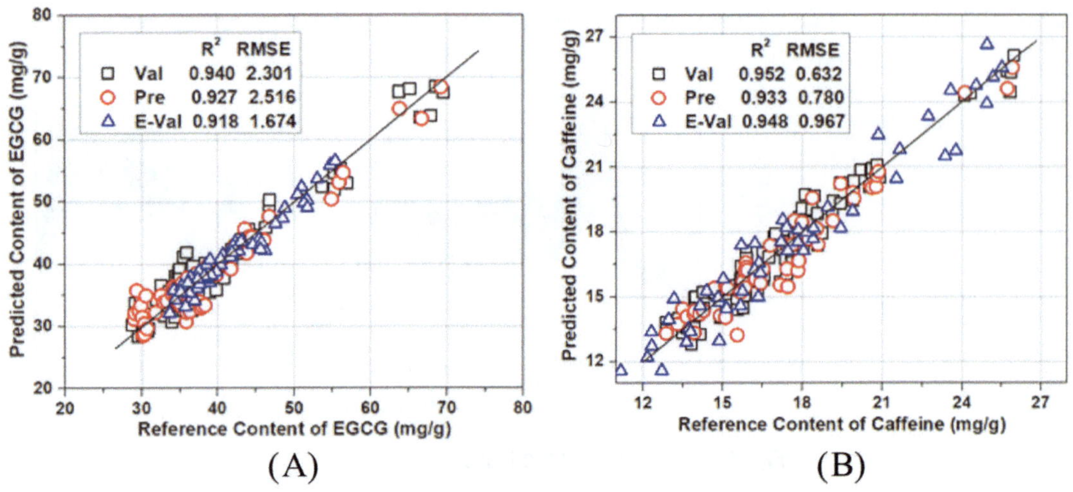

Fig. 2 Determination models of (**a**) catechins and (**b**) caffeine content

12. After collecting the spectrum, removing the sample cell, loosing the screws, putting on the fingertips, and carefully remove the salt sheet. First, wipe the liquid with soft paper, drop anhydrous ethanol, and wash off the sample. Avoid washing with water. Then, polishing with talcum powder and anhydrous ethanol under an infrared lamp. Finally, washing the surface with anhydrous ethanol, wiping it dry, drying it, wiping the mold, sample holder, etc. as required, and storing the two salt tablets in a dryer.
13. Choose File>Exit to exit the program.
14. Selecting "Shut down" from the "Start" menu on the computer desktop and a safe shutdown prompt will appear.
15. Turning off computer power, turning off the FT-IR spectrometer power, and turning off the regulated power supply.
16. Determination models for catechins and caffeine were constructed based on FT-IR spectra data and the chemical value of HPLC (*see* **Note 13**), shown in Fig. 2.
17. Appropriate preprocessing methods can improve the performance of the model (*see* **Note 14**).
18. Feature wavenumber extraction can improve the running speed of the model while ensuring accuracy, and making the model easier to develop portable equipment (*see* **Note 15**).

3.5 Constructing Determination Models Based on Hyperspectral

1. Turning on the halogen lamp, and the power of mobile platform and spectrometer.
2. Turning on the computer and running the software of "Isuzu Optics Crop".

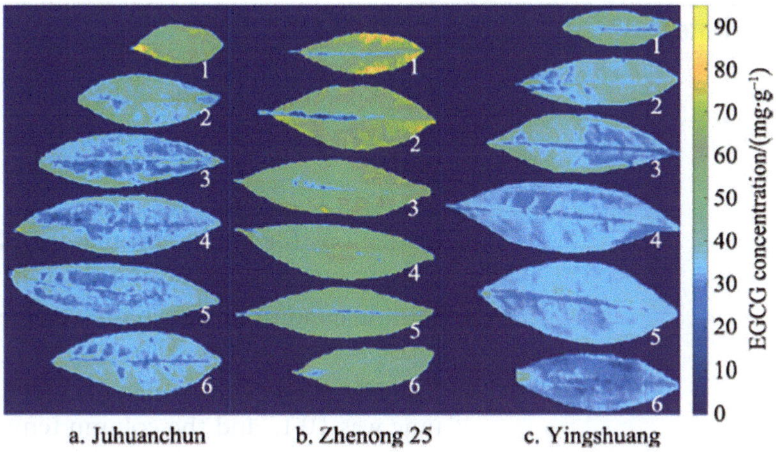

Fig. 3 EGCG concentration distribution diagram from 1st to 6th leaves of different cultivars

3. Setting the exposure time of the hyperspectral camera, the speed of the moving platform, and the distance between the lens and the sample to ensure that the collected hyperspectral images do not produce distortion.
4. The hyperspectral image of tea samples was captured through the movement of the conveyor belt.
5. White reference images (I_{white}) were obtained by collecting spectra of Teflon white board with 99% reflectance.
6. Dark reference images (I_{dark}) were obtained by closing the tungsten halogen lamps.
7. Tea leaves were placed flat on a black nylon baseplate for scanning to acquiring hyperspectral image of tea sample (R_{raw}).
8. Images of tea leaves were calibrated (R_{cal}) by the equation of $R_{cal} = (R_{raw} - I_{dark})/(I_{white} - I_{dark})$.
9. Regions of interest (ROIs) is chosen for each tea leave.
10. Mean spectra of each ROI is calculated as the representative spectra of each tea leaves.
11. Determination models for catechins and caffeine were constructed based on hyperspectral data and the chemical value of HPLC (*see* **Note 13**).
12. Inputting the spectra of each pixel into the determination model, the distribution of catechins (*see* Fig. 3) and caffeine in the entire leaves can be obtained.
13. Appropriate preprocessing methods are always welcome and can improve the performance of the model (*see* **Note 14**).
14. Feature band extraction can make the model easier to be used in portable equipment by improving the running speed of the model (*see* **Note 15**).

4 Notes

1. The volumes of mobile phases to be prepared depends on the number of extracted samples to be analyzed in each specific study.
2. The HPLC system was calibrated according to the verification regulations for liquid chromatographs (JJG 705-2014) when you are going to start using the HPLC system.
3. Elution conditions were set to change the proportion of mobile phase B from 20% to 65% in 35 min.
4. The UV detection wavelength was 280 nm, the injection volume was 10 L, and the column temperature was set at 35 °C.
5. Spectra manager CFR software was used for spectra acquisition and analysis. Spectra with high noise are usually removed in analysis.
6. The parameters of the two commonly used hyperspectrometer (ImSpector V10E and ImSpector N17E) are shown in Table 1.
7. The probe is positioned above the surface of the sample at a distance of 15 mm and the sample view angle of the probe is 30°. The light source is positioned at a distance of 150 mm and at an angle of 30°from the surface of the sample.
8. Scrubbing the salt tablets of FT-IR spectrometer from the inside out, and avoid getting too much of organic solvents, such as acetone.
9. Before the experiment, potassium bromide (KBr; SCR, Shanghai, China) was dried in an oven at 105 °C for 4 h and cooled in a desiccator.

Table 1
The parameters of ImSpector V10E and ImSpector N17E

Parameter	ImSpector V10E	ImSpector N17E
Spectral range	400–1000 nm	900–1700 nm
Spectral resolution	2.8 nm	5 nm
CCD pixel	672 (Front length)×512 (Band number)	320 (Front length)×256 (Band number)
Exposure time	0.01–1000 ms	0.01–500 ms
Sampling interval	0.59 nm	0.65 nm
Effective slit length	14.2 nm	14.2 nm
Focal length	23 mm	22 mm
Slit width	30 μm	30 μm

10. In theory, the particle size of the grinding of FT-IR sample should be smaller than the wavelength of its infrared light in order to avoid the generation of dispersion spectra. Pay attention not to absorb moisture during the grinding process and do not exhale towards the sample.
11. When making FT-IR samples, pay attention to opening and closing the sample chamber gently, and do not exhale facing the sample chamber, as it can make the background absorb well.
12. Hand washing and drying are important when collecting the FT-IR spectra of tea samples.
13. Some regression algorithms are usually applied to establish determination models: partial least squares (PLS), radial basis function neural network (RBF-NN), least-squares support vector machine (LS-SVM), and Gaussian process regression (GPR) [15, 16]. PLS is one of the reliable analytical tools for modeling and has a wide application in multivariable data analysis and regression [17]. RBF-NN has a strong ability for nonlinear fitting, which can be applied to approximate functions. LS-SVM is an optimized algorithm based on standard SVM, and has the capability for linear and nonlinear multivariate calibration and solves the multivariate calibration problems in a relatively fast way [18]. GPR is a machine learning method based on Bayesian theory and statistical learning theory.
14. Spectra often contain some undesired effects (sample physical characteristics, instrument response and working environment) which are irrelevant to the substance studied and would decrease the accuracy of the calibration models. Thus, mathematical treatments are often applied to the spectra to remove the interferences of these disturbing information before modeling [19, 20]. These pretreatments were baseline correction (BC), standard normal variate (SNV), multiplicative scatter correction (MSC), smoothing (SM), first derivate (1st DER), second derivate (2nd DER), Savitzky-Golay (SG) smoothing, wavelet transform (WT), detrending (DET), and normalization (NOR). These methods are widely used in preprocessing of spectral data.

    Three typical preprocessing methods are introduced:

    BC is mainly used to eliminate the interference of instrument background or drift on spectral signals. Existing baseline correction algorithms include polynomial fitting, segmented polynomial fitting, wavelet transform, etc. Among them, the step of piecewise polynomial fitting baseline correction is to divide the spectral curve into several intervals, and the width of the interval is used as the window width for analysis. The spectral curve characteristics of each window are processed separately. Usually, the minimum value of the spectral curve of each window is selected as the feature point, and then the feature points in each window are connected with straight lines, and the connected broken lines are fitted into an approximate linear relationship. The final fitted curve is considered as the baseline, and the corrected spectral data is obtained by subtracting the baseline from the original spectral curve.

SNV is used to correct spectral errors caused by scattering between samples, and is widely used in the preprocessing of reflection spectral data of solid agricultural products. The basic calculation idea of the algorithm is: assuming that the reflection intensity of each wavelength point in each spectrum should conform to a certain distribution (such as normal distribution), based on this assumption, each original spectral data will be standardized and normalized.

MSC is similar to SNV, which eliminates the effects of uneven particle distribution and particle size by correcting the scattering of each spectrum. MSC has a wide range of applications in solid-state diffuse reflectance spectroscopy. This algorithm assumes that each spectrum has a linear relationship with the "ideal" spectrum, but the true "ideal" spectrum does not exist. Therefore, the average spectrum of the sample is considered as the "ideal" spectrum.

15. Spectral data have many variables, among which there is only a small amount of useful information and most of them are irrelevant and collinear variables. The commonly used approach to solve this problem is to select useful variables from many variables through the characteristic wavelength selection methods. There are some selection algorithm of feature wavelength, including competitive adaptive reweighted sampling (CARS), least square support vector regression (LSSVR), shuffled frog leaping algorithm (SFLA), successive projections algorithm (SPA), which are used to find the variables with the lowest redundancy from datasets.

    CARS algorithm is a common method for selecting spectral characteristic variables. It uses MSC method to randomly select a certain proportion (generally 80%) of all samples as the modeling set, establishes the PLSR model, then uses Exponential decay function to remove variables with smaller weight values, and uses adaptive reweighting sampling algorithm to filter variables with larger weight, and finally selects the subset of the PLSR model with the smallest root mean square error of cross validation (RMSECV).

    Least square support vector machine (LSSVM) is a data analysis method combining least squares and support vector machine. Similar to support vector machine, the basic idea of this method is to map low-dimensional raw data nonlinearly to high-dimensional space, but the difference is that inequality constraints are used instead of equality constraints. In high-dimensional linear space, minimizing the loss function to obtain a linear fitting function. The parameters of this fitting function are obtained by solving a finite dimensional quadratic programming problem in the dual space. Therefore, this method is more efficient than support vector machines.

SFLA is a heuristic population evolutionary algorithm with high computing efficiency and excellent global search ability. The algorithm calculates the selected probability of each variable by simulating a Markov chain obeying the steady distribution in the model space, so as to select the characteristic variable. SFLA is used in conjunction with the PLSR method when selecting feature bands for spectra, to determine whether the variable needs to be removed during each iteration based on the absolute value of the regression coefficient in the PLSR model.

SPA is a forward variable selection algorithm that minimizes the collinearity of variable space. This method has a direct idea for the selection of characteristic bands, that is, starting from a band, forward cycling step by step. Each cycle calculates the projection of the selected band vector on the unselected band vector, and introduces the band with the largest projection distance to the set of characteristic bands.

References

1. Wu D, Yang HQ, Chen XJ, He Y, Li XL (2008) Application of image texture for the sorting of tea categories using multi-spectral imaging technique and support vector machine. J Food Eng 88:474–483
2. Zuo H, Si XY, Li P, Li J, Chen ZH, Li PH, Chen CS, Liu ZH, Zhao J (2023) Dynamic change of tea (*Camellia sinensis*) leaf cuticular wax in white tea processing for contribution to tea flavor formation. Food Res Int 163: 112182
3. Zhou RQ, Li XL, He Y, Jin JJ (2018) Determination of catechins and caffeine content in tea (*Camellia sinensis* L.) leaves at different positions by Fourier-transform infrared spectroscopy. Trans ASABE 61(4):1221–1230
4. Eremina AV, Reshetnyak VY, Vezirishvili MO (2004) HPLC determination of polyphenols in dry grape stem extract. Pharm Chem J 38(3): 143–145
5. Gatea F, Teodor ED, Matei AO, Badea GI, Radu GL (2015) Capillary electrophoresis method for 20 polyphenols separation in propolis and plant extracts. Food Anal Methods 8(5):1197–1206
6. Wade P, Hannen J (2010) Iodometric determination of caffeine. J Sci Food Agric 1(6): 177–178
7. Wanyika HN, Gatebe EG, Gitu LM, Ngumba EK, Maritim CW (2010) Determination of caffeine content of tea and instant coffee brands found in the Kenyan market. Afr J Food Sci 4(6):353–358
8. Li XL, Jin JJ, Sun CJ, Ye DP, Liu YF (2019) Simultaneous determination of six main types of lipid-soluble pigments in green tea by visible and near-infrared spectroscopy. Food Chem 270:236–242
9. Yu KQ, Zhao YR, Li XL, Shao YN, Zhu FL, He Y (2014) Identification of crack features in fresh jujube using Vis/NIR hyperspectral imaging combined with image processing. Comput Electron Agric 103:1–10
10. Gao JF, Li XL, Zhu FL, He Y (2013) Application of hyperspectral imaging technology to discriminate different geographical origins of *Jatropha curcas* L. seeds. Comput Electron Agric 99:186–193
11. Su WH, Sun DW (2018) Fourier transform infrared and Raman and hyperspectral imaging

techniques for quality determinations of powdery foods: a review. Compr Rev Food Sci Food Safety 17(1):104–122
12. Paradkar MM, Irudayaraj J (2010) A rapid FTIR spectroscopic method for estimation of caffeine in soft drinks and total methylxanthines in tea and coffee. J Food Sci 67(7): 2507–2511
13. Cui XX, Zhang XL, Tang HW, Zhang LP, Xu CH, Sun SQ (2012) Study on extracts of active substances from larch bark by FTIR spectroscopy. Spectrosc Spectral Anal 32(7): 1810–1814
14. Park YS, Lee MK, Heo BG, Ham KS, Kang SG, Cho JY, Gorinstein S (2010) Comparison of the nutrient and chemical contents of traditional Korean Chungtaejeon and green teas. Plant Foods Hum Nutr 65(2):186–191
15. Rasmussen CE, Nickisch H (2010) Gaussian processes for machine learning (GPML) toolbox. J Mach Learn Res 11(6):3011–3015
16. Schilling RJ, Carroll JJ, Al-Ajlouni AF (2001) Approximation of nonlinear systems with radial basis function neural networks. IEEE Trans Neural Netw 12(1):1–15
17. Li F, Mistele B, Hu Y, Chen X, Schmidhalter U (2014) Reflectance estimation of canopy nitrogen content in winter wheat using optimised hyperspectral spectral indices and partial least-squares regression. Eur J Agron 52:198–209
18. Wu D, He Y, Feng SJ, Sun DW (2008) Study on infrared spectroscopy technique for fast measurement of protein content in milk powder based on LS-SVM. J Food Eng 84(1): 124–131
19. Barbin DF, Elmasry G, Sun DW, Allen P (2012) Predicting quality and sensory attributes of pork using near-infrared hyperspectral imaging. Anal Chim Acta 719:30–42
20. Huang YF, Dong WT, Sanaeifar A, Wang XM, Luo W, Zhan BS, Liu XM, Li RL, Zhang HL, Li XL (2020) Development of simple identification models for four main catechins and caffeine in fresh green tea leaf based on visible and near-infrared spectroscopy. Comput Electron Agric 173:105388

Chapter 2

Comprehensive Analysis of Pistachio (*Pistacia vera* L.) Hull Phenolics by High-Performance Liquid Chromatography-Diode Array Detection-Tandem Mass Spectrometry

Sevcan Erşan

Abstract

Pistachio hull, the by-product of pistachio processing, is a plant source rich in phenolic compounds with diverse biological activities that may promote human health. Here, we describe the high-performance liquid chromatography (HPLC) coupled with diode array detection (DAD) and tandem mass spectrometry (MS/MS)-based procedure for comprehensive analysis of phenolic compounds from pistachio hull. The analytical approach encompasses sample preparation, phenolic extraction, HPLC separation, DAD and MS detections, data analysis, and interpretation. Additionally, method validation parameters are presented, ensuring the accuracy and reliability of the analytical process. The described methodology enables quantitative determination of a wide range of phenolic compounds from pistachio hull, including flavonoids, anacardic acids, and gallotannins.

Key words Nut, Waste, Polyphenol, Gallic acid, Tannin, Bioactive, HPLC-DAD-MS/MS, Mass spectrometry profiling, Fragmentation

1 Introduction

Pistachio hulls are rich in phenolic compounds associated with various potential health-promoting effects [1]. The pistachio hull phenolics were linked to anti-inflammatory [2], anti-obesity [3], and anti-cancer properties [4, 5], suggesting their potential in preventing certain diseases. Pistachio hull and their extracts were considered functional ingredients in various food products [6, 7]. They have been explored as natural alternatives for synthetic antioxidants commonly used in the food industry [8–12], highlighting their potential as techno-functional food ingredients. Hence, pistachio hull and their extracts have immense potential for use in the food, pharmaceutical, and biomedical industries.

Mostafa Gouda et al. (eds.), *Plant Chemical Compositions and Bioactivities*, Methods and Protocols in Food Science,
https://doi.org/10.1007/978-1-0716-3938-2_2,

Further research is required to fully understand the extent of these health benefits and technological functions and their association with pistachio hull phenolics. A comprehensive phenolic analysis is essential to elucidate the contribution of individual phenolic components to biological activities, ascertain optimal dosages for therapeutic effects, and standardize the phenolic compositions of extracts. This is particularly crucial as poorly characterized extracts often lead to highly variable, unpredictable, and sometimes even partly controversial biological activities.

The phenolic compositions can be accurately assessed with high specificity using high-performance liquid chromatography (HPLC) coupled with a diode array detector (DAD) and mass spectrometry (MS). This approach overcomes the limitations of commonly used nonspecific spectrophotometric methods, such as the Folin-Ciocalteu and activity-based assays. In the HPLC-DAD-MS/MS system, HPLC-separated analytes are ionized in an electrospray ionization (ESI). Subsequently, MS detects the charge-to-mass ratio (*m/z*) of precursor or fragment ions. The additional DAD provides ultraviolet-visible light (UV/Vis) spectra characteristic of each class of phenolic compounds. Consequently, phenolic compound identification can be reliably accomplished by comparing the LC retention times, UV/Vis spectra, and MS spectra with those of standard compounds [13].

When standard compounds are commercially unavailable, and their preparation at high purity is impractical, phenolic compound identification can be achieved by comparing the HPLC-DAD-MS/MS data with that from earlier research and evaluating UV and MS data and the expected fragmentation patterns. In our previous comprehensive study, we employed the HPLC-DAD-MS/MS approach following the described methodology and provided detailed insights into 66 individual phenolic compounds from pistachio hull with minimal reliance on standard compounds [13]. Among the three major phenolic classes present in pistachio hull, namely gallotannins, flavonoids, and anacardic acids, the most prevalent compounds were anacardic acids, gallic acid, monogalloyl glucose, pentagalloyl glucose, quercetin galactoside, and quercetin glucoside (Fig. 1). Subsequently, 58 phenolic compounds quantitated in the pistachio hull extracts utilizing six standard compounds [14], considering that structurally related phenolic compounds exhibit similar UV/Vis absorbances at their specific detection wavelengths when aromatic, or aglycon, parts of the phenolic structure resemble each other or remain the same [15]. As a result, the phenolic composition of pistachio hulls, determined by HPLC-DAD-MS/MS, ranged from 61.2 to 100 g/kg dry matter (DM), underscoring them as good sources of various individual phenolic compounds [14].

This chapter outlines a workflow for HPLC-DAD-MS/MS analysis of the pistachio hull phenolics (Fig. 2), extending from

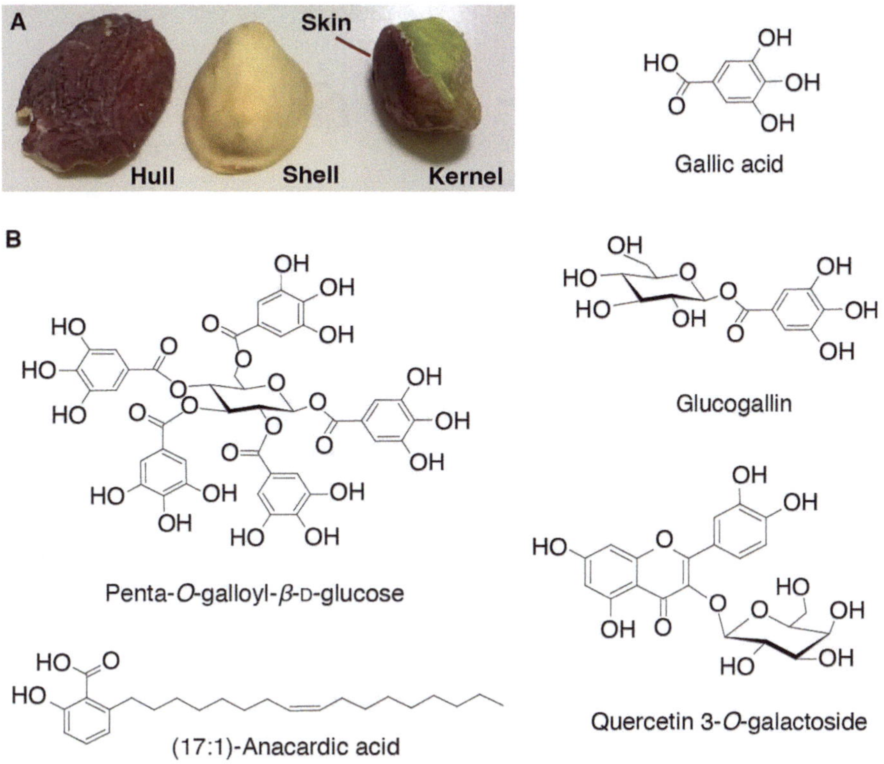

Fig. 1 Pistachio hull, the primary by-product of pistachio processing, constitutes the outer layer of pistachio drupe. The remaining components of the pistachio drupe include lignocellulosic hard shell and edible green kernel covered by red skin (**a**). Representative phenolic structures from pistachio hull (**b**)

our studies [13, 14]. The presented method warrants future research into the phenolic composition of pistachio hulls, including investigations across different varieties cultivated under diverse environmental conditions, samples collected from technologically differing processing units, and for the standardization of derived extracts. This approach allows a more comprehensive understanding of the biological activities linked to pistachio hull phenolics, thereby promoting their potential applications.

2 Materials

2.1 Equipments

1. Analytical balance, 0.1 mg sensitivity.
2. Calibrated pipettes, 1 mL, and 100 or 300 μL volumes.
3. Freeze-dryer (*see* **Note 1**).
4. Analytical grinding mill.
5. Infrared moisture analyzer (*see* **Note 2**).
6. Mili-Q water system.

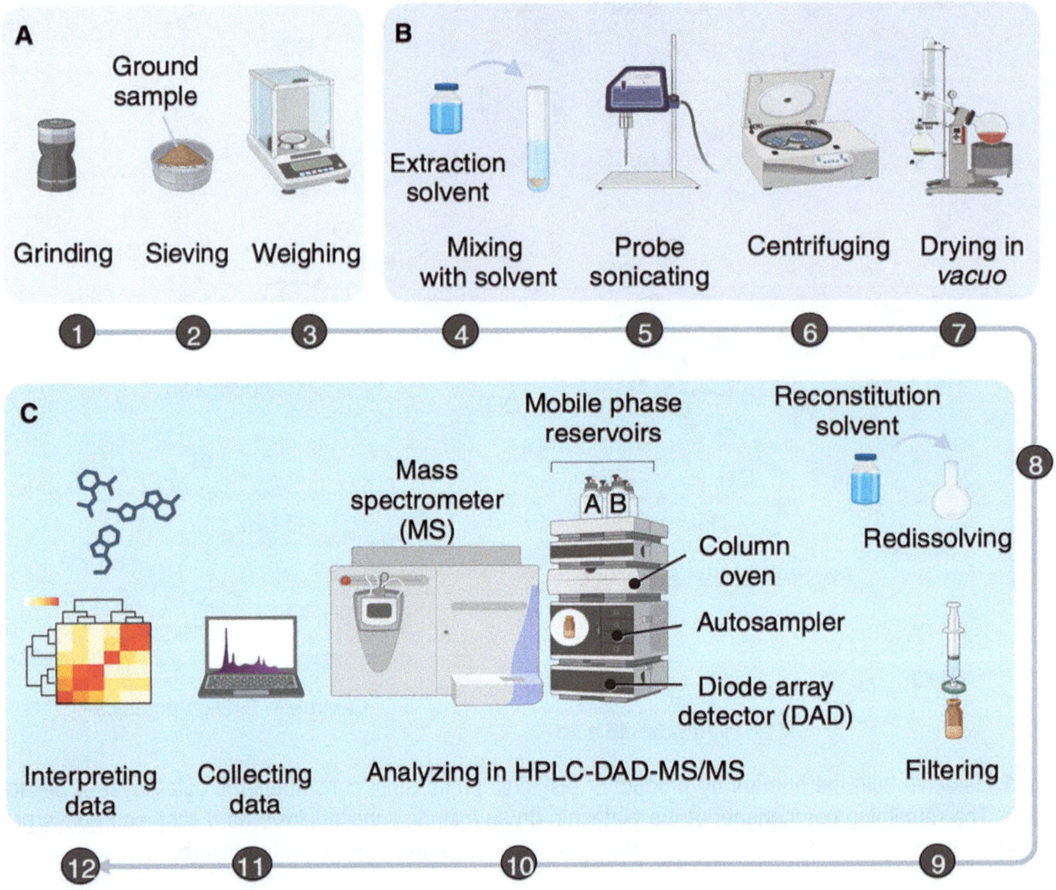

Fig. 2 Workflow for the analysis of pistachio hull phenolics. (**a**) Sample preparation; (**b**) Phenolic extraction; (**c**) HPLC-DAD-M/MS analysis and data processing. (Created with BioRender.com)

7. Vortex mixer.
8. Ultrasonic probe sonicator.
9. Ultrasonic bath (optional) (*see* **Note 3**).
10. Magnetic stirrer plate.
11. Centrifuge, compatible with available glass test tubes.
12. Rotary evaporator.
13. HPLC system (with a quaternary pump, an autosampler, and a vacuum degasser) coupled with a diode array detector (DAD) and a mass spectrometer (e.g., Esquire 3000+ ion-trap mass spectrometer, Bruker Daltonics, Bremen, Germany) fitted with an electrospray ion (ESI) source. Instrument control and data acquisition softwares (e.g., Chromquest and Esquire Control).
14. (Optional) A high-resolution mass spectrometer (e.g., micrOTOF-Q mass spectrometer, Bruker Daltonics, Bremen, Germany or Q Exactive Plus high-resolution mass spectrometer, Thermo Fisher Scientific, Bremen, Germany) (*see* **Note 4**).

2.2 Reagents

1. Water, LC-MS grade.
2. Methanol, LC-MS grade.
3. Formic acid, 98–100%.
4. Hydrochloric acid (HCl), 37%.
5. Liquid nitrogen.
6. Nitrogen gas (*see* **Note 5**).
7. Analytical standard compounds at the highest available purity (*see* **Note 6**): Gallic acid; Protocatechuic acid; *β*-Glucogallin (1-*O*-galloyl-*β*-D-glucopyranose); Penta-*O*-galloyl-*β*-D-glucose; (15:0)-Anacardic acid; Quercetin 3-*O*-galactoside; Quercetin 3-*O*-glucoside; Quercetin 3-*O*-glucuronide; Myricetin 3-*O*-galactoside; Cyanidin 3-*O*-*β*-D-galactopyranoside.

2.3 Pistachio Hull Samples

1. Pistachio hull: fresh or dried whole pistachio drupes containing hull, shell, skin, and kernel (Fig. 1) or waste stream collected from the processing unit (*see* **Note 7**).
2. Store pistachio hull samples at −20 °C until analysis.

2.4 Supplies

1. Sample storage bottles.
2. Weighing paper.
3. Micro spatulas.
4. Lab spoon or spatula.
5. Microcentrifuge tubes, 1.5 or 2 mL volume.
6. Pipette tips, compatible with calibrated pipettes.
7. Glass reagent bottles, 100 mL volume.
8. Glass test tubes (5 or 10 mL volume) with stoppers or caps, compatible with available centrifuge setup.
9. Glass pipettes, 1 and 10 mL volumes.
10. Pipette pumps, compatible with glass pipettes (**item 9**).
11. Magnetic stirring bars (*see* **Note 8**).
12. Pasteur pipettes and compatible pumps (*see* **Note 9**).
13. Round bottom evaporation flasks, 25 or 50 mL, and flask supports (*see* **Note 10**).
14. Glass graduated cylinders, 50, 100, and 1000 mL volumes.
15. LC solvent bottles, 4 of 1 L volume (*see* **Note 11**).
16. Amber LC glass vials with caps, 2 mL volume (*see* **Note 12**).
17. Vial inserts, 100 or 300 μL volume, compatible with glass vials (**item 16**).
18. Plastic or glass syringe, 1 mL volume.
19. Regenerated cellulose syringe filters, 0.45 μm pore size.

20. Supplies necessary for infrared moisture analyzer (*see* **Note 13**).
21. HPLC column: Kinetex C_{18} core-shell reversed-phase column (250 mm × 4.6 mm i.d., 5 μm particle size; Phenomenex, Aschaffenburg, Germany). SecurityGuard Ultra C_{18} guard column holder and cartridges (4.6 mm i.d.; Phenomenex, Aschaffenburg, Germany) (*see* **Note 14**).

3 Methods

Prepare all solutions using LC-MS grade water and at least analytical-grade reagents. Store all solutions at room temperature. Check material safety data sheets (MSDS) online or provided by the supplier for instructions on how to handle chemicals and precautions to minimize exposure and risks. All steps involving organic solvents should be conducted under the fume hood. Adhere to the waste disposal regulations of the institutions.

3.1 Solvent Preparation

1. Sample extraction and reconstitution solvent: Methanol/water/formic acid, 80/19/1 (v/v/v). Transfer 19 mL of ultrapure water into a 100 mL solvent bottle using a 50 mL graduated cylinder. Add 1 mL of formic acid using a glass 1 mL pipette. Thoroughly mix by swirling the solvent bottle (**Warning!** *see* **Note 15**). Add 80 mL of methanol using a 100 mL graduated cylinder. Thoroughly mix.
2. Standard preparation solvent for cyanidin 3-*O*-*β*-D-galactopyranoside: HCl 0.1% (v/v) in methanol. Transfer 100 mL methanol to a 100 mL solvent bottle using a 100 mL graduated cylinder. Add 0.1 mL of concentrated HCl using a 1 mL glass pipette. Thoroughly mix by swirling the solvent bottle.
3. Mobile phase A: Formic acid 1% (v/v) in water (**Warning!** *see* **Notes 15** and **16**). Measure 990 mL water using a 1 L graduated cylinder. Transfer a portion (200–300 mL) of water to a 1 L LC solvent bottle. Add 10 mL of formic acid using a 10 mL glass pipette. Swirl the bottle thoroughly to mix. Add the remaining water from the graduated cylinder to the solvent bottle.
4. Mobile phase B: Formic acid 1% (v/v) in methanol. Transfer 990 mL methanol to a 1 L LC solvent bottle using a 1 L graduated cylinder. Add 10 mL of formic acid using a 10 mL glass pipette. Mix content thoroughly (*see* **Note 17**).
5. Column wash solvent C: Water (*see* **Note 16**).
6. Column wash solvent D: Methanol.

3.2 Standard Preparation

3.2.1 Preparation of Stock Solutions of Standard Compounds

1. Accurately weigh 1–10 mg of standard compound using a weighing paper and an analytical balance (*see* Table 1 for recommended concentrations and **Note 18**).
2. Transfer the weighed standard compounds to a test tube. Note the weight for later concentration calculations (*see* **Note 19**).
3. Add an organic solvent (e.g., 1–10 mL methanol) to achieve a standard solution at a desired concentration (*see* Table 1 for recommended concentrations). Use HCl 0.1% (v/v) in methanol for cyanidin 3-*O*-*β*-D galactopyranoside (*see* **Note 20**).
4. Vortex standard solutions thoroughly until complete dissolution. Alternatively, briefly immerse the test tube into an ultrasonic bath to aid in dissolution (*see* **Note 21**).
5. Aliquot the prepared standard solutions into small portions, such as 100 μL each, transfer them into sample vials or microcentrifuge tubes, and store them at −20 or −80 °C (*see* **Note 22**).

3.2.2 Preparation of Standard Solutions for Calibration Curves

1. Prepare a working standard solution: transfer 100 μL of stock standard solution to a microcentrifuge tube. Add 900 μL of extraction solvent and thoroughly mix the solution by pipetting or vortex (*see* Table 1, **Note 23**).

Table 1
Recommended concentrations for the preparation of standard solutions

Standard compound	Stock solution concentration (mg/mL)	Working solution concentration (mg/mL)	Calibration curve concentration range (μg/mL)
Gallic acid	10	1	500–10
β-Glucogallin	10	1	500–10
Penta-*O*-galloyl-*β*-D-glucose	10	1	500–10
Quercetin 3-*O*-glucoside	2	0.2	100–5
Quercetin 3-*O*-galactoside	2	0.2	60–5
Quercetin 3-*O*-glucuronide	2	0.2	60–5
Myricetin 3-*O*-galactoside	1	0.1	60–5
Protocatechuic acid	1	0.1	60–5
Cyanidin 3-*O*-*β*-D-galactopyranoside	1	0.1	25–1
(15:0)-Anacardic acid	10	1	500–10

2. Prepare a series of dilutions from working standard solutions using extraction solvent. Aim for at least five or seven, preferably equally spaced data points for each standard. Prepare three independent replicates for each standard and data point (*see* **Note 24**). The concentrations of standard solutions used for establishing calibration curves should cover the expected range of concentrations for the compounds of interest (*see* Table 1 and Fig. 3 for suggested ranges of concentrations).

3.3 Sample Preparation

1. Separate pistachio hulls manually from the remaining part of the pistachio drupe. Skip to the next step if pistachio hulls are collected from the processing line as waste materials without presence of other parts.
2. Freeze-dry pistachio hulls until reaching constant weight or moisture content <5% if samples are fresh or wet (*see* **Note 1**).
3. Submerge samples in liquid nitrogen, usually in a wire container, then transfer the sample to an analytical laboratory mill, ensuring no liquid remains, and grind (*see* **Note 25**).
4. Sieve ground pistachio hull, typically using a sieve shaker, and collect fractions having particle sizes between 0.5 μm and 1 mm (*see* **Note 26**).
5. Determine moisture contents by drying a portion of the sample (e.g., ~1 g) in a rapid moisture analyzer (*see* **Note 2**).
6. Store ground samples in a sample bottle away from light at −20 °C.

3.4 Phenolic Extraction

Extract the phenolic content of pistachio hulls using a probe sonicator or magnetic stirrer, depending on the availability of the instrumentation.

3.4.1 Ultrasound-Assisted Solvent Extraction

1. Accurately weigh ~250 mg of ground pistachio hull in a glass tube.
2. Prepare sample blanks as above by omitting the pistachio hull sample.
3. Add 5 mL of extraction solvent.
4. Immerse the ultrasonic probe into the liquid in the test tube (*see* **Note 27**). Place the glass tube in an ice-filled bucket to avoid heating during ultrasonication.
5. Probe sonicate for 30 s at 70% amplitude.
6. Centrifuge at 1000 x *g* at a higher speed for 5 min.
7. Using a Pasteur pipette, transfer supernatant to a round-bottom evaporation flask while being cautious not to remove any solid residue (*see* **Note 9**).

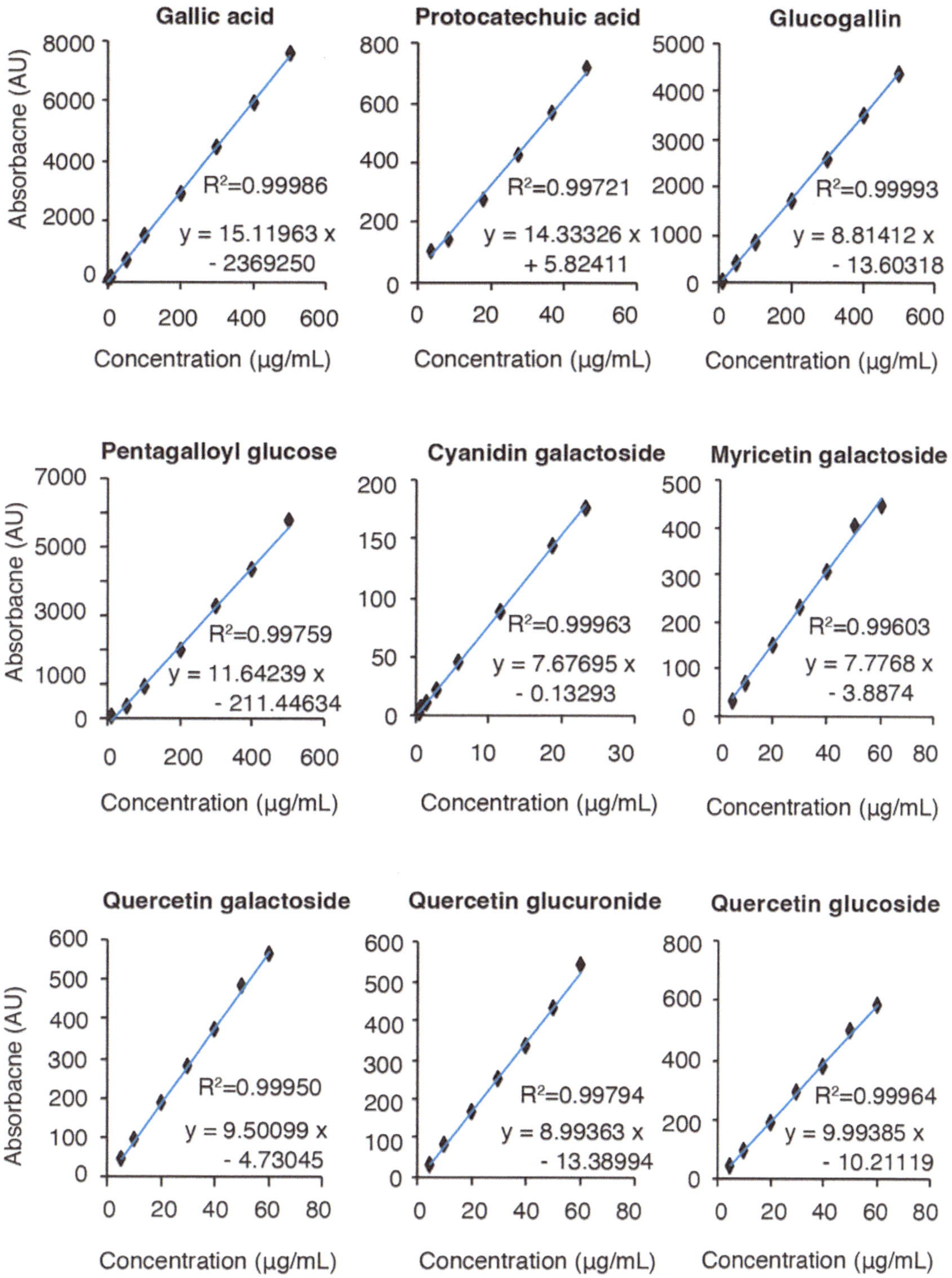

Fig. 3 Representative calibration curves of phenolic compounds

8. Re-extract the solid residues three more times, as mentioned in **steps 2–7**, and combine the supernatants (*see* **Note 28**).
9. Evaporate to dryness in *vacuo* at 30 °C (*see* **Note 29**).
10. Store at −20 °C until analysis.

3.4.2 Magnetic Stirring-Based Solvent Extraction

1. Accurately weigh ~250 mg of ground pistachio hull in a glass tube.
2. Prepare sample blanks as above by omitting the pistachio hull sample.
3. Add 5 mL of extraction solvent.
4. Put magnetic stirring fish (*see* **Note 8**).
5. (optional) Flush the headspace of the extraction flask with nitrogen to prevent oxidation during prolonged extraction times.
6. Stir at ~600 rpm for 30 min in a magnetic stirring plate.
7. Centrifuge at 1000 x *g* or a higher speed for 5 min.
8. Using a Pasteur pipette, transfer supernatant to a round-bottom evaporation flask while being cautious not to remove any solid residue (*see* **Note 9**).
9. Re-extract the solid residues three more times, as mentioned in **steps 2–7**, and combine the supernatants (*see* **Note 28**).
10. Evaporate to dryness in *vacuo* at 30 °C (*see* **Note 29**).
11. Store at −20 °C until analysis.

3.5 HPLC-DAD-MS/MS Analysis

1. Re-dissolve dried extracts and sample blanks in 1 mL reconstitution solvent and vortex well until complete dissolution (*see* **Note 30**).
2. Transfer the sample to be analyzed to a 2 mL syringe, filtrate through a 0.45 μm pore size regenerated cellulose membrane filter, and collect filtrate into an amber glass vial. Apply the same procedure separately to all reconstituted extracts, blanks, and standards.
3. Put sample, sample blank, and standard compound vials into the autosampler (Fig. 2).
4. Assemble analytical and guard columns.
5. Set column temperature at 35 °C (*see* **Note 31**).
6. Place mobile phases in HPLC solvent lines (*see* **Notes 16** and **17**).
7. Purge and prime HPLC with mobile phases (*see* **Note 32**).
8. Program mobile phase gradient as in Table 2.
9. Set injection volume as 3 μL (*see* **Note 33**).

Table 2
HPLC gradient profile

t (min)	Mobile phase A (%)	Mobile phase B (%)
0	98	2
10	98	2
37	63	37
42	63	37
60	60	40
70	40	60
90	0	100
104	0	100
105	98	2
112	98	2

10. Set DAD to scan between 200 and 600 nm with specific monitoring data collected at 280, 310, 350, and 520 nm.
11. Tune and calibrate MS using an MS calibration solution.
12. Set the ESI and MS parameters: Drying and nebulizing gas at a flow rate of 11 mL/min and a pressure of 60 psi, respectively. Nebulizer temperature at 365 °C. The potential on the capillary at ±2287 V, alternating positive and negative modes (*see* **Note 34**). Fragmentation amplitude of 1.2 V.
13. Set MS to collect spectra in scanning mode from *m/z* 100 to *m/z* –2000 and for MS/MS analysis.
14. Set HPLC to a slow flow rate (typically 0.1 mL/min) with 2% mobile phase B and 98% mobile phase A and gradually increase the flow rate to 1 mL/min over 5 min.
15. Condition the column at a 1 mL/min flow rate and 2% mobile phase B and 98% mobile phase A until DAD and MS baselines are stable.
16. Run all samples, standards, and sample blanks, ideally in random order (*see* **Note 35**). Start analysis with two instrumental blank injections from vials containing only extraction solvent.
17. After completing the HPLC-DAD-MS/MS run, flush the system with 50:50 mobile phase C: mobile phase D.
18. Flush the column with storage solvents (e.g., 50:50 mobile phase C: mobile phase D), remove and seal the column, and store it as necessary (*see* **Note 36**).

3.6 Data Processing

1. Process UV and MS data using the available software: Examples are ChromQuest for DAD data and Esquire Control for MS. The *m/z* values and UV/Vis maxima monitored for each compound peak are listed in Table 3. Extract data for each chromatographic peak's retention time, maximum UV/Vis wavelength, *m/z* value for the highest intensity peak after eliminating contaminant, background and artifact ions (e.g., dimers, sodium adducts), if any present, in MS^1 scan and corresponding MS^2 spectra.
2. Identify compounds by comparing the detected peaks' retention times, UV, and mass spectra with those of standard compounds and with data presented in Fig. 4 and Table 3 (*see* **Note 37**).
3. Extract absorbances for each compound peak separately at their UV/Vis absorbance maxima (Table 3), typically 280 nm (gallotannins), 310 nm (anacardic acids), 350 nm (flavonols), and 520 nm (anthocyanins) for quantification (*see* **Note 38**).
4. Create calibration curves by plotting the graph of absorbance on the *y*-axis versus concentration on the *x*-axis using the analytical data collected from DAD for the serially diluted standard compounds. Add the best-fit linear regression line, calibration equation, and R^2 value or coefficient of determination of the regression line (*see* **Note 39**). These can be done using programming languages such as MS Excel, Phyton, or MATLAB. Representative calibration curves are given in Fig. 3.
5. Calculate the contents using calibration equations of respective standard compounds (Eq. 1). When standard compounds are not available, use calibration equation of structurally related substances (e.g., gallic acid for gallic acid derivatives, quercetin 3-*O*-galactoside for flavonols, and (15:0)-anacardic acid for other anacardic acids) for quantitation. Ensure that molecular-weight-correction factors (MWCF) are included.

$$C = \frac{A - b}{a} \times \frac{V}{m} \times \frac{100}{\text{DM}} \times \text{MWCF} \quad (1)$$

where C is concentration in g/kg dry matter (DM), A is absorbance at their respective UV/Vis absorbance maxima in AU (arbitrary units), a and b are slope and intercept, respectively, of the calibration curve, V is volume of solvent in L used to dissolve the dry extract, w is the weight of pistachio hull sample extracted in g, DM is percent dry matter in g/g and calculated as in Eq. 2, and MWCF is molecular weight correction factor calculated using Eq. 3.

$$\text{DM} = 100 - \text{MC} \quad (2)$$

where DM is percent dry matter in g/g and MC is percent moisture content.

Table 3
HPLC retention times, UV/Vis spectra, and MS data of pistachio (*Pistacia vera* L.) hull phenolics

No.	t_R[a] (min)	Compound identity	HPLC-UV/Vis abs. max (nm)	HR-ESI(-)-MS $[M-H]^-$ *m/z* exp. (theo.)	Molecular formula	HPLC-ESI-MS/MS experiment *m/z*
1	6.0	1-*O*-galloyl *β*-D-glucopyranose (*β*-glucogallin)[b]	274	331.0684 (331.0671)	$C_{13}H_{16}O_{10}$	[331]: 169(100), 271(56), 193(31), 211(31), 183(25), 315(10) [331→169]:125(100)
2	6.8	Gallic acid[b]	272	na[c]	na	[169]: 125(100)
3	10.0	Gallic acid derivative (1)	276	296.0788 (296.0789)	na	[296]:169(100), 125(15), 195(5), 107(5), 171(2) [296→169]:125(100)
4	11.2	Galloyl quinic acid	275	343.0689 (343,0671)	$C_{14}H_{16}O_{10}$	[343]: 191(100), 169(17), 125(12)
5	13.4	Protocatechuic acid[b]	259	na	na	[152]: 109(100)
6	14.5	Galloyl dihexose (1)	274	493.1195 (493.1194)	$C_{19}H_{26}O_{15}$	[493]: 313(100), 169(39), 283(29), 331(10) [493→313]: 283(100), 169(44), 223(43), 135(39), 241(15)
7	16.5	Galloyl dihexose (2)	274	493.1209 (493.1194)	$C_{19}H_{26}O_{15}$	[493]: 271(100), 313(25), 211(20), 331(20), 169(7) [493→271]: 211(100), 125(4)
8	17.2	Galloyl dihexose (3)	274	493.1205 (493.1199)	$C_{19}H_{26}O_{15}$	[493]: 313(100) [493→313]: 169(100), 125(92), 189(30), 242(18)
9	18.0	Galloyl shikimic acid (1)	274	325.0571 (325.0565)	$C_{14}H_{14}O_9$	[325]: 169(100), 125(22), 139(9), 193(8) [325→169]: 125(100)
10	19.6	Galloyl shikimic acid (2)	274	325.0568 (325.0565)	$C_{14}H_{14}O_9$	[325]: 169(100), 125(15), 281(7), 111(4), 173(3) [325→169]: 124(100), 125(28)
11	20.2	Gallic acid derivative (2)	276	na	na	[571]: 285(100), 169(7) [571→285]: 133(100), 169(49), 170(10)

(continued)

Table 3
(continued)

No.	t_R[a] (min)	Compound identity	HPLC-UV/Vis abs. max (nm)	HR-ESI(-)-MS $[M-H]^-$ *m/z* exp. (theo.)	Molecular formula	HPLC-ESI-MS/MS experiment *m/z*
12	20.8	Galloyl shikimic acid (3)	274	325.0576 (325.0565)	$C_{14}H_{14}O_9$	[325]: 169(100), 125(15), 137(3), 173(3) [325→169]: 125(100)
13	21.1	Digallic acid (1)	280	321.0261 (321.0252)	$C_{14}H_{10}O_9$	[321]: 169(100) [321→169]: 125(100)
14	21.5	Digalloyl hexose (1)	280	483.0805 (483.0780)	$C_{20}H_{20}O_{14}$	[483]: 331(100), 271[(16), 169(15) [483→331]: 169(100), 313(60), 271(51), 193(25), 123(15), 241(12), 195(12), 211(10)
15	22.3	Methyl gallate	272	na	na	[183]: 168(100), 124(10) [183→168]: 124(100)
16	23.8	Digalloyl hexose (2)	288	483.0794 (483.0780)	$C_{20}H_{20}O_{14}$	[483]: 331(100), 169(27), 332(17), 271(11) [483→331]: 169(100), 241(53), 125(15)
17	25.0	Gallic acid derivative (3)	274	509.0972 (509.0996)	$C_{15}H_{26}O_{19}$	[509]: 267(100), 429(11) [509→267]: 139(100)
18	26.0	Gallic acid derivative (4)	274	423.0931 (423.0933)	$C_{19}H_{20}O_{11}$	[423]: 313(100), 169(70), 125(36), 211(24) [423→313]: 169(100), 313(39), 295(17)
19	26.4	Digalloyl quinic acid	274	495.0776 (495.0780)	$C_{21}H_{20}O_{14}$	[495]: 343(100), 191(29), 344(16), 271(4) 169(3) [495→343]: 191(100)
20	27.3	Digallic acid (2)	275	321.0246 (321.0252)	$C_{14}H_{10}O_9$	[321]: 169(100), 125(16) [321→169]: 125(100)
21	28.1	Gallic acid derivative (5)	268	403.1254 (403.1246)	$C_{17}H_{24}O_{11}$	[403]: 169(100), 151(55), 313(37), 125(36), 271(21), 211(18), 179(15) [403→169]: 124(100), 125(34), 107(34)

22	29.5	Gallic acid derivative (6)	274	467.1197 (467.1195)	$C_{21}H_{24}O_{12}$	[467]: 313(100), 169(17), 295(7) [467→313]: 169(100), 153(10), 191(5)
23	30.6	Digalloyl shikimic acid (1)	275	477.0688 (477.0675)	$C_{21}H_{18}O_{13}$	[477]: 325(100), 169(47), 326(33) [477→325]: 169(100)
24	31.0	Digalloyl shikimic acid (2)	276	477.0703 (477.0675)	$C_{21}H_{18}O_{13}$	[477]: 325(100), 169(47) [477→325]: 169(100), 125(11), 281(10)
25	31.6	Cyanidin 3-O-β-D-galactopyranoside[b]	278, 517	na	na	[d][449]: 287(100) [449→287]: 137(100)
26	33.2	Luteic acid	278	319.0133 (319.0096)	$C_{14}H_8O_9$	[319]: 239(100), 139(15), 240(14) [319→239]: 139(100), 124(12)
27	33.7	Cyanidin pentoside	520	na	na	[d][419]: 287(100)
28	33.8	Trigalloyl quinic acid	273	647.0873 (647.0890)	$C_{28}H_{24}O_{18}$	[647]: 495(100), 343(22), 191(2) [647→495]: 343(100), 191(46)
29	34.5	Quinic acid derivative (1)	276	523.1455 (523.1157)	$C_{24}H_{28}O_{13}$	[523]: 209(100), 371(45), 505(30), 313(19), 169(14), 191(14) [523→209]: 191(100), 165(21), 151(18), 123(2)
30	35.3	Myricetin galloyl hexoside	270, 354	631.0934 (631.0941)	$C_{28}H_{24}O_{17}$	[631]: 479(100), 317(10) [631→479]: 316(100), 317(69), 325(15), 271(11)
31	35.8	Tetragalloyl hexose	278	787.1010 (787.0999)	$C_{34}H_{28}O_{22}$	[787]: 617(100) [787→617]: 465(100), 589(55), 221(41), 277(34), 449(30), 600(26), 296(20), 235(18), 466(12)
32	36.7	Trigallic acid	274	473.0373 (473.0362)	$C_{21}H_{14}O_{13}$	[473]: 321(100), 169(25) [473→321]: 169(100), 125(6)
33	36.9	Myricetin hexuronide	357, 252	493.0640 (493.0624)	$C_{21}H_{18}O_{14}$	[493]: 317(100), 299(4), 151(3), 137(2) [493→317]: 227(100), 151(23), 137(18)
34	37.1	Myricetin 3-*O*-galactoside	359, 252	479.0848 (479.0831)	$C_{21}H_{20}O_{13}$	[479]: 317(100), 271(20), 179(19), 287(15) [479→317]: 215(100), 271(99), 164(24), 270(22), 287(22), 242(17)

(continued)

Table 3
(continued)

No.	t_R[a] (min)	Compound identity	HPLC-UV/Vis abs. max (nm)	HR-ESI(-)-MS $[M-H]^-$ *m/z* exp. (theo.)	Molecular formula	HPLC-ESI-MS/MS experiment *m/z*
35	37.4	Myricetin hexoside	357, 252	479.0884 (479.0831)	$C_{21}H_{20}O_{13}$	[479]: 316(100), 317(57), 271(12) [479→316]: 271(100), 179(40), 317(35), 180(29), 255(26), 137(23)
36	37.9	Methyl-digallate	274	335.0414 (335.0409)	$C_{15}H_{12}O_9$	[335]: 183(100), 253(5) [335→183]: 168(100)
37	38.3	Penta-*O*-galloyl-*β*-D glucose[b]	280	939.1130 (939.1109)	$C_{41}H_{32}O_{26}$	[939]: 769(100), 617(18), 787(12) [939→769]: 617(100), 387(43), 601(40), 465(28), 323(21), 259(19), 403(19), 725(17), 245(12), 573(12), 386(11)
38	38.9	Quercetin galloyl hexoside (1)	257, 354	615.1031 (615.0992)	$C_{28}H_{24}O_{16}$	[615]: 463(100), 301(32) [615→463]: 301(100), 179(8), 229(7), 253(6), 272(6), 151(2)
39	40.6	Hexagalloyl hexose	263, 353	1091.1423 (1091.1213)	$C_{48}H_{36}O_{30}$	[1091]: 939(100), 769(16) [1091→939]: 769(100), 770(13), 617(10)
40	41.0	Quercetin 3-*O*-galactoside[b]	252, 352	463.0922 (463.0882)	$C_{21}H_{20}O_{12}$	[463]: 301(100), 179(8) [463→301]: 151(100), 229(58), 271(43), 343(30), 254(26), 258(24), 181(16)
41	41.2	Quercetin 3-*O*-glucuronide[b]	255, 355	477.0715 (477.0675)	$C_{21}H_{18}O_{13}$	[477]: 301(100), 179(7), 151(2) [477→301]: 179(100), 152(20), 180(11), 121(8), 256(7), 273(6), 229(6)
42	41.6	Quercetin 3-*O*-glucoside[b]	260, 355	463.0915 (463.0877)	$C_{21}H_{20}O_{12}$	[463]: 301(100) [463→301]: 179(100), 271(52), 152(39), 272(32), 255(31), 151(27), 203(21)

43	42.1	Quercetin galloyl hexoside (2)	254, 358	615.1031 (615.0986)	$C_{28}H_{24}O_{16}$	[615]: 301(100), 313(15), 315(8), 273(6), 463(3) [615→301]: 179(100), 151(99)
44	42.7	Gallic acid derivative (7)	279	347.0409 (347.0406)	$C_{16}H_{12}O_9$	[347]: 267(100), 139(5), 249(4), 269(4), 124(2) [347→267]: 139(100), 249(28), 83(16), 140(10), 205(8)
45	42.9	Quercetin galloyl hexoside (3)	276, 358	615.1098 (615.0986)	$C_{28}H_{24}O_{16}$	[615]: 301(100), 179(5), 463(3) [615→301]: 151(100), 170(32), 107(21), 179(20)
46	43.3	Quercetin galloyl deoxyhexose	274, 355	599.1075 (599.1037)	$C_{28}H_{24}O_{15}$	[599]: 463(100), 301(36) [599→463]: 301(100), 255(7)
47	43.7	Quercetin pentoside	254, 355	433.0810 (433.0771)	$C_{20}H_{18}O_{11}$	[433]: 301(100), 255(6), 242(4), 193(4) [433→301]: 271(100), 179(42), 151(41), 255(11), 243(9)
48	44.1	Unknown (1)	275, 357	na	na	[629]: 327(100), 459(15), 477(15), 328(11) [629→327]: 113(100), 169(49), 283(40), 175(38), 170(37), 177(35)
49	45.0	Kaempferol hexoside	272, 350	447.0963 (447.0927)	$C_{21}H_{20}O_{11}$	[447]: 285(100), 255(28), 327(19) [447→285]: 255(100), 223(6), 256(5)
50	45.4	Quercetin galloyl hexuronide	276, 352	629.0852 (629.0779)	$C_{28}H_{22}O_{17}$	[629]: 477(100), 301(26) [629→477]: 301(100), 323(5), 175(3)
51	46.3	Quercetin galloyl pentoside	277, 352	585.0910 (585.0881)	$C_{27}H_{22}O_{15}$	[585]: 301(100) [585→301]: 151(100), 179(45), 165(44), 121(9)
52	46.9	Kaempferol hexoside	270, 350	447.0952 (447.0927)	$C_{21}H_{20}O_{11}$	[447]: 285(100), 284(92), 255(63), 163(18), 151(17) [447→285]: 255(100), 197(13), 240(10), 198(8), 213(7), 227(6), 169(5)
53	48.9	Kaempferol pentoside	270, 350	417.0907 (417.0822)	$C_{20}H_{18}O_{10}$	[417]: 284(100), 255(37) [417→284]: 255(100), 160(5), 165(4), 227(4), 195(3), 151(3)

(continued)

Table 3
(continued)

No.	t_R[a] (min)	Compound identity	HPLC-UV/Vis abs. max (nm)	HR-ESI(-)-MS $[M-H]^-$ *m/z* exp. (theo.)	Molecular formula	HPLC-ESI-MS/MS experiment *m/z*
54	88.4	(16:1)-Anacardic acid	238, 310	359.2521 (359.2586)	$C_{23}H_{36}O_3$	[359]: 315(100), 341(43), 161(21), 315(17), 107(13), 343(13), 317(11) [359→315]: 108(100)
55	88.5	(13:2)-Aanacardic acid	238, 310	315.1968 (315.1966)	$C_{20}H_{28}O_3$	[315]: 271(100), 107(6) [315→271]: 107(100)
56	88.9	(11:0)-Anacardic acid	238, 308	291.1960 (291.1966)	$C_{18}H_{28}O_3$	[291]: 247(100) [291→247]: 106(100)
57	89.1	(15:3)-Anacardic acid	238, 310	341.2112 (341.2122)	$C_{22}H_{30}O_3$	[341]: 297(100) [341→297]:149(100)
58	89.6	(13:1)-Anacardic acid	246, 310	317.2115 (317.2117)	$C_{20}H_{30}O_3$	[317]: 273(100) [317→273]: 107(100)
59	90.7	(13:0)-Anacardic acid	248, 312	319.2289 (319.2279)	$C_{20}H_{32}O_3$	[319]: 275(100) [319→275]: 106(100)
60	91.0	(15:1)-Anacardic acid	240, 310	345.2429 (345.2434)	$C_{22}H_{34}O_3$	[345]: 301(100), 119(4) [345→301]: 119(100)
61	91.2	Unknown (2)	260	455.3510 (455.3405)	$C_{30}H_{48}O_3$	[455]: 418(100) [455→418]: 434(100), 395(66), 375(60), 399(15)
62	91.4	(17:2)-Anacardic acid	238, 310	371.2599 (371.2592)	$C_{24}H_{36}O_3$	[371]: 327(100) [371→327]: 327(100), 119(18), 107(10)

63	92.0	(15:0)-Anacardic acid[b]	242, 311	347.2623 (347.2592)	$C_{22}H_{36}O_3$	[347]: 303(100) [347→303]: 106(100)
64	92.2	(17:1)-Anacardic acid	248, 312	373.2743 (373.2743)	$C_{24}H_{38}O_3$	[373]: 329(100) [373→329]: 106(100)
65	92.4	Merulinic acid	260	373.2748 (373.2748)	$C_{24}H_{38}O_3$	[373]: 329(100) [373→329]: 119(100)
66	93.1	(17:0)-Anacardic acid	238, 312	375.2951 (375.2977)	$C_{24}H_{40}O_3$	[375]: 331(100) [375→331]: 106(100)

[a] t_R retention time

[b] Verified by standard compounds

[c] *na* not available

[d] Positive ionization mode was used for the identification of anthocyanins

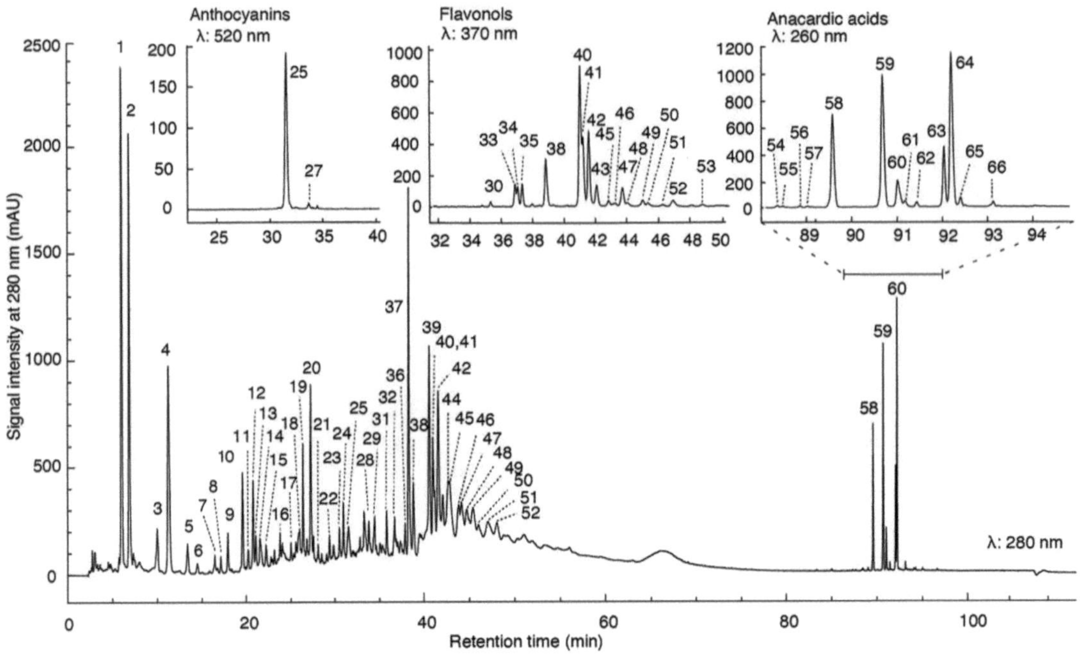

Fig. 4 HPLC separation of phenolic compounds from red *Pistacia vera* L. hulls monitored at different wavelengths. Peak assignments are shown in Table 3. (Adapted with permission from ref. [13], Copyright 2016 American Chemical Society)

$$MWCF = \frac{MW_1}{MW_2} \qquad (3)$$

where MW_1 is the molecular weight of the compound of interest in g/mol, and MW_2 is the molecular weight of the respective standard compound in g/mol.

6. Report data as a mean ± standard deviation of at least three independent replicates.

3.7 Method Validation

1. Calibration linearity: Evaluate R^2 of calibration curves (*see* **Note 39**).
2. Limit of detection (LOD) and limit of quantitation (LOQ): Calculate the standard deviation of the y-intercept (σ) and the average slope of regression lines (S) from three independent replicates of calibration curves (*see* **Note 40**). Calculate LOD and LOQ using Eqs. 4 and 5. The results can be reported as "not detected (nd)" when below than LOD and "trace (tr)" when lower than LOQ.

$$LOD = \frac{3.3 \times \sigma}{S} \qquad (4)$$

$$LOQ = \frac{10 \times \sigma}{S} \qquad (5)$$

3. Extraction recovery: Transfer a methanolic solution of standard compound(s) at concentrations of expected high, middle, and low levels in the pistachio hull to an extraction tube (*see* **Note 41**). Evaporate solvent under a gentle nitrogen stream. Weigh the pistachio hull sample into the same extraction tube. Prepare blank samples by omitting standard compound(s) but including the pistachio hull sample. Perform extraction and analysis following the above procedure. Calculate recovery using Eq. 6.

$$\text{Recovery\%} = \frac{\text{Measured content}_{\text{spiked}} - \text{Measured content}_{\text{control}}}{\text{Added content}_{\text{spiked}}} \times 100 \quad (6)$$

4. Intra-day repeatability: Analyze freshly prepared six independent sample extracts within one day. Calculate the coefficient of variation, CV%.
5. Inter-day repeatability: Analyze freshly prepared six independent sample extracts on four different days within a month, ideally analyzing once a week. Calculate CV% based on the analysis performed on different days.
6. Interpret the results to evaluate how consistent the method is within the same and different days. A low CV below 5% is an indication of good repeatability.

4 Notes

1. Freeze-drying is necessary only when working with fresh samples with moisture content above 5%. Native enzymes, such as polyphenol oxidase, are released when the matrix is disturbed and degrade phenolic compounds in high moisture content samples. Immediate submerging in liquid nitrogen followed by freeze-drying is recommended for high-moisture containing samples.
2. An infrared moisture analyzer is rapid and ideal for determining moisture contents. However, other drying methods, such as conventional oven drying, can be employed depending on equipment availability.
3. An ultrasonic bath for general purposes is sufficient as it will be used for dissolution whenever necessary (*see* **Note 21**).
4. High-resolution mass spectrometry can provide more certainty for identifying compounds, especially in complex mixtures obtained from pistachio hull (Table 3). It is particularly useful for reliable identifications if there is a deviation from the presented method, such as using a different column, or if samples exhibit different phenolic profiles due to variations in factors like variety or cultivating conditions. Otherwise, a

low-resolution mass spectrometer might be sufficient for compound identification by following peak elution orders and peak distributions illustrated in Fig. 4 and Table 3.

5. A nitrogen gas tube connected to a pressure regulator and tubing for gas delivery will suffice if a dedicated nitrogen drying setup is unavailable.
6. High purity standards (at least 98–99%) are preferred whenever available. In the case where only 90–95% purity standards are also available, they are also acceptable with the understanding that corrections for purity are necessary when determining concentrations. Lower purity standards are recommended solely for compound identification purposes.
7. The physical characteristics of waste streams collected from processing units may vary based on the technology of the processing unit and at which processing stage they are collected. Consequently, these waste streams may contain other parts of pistachio drupe, which require manual separation. They can also be wet; in which case they must be submerged in liquid nitrogen for snap freezing and stored at least at -20 °C until analysis.
8. Magnetic stirring bars should be small, allowing them to mix in a glass test tube effectively. Alternatively, samples can be extracted in a 10 mL bottle. However, this will require their transfer to a test tube for centrifugal separation after extraction.
9. Single-use plastic Pasteur pipettes do not need pumps, while glass Pasteur pipettes require a pump for proper use.
10. Pear-shaped evaporation flasks are preferred as dry extracts are collected efficiently.
11. Ideally, only use borosilicate glass bottles for LC-MS to avoid high background noise. Do not use detergents as they can contaminate bottles unless completely necessary. Rinse the bottle with distilled water and the desired solvent before refilling it.
12. Amber vials are recommended as phenolic compounds, especially coloring anthocyanins, are light-sensitive.
13. These include drying pans, desiccators, silica gel, and a drying oven for pre-drying.
14. The column presented here produced satisfactory separation for a wide spectrum of phenolic compounds from pistachio hull (Fig. 4). Alternatively, other readily available C_{18} columns can be employed, however, with anticipated variations in retention times and peak separation capabilities. Identification can still be accomplished by comparing DAD and MS data with those presented in Table 3.

15. Always slowly add acid to the water rather than the other way around to avoid splashing due to the exothermic nature of their mixing.
16. Ideally, discard and prepare water-based solutions (mobile phases A and C) everyday due to the risk of microbial growth that can block LC-system.
17. The organic mobile phase solvent B is ideally changed every second day to prevent alterations in its concentration due to evaporation, which results in shifts in retention times and distortion of peak shapes. Change seal wash and injection solvents every week.
18. Accuracy and precision of weighing in analytical balances can be compromised due to the accumulation of static electricity, resulting in unstable readings. Avoid anything that can generate static electricity, such as rubbing rubber-gloved hands. If an item to be weighed has static electricity, remove it and weigh it again. Optionally, use static electric removal tools.
19. Note the exact weighing of standards and calculate the exact concentration using Eq. 7.

$$C = \frac{m}{V} \times \mathrm{CF} \tag{7}$$

where C is the concentration of the standard solution in mg/mL, m is the exact mass of the standard compounds weighted in mg, V is the volume of organic solvent in mL where the standard compound is dissolved, and CF is the correction factor for molecular weight calculated using Eq. 8.

Some phenolic standards are supplied as hydrates, like gallic acid monohydrate, or in salt form, such as cyanidin galactoside chloride. For standard compounds such as penta-*O*-galloyl-*β*-D-glucose hydrate, the hydration degree can vary depending on the production batch. Check the hydration degree from the analysis certification that comes with the order or available online via the supplier's webpage. When calculating the concentration of standard solutions, account for the hydration degree or salt, that will be released to the solution and will not be detected in mass spectrometer as the part of the molecule, as in Eq. 8.

$$\mathrm{CF} = \frac{\mathrm{MW}_1}{\mathrm{MW}_2} \tag{8}$$

where MW_1 is the molecular weight of a compound without hydration or salt addition in g/mol, and MW_2 is the molecular weight of standard compounds as supplied (hydrated or in salt form) in g/mol.

20. Stock solutions of phenolic standards can be prepared in pure methanol. Anthocyanins, such as cyanidin 3-*O*-*β*-D

galactopyranoside, should be prepared in acidic solvents to enhance their stability. Also, consider suppliers' recommendations for the choice of organic solvents.

21. Certain phenolic compounds, like quercetin derivatives, might not readily be dissolved in organic solvents, even within the compound solubility limits. In such cases, extended vortex time or ultrasonication can aid in dissolution. To ensure complete dissolution, centrifuge the standard solution and subsequently check for the presence of any solid residue. Quantitatively dissolved solutions should not contain any solid residue after centrifugation.
22. Concentrated stock solutions of standards can be stored for extended periods of up to 12 months, either at −20 °C or ideally at −80 °C.
23. Individual standards can be prepared and analyzed separately. Alternatively, a standard mix can be prepared to save analysis time.
24. As DAD-based compound quantitation can be done with structurally similar compounds (*see* Subheading 3.6), calibration curves prepared for a selection of compounds, such as gallic acid, protocatechuic acid, penta-*O*-galloyl-*β*-D-glucose, quercetin 3-*O*-galactoside, cyanidin 3-*O*-*β*-D-galactopyranoside, and (15:0)-anacardic acid, will suffice.
25. Immersion in liquid nitrogen before grinding is necessary to avoid overheating that will results in decomposition of phenolic compounds during grinding. The exclusion of this step will compromise the accuracy and repeatability of results. Apply short pulses of grinding (e.g., 10 s) to avoid sample heating.
26. Sieving the ground sample is recommended for achieving a homogeneous sample, ensuring exhaustive extraction with the described extraction protocol, and for repeatable results. However, this step can be eliminated if the equipment is unavailable. Finely ground samples are also fine but may lead to filtration issues due to clogging.
27. Various sizes of probes might be available. Ensure that the same probe size, compatible with the glass tubes used for extraction, is used consistently.
28. Re-extraction is necessary for the exhaustive recovery of phenolic compounds. We found that ≥90% of phenolic compounds were recovered with four extraction cycles [14].
29. Use a rotary evaporator with a bath temperature of 30 °C with a cooling circulating water. Alternatively, nitrogen gas can be employed for drying.
30. Extracts are ideally analyzed on the same day of preparation. If this is not possible, the headspace of dry extracts can be flushed

with nitrogen, and they can be stored for analysis on subsequent days. Re-dissolve the extracts on the day of analysis.

31. Wait until the column and DAD lamps get warm, which may take around half an hour, depending on the efficiency of the column heater and DAD.
32. Purge each line to get rid of any visible bubbles.
33. The 3 μL injection volume is ideal for the presented method. Larger injection volumes can lead to band broadening, as the injection solvent (containing 80% methanol) is stronger than the mobile phase at initial HPLC gradient conditions (containing 2% methanol). If desired, the injection volume can be increased up to 10 μL with this column after diluting extracts in water-based solutions.
34. ESI source operated in positive ion mode for anthocyanins and negative ion mode for detecting all other phenolic compounds.
35. The extracts are stable for up to 48 h at room temperature.
36. Check column care guidelines for recommended column storage solvents.
37. Retention times may vary compared to those presented in Table 3. Consider retention time shifts by comparing the retention times of standard compounds with those listed in Table 3 and adjust retention times of the remaining compound peaks, as necessary.
38. Be aware that DAD response may change within the lifespan of DAD lamps, potentially leading to differences on absorbances for the samples and standards analyzed at different times.
39. R^2 value equal to 1 represents the best fit, while R^2 values around 0.99–0.98 are generally acceptable.
40. Given the linear regression equation is $y = ax + b$, a and b represent constants for slope and intercept, respectively.
41. Expected concentrations for use in recovery studies can be determined based on the results obtained for a particular set of samples analyzed. Alternatively, refer to ref. 14 for expected ranges.

References

1. Arjeh E, Akhavan H-R, Barzegar M, Carbonell-Carrachina Á (2020) Bio-active compounds and functional properties of pistachio hull: a review. Trends Food Sci Technol 97:55–64
2. Grace MH, Esposito D, Timmers MA, Xiong J, Yousef G, Komarnytsky S, Lila MA (2016) Chemical composition, antioxidant and anti-inflammatory properties of pistachio hull extracts. Food Chem 210:85–95
3. Noorolahi Z, Sahari MA, Barzegar M, Gavlighi HA (2020) Tannin fraction of pistachio green hull extract with pancreatic lipase inhibitory and antioxidant activity. J Food Biochem 44:e13208

4. Karaoglu EC, Tarhan L (2022) Pistachio (*Pistacia vera* L.) hull samples from Turkey: phenolic compounds, antioxidant properties, and cytotoxic activities against HeLa, MCF-7, OE-33, and ACC-201 cancer cell lines. Food Meas 16:2300–2313
5. Seifaddinipour M, Farghadani R, Namvar F, Bin Mohamad J, Muhamad NA (2020) In vitro and in vivo anticancer activity of the most cytotoxic fraction of pistachio hull extract in breast cancer. Molecules 25:1776
6. Nateghi L, Hosseini E (2023) Investigating the oxidative stability of grape seed oil using aqueous extract of pistachio green hull. J Food Meas Charact 17:4434–4447
7. Zandi-Darehgharibi F, Haddadi H, Rafieian-Kopaei M, Fallah AA (2023) Effects of pistachio green hull crude extract and its polyphenol fraction on oxidative stability of sunflower oil during accelerated storage. Biomass Conv Bioref 13:6973–6980
8. Noorolahi Z, Sahari MA, Gavlighi HA, Barzegar M (2022) Pistachio green hull extract as natural antioxidant incorporated to omega-3 rich kappa-carrageenan oleogel in dry fermented sausage. Food Biosci 50:101986
9. Elhadef K, Ennouri K, Fourati M, Ben Hlima H, Akermi S, Mellouli L, Smaoui S (2021) Pistachio hull extract as a practical strategy to extend the shelf life of raw minced beef: chemometrics in quality evaluation. Evid Based Complement Alternat 16:2429766
10. Sadeghinejad N, Sarteshnizi RA, Gavlighi HA, Barzegar M (2019) Pistachio green hull extract as a natural antioxidant in beef patties: effect on lipid and protein oxidation, color deterioration, and microbial stability during chilled storage. LWT 102:393–402
11. Koç B, Kayabaşi GA (2023) Enrichment of white wheat bread with pistachio hulls and grape seeds: effect on bread quality characteristics. Appl Sci 13:3431
12. Lalegani S, Ahmadi Gavlighi H, Azizi MH, Amini Sarteshnizi R (2018) Inhibitory activity of phenolic-rich pistachio green hull extract-enriched pasta on key type 2 diabetes relevant enzymes and glycemic index. Food Res Int 105:94–101
13. Erşan S, Güçlü Üstündağ Ö, Carle R, Schweiggert RM (2016) Identification of phenolic compounds in red and green pistachio (*Pistacia vera* L.) hulls (Exo- and Mesocarp) by HPLC-DAD-ESI-(HR)-MS^n. J Agric Food Chem 64:5334–5344
14. Erşan S, Güçlü Üstündağ Ö, Carle R, Schweiggert RM (2017) Determination of pistachio (*Pistacia vera* L.) hull (exo- and mesocarp) phenolics by HPLC-DAD-ESI/MS^n and UHPLC-DAD-ELSD after ultrasound-assisted extraction. J Food Compos Anal 62:103–114
15. Chandra A, Rana J, Li Y (2001) Separation, identification, quantification, and method validation of anthocyanins in botanical supplement raw materials by HPLC and HPLC-MS. J Agric Food Chem 49(8):3515–3521

Chapter 3

Authentication of Microalgae Edible Lipids Using Non-Targeted C12-NMR Spectroscopy

Ashfaq Ahmad, Syed Salman Ashraf, and Hina Zaman

Abstract

Microalgae-derived lipids have been established as potential sustainable sources of edible oils and bioactive compounds, and non-targeted C12-NMR spectroscopy is a reliable method for verifying their quality and authenticity. This chapter provides an in-depth investigation of the principles, applications, and recent advancements in non-targeted C12-NMR spectroscopy for the authentication of microalgae-derived edible lipids. It highlights the benefits, challenges, and future directions of the technique in this dynamic field, and emphasizes the application of Nuclear Magnetic Resonance (NMR) technology to the characterization of metabolites and lipids in microalgae, emphasizing its qualitative and quantitative capabilities. This chapter also introduces metabolomics and lipidomics within the context of NMR technology, outlines protocols for metabolite isolation and measurement in microalgae, and emphasizes their potential for resource recovery and sustainable value-added product development.

Key words Microalgae, Lipid, Authenticity, C12-NMR spectroscopy, Bioactive compounds

1 Introduction

Microalgae have become increasingly recognized as promising sources of edible lipids owing to their exceptional nutritional value and sustainability. Microalgae provide a supply of nutrient-rich edible lipids that contain essential minerals, vitamins, and proteins as well as a variety of essential fatty acids, including alpha-linoleic acid (ALA), eicosapentaenoic acid (EPA), docosahexaenoic acid (DHA), and omega-6 fatty acids. It has been noted that the authenticity and quality of microalgae-derived edible lipids have been a subject of concern due to instances of adulteration, mislabeling, and the use of low-quality lipid implants. Traditional analytical methods have limited ability to accurately assess the composition and origin of lipids, leading to the development of advanced techniques that are more effective than traditional methods [1–4]. The verification of edible lipids derived from microalgae

Mostafa Gouda et al. (eds.), *Plant Chemical Compositions and Bioactivities*, Methods and Protocols in Food Science,
https://doi.org/10.1007/978-1-0716-3938-2_3,

is crucial to guarantee the quality, safety, and compliance with regulatory standards of the product. Non-targeted C12-NMR spectroscopy has emerged as a powerful and comprehensive tool to analyze the composition and authenticity of edible microalgae-derived lipids. However, it is crucial to verify the authenticity and quality of these lipids in various foods and nutraceuticals. A C12-NMR technique provides valuable information about the composition of microalgae-derived lipids. The technique is non-destructive and comprehensive. The advancement of microalgal biotechnology has been a notable contributor to challenges since the mid-nineteenth century [5]. The diverse array of bioactive compounds produced by microalgae and their potential applications have led to the emergence of microalgal biotechnology, which facilitates the large-scale commercial production of microalgae. The economic potential of microalgae can be improved through several factors, including the cost-effective production of algal biomass feedstocks, the extraction of unique metabolites, and the utilization of saline water, such as seawater and groundwater, due to their increased nutrient content [6]. Aquatic microalgae are crucial components of the food chain and have the potential to increase aquaculture yields by affecting the causal dynamics of the food web. Many marine microalgae, including diatoms, flagellates, and dinoflagellates, produce remarkable secondary metabolites known as High-Value Molecules (HVMs) such as carotenoids, polyunsaturated fatty acids (PUFA), polysaccharides, and glycolipids. These microalgae have evolved to adapt to rapid changes in their living environment, leading to the production of valuable compounds [7–9]. Several challenges must be overcome to increase the demand for environmentally sustainable materials, including concerns regarding sustainability, biodegradability, environment, economics, and compatibility with biological systems [10, 11]. Microalgae offer an exceptional ability to provide ecological services as well as address sustainability concerns at the same time. Microalgae can establish biorefineries that are sustainable and generate multiple benefits while promoting biodiversity in the bio-economy [12].

The application of genetic engineering has resulted in the significant improvement of microalgae species, enhancing their competitiveness. Algal-omics technology has provided an extensive amount of information on the metabolic responses, growth, and functions of algal communities. This technology has also facilitated the development of more efficient bioremediation processes and a more comprehensive understanding of environmental toxicity hazards, making these processes more accessible to a wider range of communities [13, 14]. Microalgae produce biochemically active compounds with potential applications in the food, drug, and biofuel industries, which have been thoroughly reviewed [15–18]. Advancements in recent years have opened up the possibility of using microalgae to make sustainable biorefineries that can capture

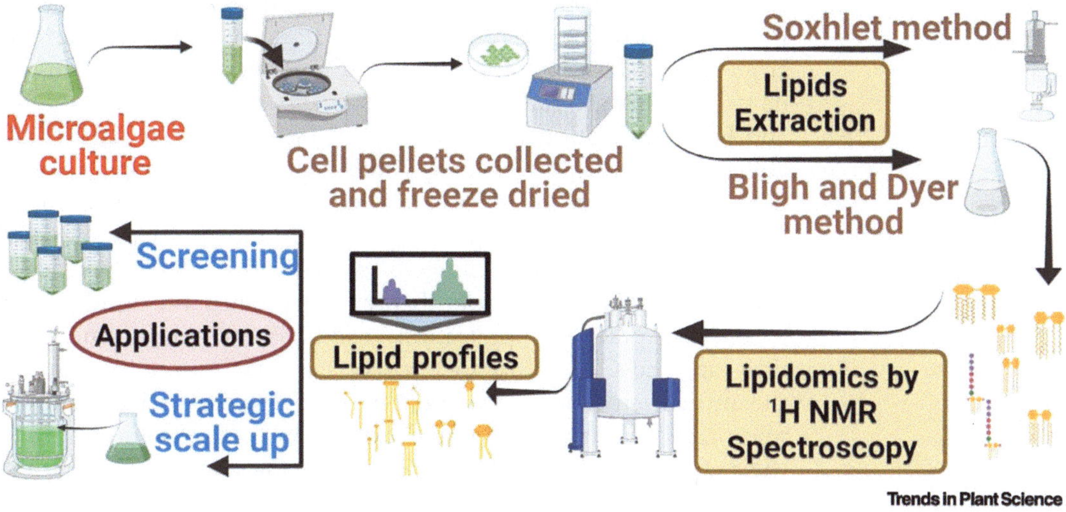

Fig. 1 Workflow of microalgal cultivation, lipid extraction, and lipid profiling via NMR [23]

and convert atmospheric carbon dioxide (CO_2) as well as transform waste materials into valuable products such as natural ingredients, health supplements, cosmetics, and personal care items [19–22]. NMR has been refined over the past two decades as a highly dependable and adaptable analytical method for the scrutiny of both liquid and solid materials. Its application in the assessment and examination of diverse substances has become prevalent. Early NMR setups had low resolution, but advancements in technology in the late twentieth century greatly increased the resolution of NMR in both liquid and solid samples. These advancements expanded its applications, such as identifying materials, understanding biological and chemical pathways, and analyzing dietary health. Figure 1 illustrates the process flow of microalgal cultivation, lipid extraction, and lipid profiling using NMR.

To accurately assess the metabolite composition, physical and chemical attributes, and functionality of various microalgae strains, scientists need reliable, non-invasive, and comprehensive analytical techniques. To maximize the potential of microalgae, the focus of metabolic measurements should shift towards resource recovery and the development of value-added products that demonstrate scalability and sustainability. The integration of NMR-based metabolomics and lipidomic techniques has increased the techno-economic and commercial feasibility of microalgal biorefineries. Over the past two decades, NMR has become one of the top three analytical methods used in metabolomics and lipidomics. Two additional methods that are commonly utilized are gas chromatography (GC), mass spectrometry (MS), and liquid chromatography-mass spectrometry (LC-MS). While these techniques have a wider detection range, up to 10–100 times higher than

NMR, there are specific instances where NMR is preferred due to its unique advantages [24]. NMR is preferred for long-term or large-scale clinical metabolomics and lipidomic studies because of its straightforward sampling process, ability to quantify metabolites and lipids, exceptional reproducibility, nondestructive nature, and ability to identify novel compounds. This spectroscopic method does not require any prior knowledge of the substances being studied, allowing for detailed examination of lipid profiles that can identify different microalgae species and detect impurities or pollutants [25]. This book chapter delves into the fundamental principles, practical applications, and recent advancements of non-targeted C12-NMR spectroscopy as a technique for ensuring the authenticity and intactness of lipids. This chapter also examines microalgal bioactive compound extraction methods and improves their usefulness in NMR-based metabolomics and lipidomic research, using a thorough and systematized approach.

2 Application of Microalgae-Derived Lipids

Microalgae are tiny photosynthetic organisms that inhabit various global environments and thrive primarily in aquatic environments [26]. Recently, microalgae have gained significant attention owing to their exceptional nutritional value and positive environmental impacts [27]. These single-celled microorganisms exhibit remarkable species diversity and have directly adapted to diverse ecological niches. Notably, microalgae are well known for their high lipid content, which makes them a valuable source of consumable lipids. These lipids, in the form of triacylglycerides (TAGs), are rich in essential fatty acids, particularly omega-3 and omega-6 [27]. These fatty acids play a crucial role in human health and influence various physiological functions such as cardiovascular well-being, brain development, and inflammation regulation. The significance of microalgae-derived consumable lipids lies in their sustainability and eco-friendly production. Unlike traditional sources such as fish oil (e.g., cod liver oil) for omega-3 fatty acids, microalgae can be deliberately cultivated under controlled conditions, relieving pressure on marine ecosystems [28]. Microalgae offer a consistent and reliable source of lipids, ensuring a stable supply for both the food and nutraceutical sectors. Microalgae-derived consumable fats have gained widespread use in food and nutraceutical industries because of their versatility and nutritional properties. Edible lipids from microalgae are used in the production of various functional foods, including dietary supplements, energy bars, and beverages [29]. They enhance the value of these products by providing essential fatty acids.

Microalgae-derived lipids also play a crucial role in the manufacture of nutraceuticals with potential health benefits. For example, omega-3 fatty acids have demonstrated the potential to reduce the risk of cardiovascular diseases and cognitive decline, making them popular dietary supplements [27]. Microalgae are essential for supplying lipids and other nutrients to farmed fish and shellfish in aquaculture feeds, promoting the growth and well-being of aquatic species, while contributing to sustainability by reducing reliance on wild fish stocks [29]. Additionally, they have applications in culinary uses, such as salad dressings, cooking oils, and condiments, thereby adding distinct flavors and essential nutrients to meals [30]. Microalgae-derived edible lipids are used to produce biofuels, bio-based plastics, and other sustainable materials, making them suitable for biotechnological processes [31].

3 Non-Targeted C12-NMR Spectroscopy: Principles and Methodology

3.1 Fundamentals of C12-NMR Spectroscopy

Non-targeted C12-NMR spectroscopy has a wide range of applications, including in the field of food and nutrition, where it is important to ensure the quality of edible lipids. This technique is useful for verifying the authenticity of the microalgae-derived lipids used in dietary supplements, functional foods, and other nutritional products. Using this method, researchers and industry professionals can identify adulteration or mislabeling, ultimately enhancing consumer trust and product integrity. Purcell and Bloch formulated the theory of nuclear magnetic resonance in 1946, which is essential for understanding the connection between the magnetic moments of atomic nuclei and their response to a magnetic field [32]. NMR is based on the principle that the nucleus of a positively charged atom behaves like a tiny bar magnet because of its spin on its own axis, creating a minuscule magnetic field. When no external magnetic field is present, nuclei remain unaligned in their ground state, resulting in no energy difference between them. As a result, no transitions occur between energy levels [33]. When a magnetic field external to a sample is present, the charged nuclei exhibit a unique motion known as processional motion, which is characterized by a particular frequency called the Larmor or Processional Frequency. This motion arises from the interaction between the nucleus's spin and the downward gravitational force, similar to gyroscopic motion. The specific Larmor frequency for each nucleus rises with the magnetic field's strength, which is a crucial factor in NMR measurements. Typically, the Larmor frequency falls within the range of 50–900 MHz [34].

NMR is an analytical technique used to characterize organic molecules by identifying their carbon-hydrogen frameworks through absorption of electromagnetic radio waves. The analysis of complex samples with edible lipids derived from microalgae is

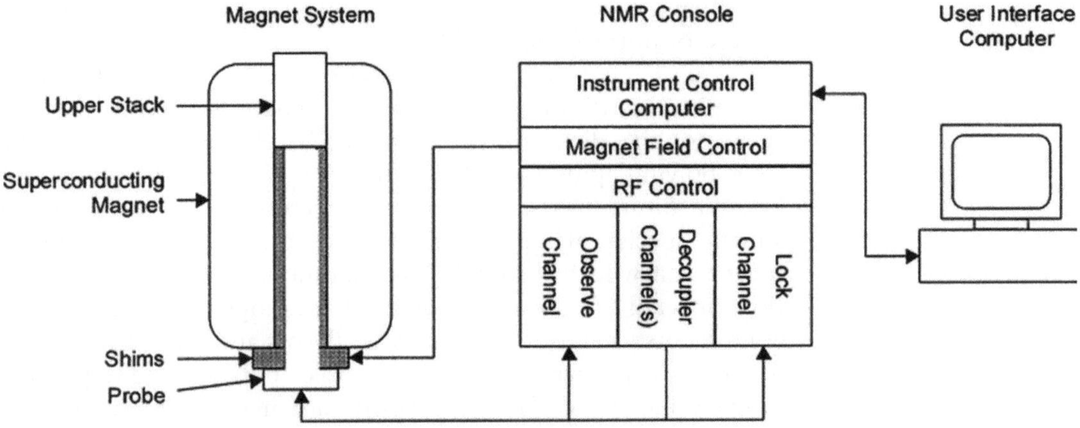

Fig. 2 Schematic representation of C12-NMR spectroscopy setup [37]

currently performed using non-targeted C12-NMR spectroscopy [35]. An illustration of the setup used in C12-NMR spectroscopy is provided in Fig. 2. NMR spectroscopy is dependent on the interaction between the magnetic field and the atomic nuclei. When atomic nuclei are subjected to a strong magnetic field, specific nuclei such as carbon-12 (C12) selectively absorb and emit electromagnetic radiation at distinct frequencies. The behavior of this phenomenon is influenced by the chemical surroundings of the nuclei, which enable the detection and measurement of various chemical compounds present in a specimen. In terms of the analysis of complex lipid mixtures in microalgae, non-targeted C12-NMR spectroscopy proves to be a beneficial technique. Unlike other methods, this method does not rely on prior knowledge of the specific compounds within a sample. Instead, it captures the entire NMR spectrum and records signals from all carbon nuclei present in the sample [36]. This approach allows for simultaneous analysis of multiple compounds, making it highly suitable for the analysis of complex lipid mixtures in microalgae.

3.2 Adaptation for Microalgae Lipid Analysis

To analyze microalgae-derived edible lipids using C12-NMR spectroscopy, three key steps must be followed such as sample preparation, data acquisition, and data processing and analysis. To obtain the desired lipids for analysis, microalgae samples must be prepared using techniques such as solvent extraction, the Bligh and Dyer method, or solid-phase extraction. The resulting lipid extract was then dissolved in a suitable NMR solvent to ensure lipid solubility and obtain accurate spectral data [38]. C12-NMR spectroscopy involves subjecting a sample to a strong magnetic field and radiofrequency pulses. During data collection, the carbon nuclei in the sample absorbed energy from these pulses and released them as NMR signals, which were then detected and recorded as a spectrum, presented as a graph showing chemical shifts on the *x*-axis and signal intensity on the *y*-axis. The process of analyzing the acquired NMR data involved correcting the baseline, phase, and integrating peaks. Software tools are commonly used to interpret and distinguish individual peaks in a spectrum that represent various lipid components [39].

3.3 Parameters and Considerations

Several parameters and considerations are crucial for the success of non-targeted C12-NMR spectroscopy in analyzing microalgae-derived edible lipids [35]. The quality of the NMR spectrum depends on its resolution, which determines the extent to which the individual peaks can be resolved. It is essential to have high spectral resolution to detect and distinguish individual lipid compounds in complex mixtures. A better sensitivity of the NMR apparatus is necessary, particularly when studying diluted samples with limited lipid components. High sensitivity enables researchers to detect even small amounts of lipid components, which is vital for understanding their characteristics and their potential impact. The consistency of the samples was crucial for obtaining reliable NMR data. If the samples are inconsistent, the spectra can be distorted, potentially leading to incorrect conclusions. Researchers have compared acquired spectra with existing databases to enhance compound identification. These databases contain reference spectra for known lipids, making it easier to identify the specific components. This approach leverages existing knowledge and spectral properties to identify compounds. Non-targeted C12-NMR spectroscopy provides comprehensive analysis of lipids from microalgae used in food. Accurate results require proper sample preparation, data collection, and the consideration of important factors (Fig. 3).

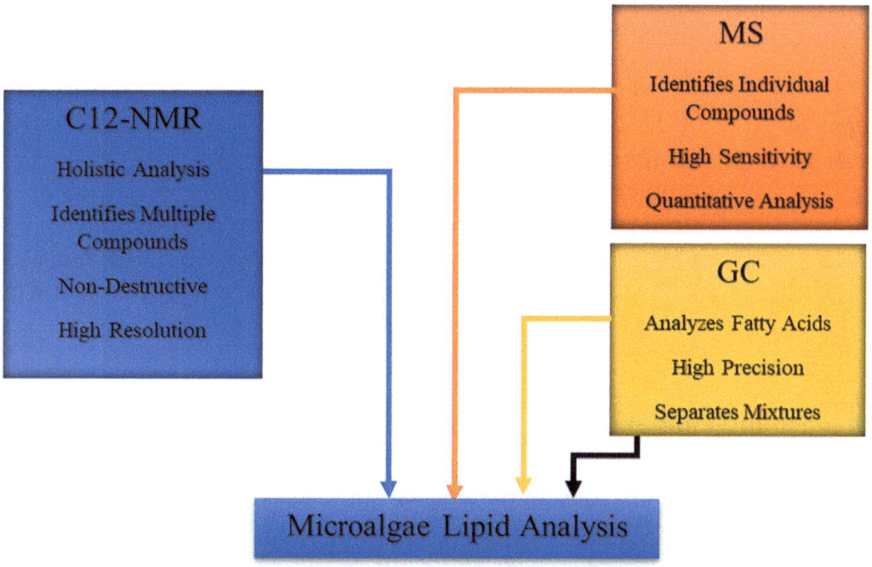

Fig. 3 Comparison of C12-NMR with other analytical techniques

4 Authentication of Microalgae Edible Lipids

Authentication is of utmost importance for maintaining the quality and genuineness of microalgae-sourced edible lipids. Using non-targeted C12-NMR spectroscopy, information about the lipid composition can be obtained. This section describes the process of verifying the authenticity of these lipids by C12-NMR spectroscopy.

4.1 Total Lipid Content Determination

NMR spectroscopy without targeting, specifically C12-NMR, can provide highly precise measurements of the lipid content in microalgae samples. This enables the accurate determination of the amount of lipids present and assessment of the nutritional value of edible lipids derived from microalgae [40]. Consequently, product labeling can be achieved with better accuracy.

4.2 Identification of Lipid Classes

NMR spectroscopy is a valuable tool for identifying the lipid components present in microalgae samples. The distinct NMR spectra of different types of lipids, such as TAGs, phospholipids, and sterols [40], allow for accurate determination of the composition of lipid classifications. By analyzing specific spectrum peaks, it is possible to ensure the accuracy of the advertised specifications of the microalgae.

4.3 Quantification of Fatty Acids

NMR is an effective technique for analyzing fatty acids present in microalgal lipids. By examining the NMR spectrum, which provides information about the peaks and chemical movements of fatty acids, it is possible to identify the specific components of these fatty acids, including omega-3 and omega-6. This information is essential for ascertaining the makeup of fatty acids and adhering to nutritional recommendations and regulatory specifications for microalgae-sourced lipid products [40].

4.4 Detection of Contaminants or Adulterants

C12-NMR spectroscopy plays a crucial role in ensuring the authenticity of edible microalgae-derived lipids by verifying their quality and origins. This technique helps identify foreign substances, maintain product quality, and enhance consumer trust. It is also used to determine lipid content, identify lipid types, measure fatty acids, and detect additives or contaminants, thereby meeting nutritional and safety standards [25] (Table 1).

5 Advantages and Limitations of C12-NMR Spectroscopy

Non-targeted C12-NMR spectroscopy is an analytical technique with several advantages. However, this also presents challenges and limitations. The following section explores both aspects to provide

Table 1
Summary of C12-NMR applications in microalgae lipid authentication

Application	Description
Total lipid content determination	Quantification of the total lipid content in microalgae samples.
Identification of lipid classes	Identification and quantification of different lipid classes.
Quantification of fatty acids	Precise quantification of individual fatty acids within lipids.
Detection of contaminants or adulterants	Identification of contaminants or adulterants in microalgae lipid samples.
Species identification	Differentiation and identification of microalgae species based on lipid profiles.
Comparisons with other analytical techniques	Evaluation of C12-NMR spectroscopy against other analytical methods like MS and GC.
Insights into microalgae lipid composition	Examination of structural features and lipid class distribution in microalgae lipids.
Quality control	Routine quality control in microalgae lipid production processes.

an informed perspective on the potential for the authentication of microalgae-derived edible lipids. This section examines the benefits of C12-NMR spectroscopy by discussing its noninvasive properties, rapid analysis time, and capacity to simultaneously analyze multiple samples. Furthermore, it considers the limitations of the technique, including the potential for spectral overlap, difficulties in data interpretation, and requirement for careful sample preparation. The benefits of using the C12-NMR spectroscopic technique are as follows:

- Analysis: C12-NMR spectroscopy provides a comprehensive view of lipid composition, enabling the simultaneous analysis of multiple components, making it an ideal method for determining the lipid profile of complex mixtures, such as microalgae-derived edible lipids.
- Nondestructive NMR spectroscopy, specifically C12-NMR, is a noninvasive method that allows sample preservation for any additional analysis or quality control checks that may be required [35].
- High-resolution C12-NMR spectroscopy can provide a high spectral resolution, allowing the discrimination of individual lipid classes and fatty acid groups [25].
- Quantitative information: NMR spectroscopy in C12 enables the determination of total lipid content and individual fatty acids, which aids in the accurate identification and nutritional evaluation of these compounds [40].

- Identification capability: C12-NMR can be used to determine lipid classes and their relative amounts in microalgae-derived edible lipids, which aids their characterization [25].
- Contaminant detection: NMR spectroscopy can detect contaminants or adulterants in microalgae-derived edible lipid samples, thereby improving product safety and authenticity [25].
- Sustainability: C12-NMR spectroscopy is an environmentally friendly and sustainable option, with the exception of some analytical techniques that require time-consuming sample preparation or dangerous chemicals [40].

6 NMR-Based Metabolomics Techniques for Microalgae

The conventional approach for investigating the chemistry of an organism involves the use of natural product chemistry, which involves a multistep process for identifying and characterizing metabolite structures [41]. The previous method is both tedious and often inadequate, because it primarily focuses on the major compounds present. Metabolomics is a more comprehensive approach that can identify a large number of metabolites from a small sample in a short period of time. This technique has multiple applications in fields, such as medicine, toxicology, nutrition, and various biological research areas, making it a versatile tool. Since its introduction in the post-genomic era, metabolomics has been widely adopted across multiple research domains [42]. The field of metabolomics is highly regarded for its ability to detect and quantify low-molecular-weight metabolites, provide valuable information about the end products of cellular regulation, and provide comprehensive insights into biological samples. NMR-based metabolomics is a powerful technique that can quickly provide valuable insights for functional genetic annotation and assessing the effects of environmental factors on organisms, even in the absence of prior knowledge [43]. Metabolites can act as indicators of natural selection during the evolution of algae and other living organisms, and they can also contribute to the development of chemical diversity [44]. The synthesis of these complex metabolites is intricately linked to associated microbiota. Metabolomics constitutes the final stage of the omics continuum, encompassing a comprehensive array of metabolites including amino acids, sugars, organic acids, and nucleotides [45, 46].

NMR-based metabolomics is regarded for its capacity to detect unbiased metabolites, determine their structures, and achieve a high level of reproducibility in quantification. This method is particularly effective for precisely quantifying a large number of metabolites and provides a broader range of time and dynamic measurements than GC/LC-MS. Although its measurements are often semi-quantitative, especially in untargeted MS-based metabolomics, the accuracy and consistency of NMR-based metabolomics research are widely recognized [42]. In a study conducted by Azizan et al. [47], the utilization of various solvent extracts of the diatom *Chaetoceros calcitrans* was examined using NMR spectroscopy to generate dynamic spectra of metabolites. A total of 29 metabolites were successfully isolated using five different solvents: acetone, chloroform, methanol, 70% ethanol, and hexane. Among these solvents, the chloroform extract demonstrated the most promising antioxidant activity by effectively neutralizing the DPPH radical and inhibiting the activity of other antioxidant compounds such as fucoxanthin, astaxanthin, violaxanthin, zeaxanthin, canthaxanthin, and lutein. These findings suggest the potential application of the chloroform extract as a valuable functional ingredient in the mariculture industry. Bustamam et al. [48] analyzed the metabolites present in *Isochrysis galbana* using solvents with varying polarities, including hexane, ethyl acetate, absolute ethanol, 50% ethanol, and an aqueous solution, using 1D 500 MHz 1HNMR spectra. A total of 21 metabolites were tentatively identified, including amino acids, carbohydrates, fatty acids, carotenoids, and chlorophyll, along with their respective chemical shift assignments, multiplicities, and coupling constant values. The aqueous extract yielded the highest amount of metabolites, while the hexane extract produced the lowest yield, suggesting that *Isochrysis galbana* is primarily composed of polar metabolites and contains relatively few nonpolar or less polar compounds. Using NMR fingerprinting, it was revealed that ethyl acetate contained the highest concentrations of fatty acids and carotenoids, while aqueous extract contained the highest concentrations of amino acids and carbohydrates. Therefore, both ethyl acetate and aqueous extracts were found to be suitable for the extraction of a wide range of compounds from these microalgae.

7 NMR-Based Lipidomics Strategies for Microalgae

Lipidomics is the comprehensive analysis of the structure and function of all lipids within a cell or organism, including their interactions with proteins and metabolites. This field of study has helped improve our understanding of health and disease by systematically quantifying the lipidome, exploring lipid metabolism kinetics, and investigating interactions with cellular proteins [49]. Microalgae are increasingly used as a source of lipids for various industrial applications because of their high lipid content [50, 51]. Rapid growth of lipid-rich microalgae on uncultivable land is crucial in this context [52]. Algal lipid composition is determined by a combination of genetic and phenotypic factors that are affected by environmental conditions and the surrounding area [53]. The utilization of marine water as a medium for microalgae cultivation presents several advantages, including elevated lipid content, enhanced lipid extraction, heightened resistance to pollution, and improved sedimentation efficiency, which can be further enhanced through floc formation and the attainment of larger cell sizes. These advantages render downstream processing of economically viable microalgae [54].

Since their potential as a viable source of third-generation biofuels, including crude oil in the form of triacylglycerols or methyl esters (biodiesel), microalgal fatty acids have seen a marked increase in interest, as well as their application in nutritional health products [55]. Certain microalgal species have a predominance of esterified lipids and are abundant in valuable PUFAs. Microalgae are particularly rich in glycolipids, such as monogalactosyl diacylglycerols (MGDGs), digalactosyl diacylglycerols (DGDGs), and sulfoquinovosyl diacylglycerols (SQDGs), which are rich in PUFAs. These glycolipids from microalgae are rich in n-3 fatty acids that have various therapeutic properties in humans, such as antifungal, antiviral, antitumor, anti-inflammatory, and antimicrobial properties [56]. Microalgae characterization, which is a comprehensive process, is hindered by the wide range of lipids present in the lipidome. Microalgae research has been limited by the intricate nature of their lipidome. This challenge has been addressed by the development of NMR-based methods as a modern technology, which offers new avenues for understanding lipidomics in an efficient manner [57, 58].

8 Challenges and Limitations

This section highlights recent advancements in non-targeted C12-NMR spectroscopy techniques, particularly the integration of chemometric approaches and high-throughput screening methods. Non-targeted C12-NMR spectroscopy is a valuable tool for verifying microalgae-derived edible lipids, offering numerous advantages, including the provision of detailed analysis in a non-invasive manner, the capacity to provide precise and quantitative information, and high-resolution capabilities. The utilization of this technique ensures accurate and reliable results. NMR spectroscopy presents both advantages and challenges in the analysis of microalgae-derived edible lipids. While MS provides highly sensitive and compound-specific information, NMR spectroscopy offers a more comprehensive picture of lipid profiles, enabling the identification of unknown compounds and unexpected contaminants. However, the cost of equipment, limitations on sample size, difficulties in data interpretation, and sensitivity issues must be considered. To address these challenges, it is important to carefully evaluate the specific needs of each analysis and select the most appropriate technique. In this study, NMR spectroscopy was compared with MS and GC for the analysis of microalgae-derived edible lipids, and NMR spectroscopy was found to be a valuable tool for the identification of unknown compounds and unexpected contaminants [40]. In a comparison of NMR spectroscopy and GC for the quantification of fatty acids in microalgae-derived edible lipids, NMR spectroscopy was found to be the superior option as it provides both structural and compositional information on various lipid classes while maintaining high levels of accuracy. NMR spectroscopy is a crucial and versatile tool in the realm of advanced analytical techniques and knowledge platforms, offering significant benefits in the fields of metabolomics and lipidomics. One of the notable benefits of NMR techniques is their capability to identify novel compounds without the requirement for chemical modification. This attribute makes NMR techniques highly suitable for the rapid screening of microalgae strains, as they provide crucial metabolite and lipid profiles that are difficult to obtain through GC-MS and LC-MS methods.

It is imperative to recognize that NMR possesses the capability to detect and quantify metabolites at exceptionally low concentrations, as low as 1 μM. However, it should be acknowledged that this capacity allows for the identification of up to 20 distinct metabolites in certain tissues (in vivo) and up to 100 metabolites in specific bio-fluids (in vitro). Notably, the quantity of discernible metabolites rises by an order of magnitude in the milieu of the metabolome spanning diverse biofluids and tissues. As a result, a noteworthy future challenge looms in devising strategies to further improve NMR's ability to address this matter [59]. The challenges and limitations associated with the technique developed in [60] are outlined as follows:

- *Equipment cost*: High-quality NMR spectrometers are expensive, making them inaccessible to smaller laboratories or research groups with limited budgets.
- *Sample size requirements*: Accurate NMR analysis often requires a substantial sample volume and concentration, which can be a constraint when working with limited or precious samples.
- *Complex data interpretation*: Interpreting NMR spectra can be intricate and requires expertise in both NMR spectroscopy and lipid chemistry, making it a technique that may require specialized knowledge.
- *Sensitivity issues*: C12-NMR spectroscopy may have limitations in detecting trace or low-abundance compounds within highly complex mixtures, which can reduce its utility in such scenarios.
- *Time-consuming data processing*: Data collection and processing for NMR analysis can be time-consuming, which may not be ideal when a more expeditious analysis is needed.

- *Limited quantification range*: The technique may struggle to accurately quantify compounds with extremely high or low concentrations, limiting its applicability in certain contexts.
- *Availability of reference data*: Reference data for specific microalgae species or lipid classes may be limited or unavailable, which can restrict the scope of the analysis.

9 Future Developments and Trends

Technology advancements and emerging trends are continually driving the field of microalgae-derived edible lipid analysis. It is anticipated that non-targeted C12-NMR spectroscopy will have a significant impact on the future of this field. Ideally, future developments should focus on improving the sensitivity of C12-NMR spectroscopy, which will enable the detection of trace amounts of lipids and minor components in microalgal samples. The utilization of microalgae-sourced edible algae in various industries such as pharmaceuticals and biotechnology can be enhanced by adopting a formal tone. Advancements in automation and data collection methods have facilitated the conduct of comprehensive analyses of microalgae-derived edible lipid samples, which has contributed to the improvement of quality control and the scaling up of production processes. Moreover, advancements in the development of smaller NMR instruments have made C12-NMR spectroscopy more accessible to a wider range of users, including smaller research laboratories and on-site testing in production facilities, as well as combining C12-NMR spectroscopy with other analytical methods, such as MS and GC [61]. Utilizing metabolomics techniques, we can improve our comprehension of the metabolic pathways involved in microalgal production of edible lipids and their reactions to environmental factors [35]. The acquisition of this knowledge is essential for enhancing lipid yield in microalgae cultivation through the utilization of advanced data analysis and bioinformatics tools to manage the extensive spectral data generated by C12-NMR spectroscopy [61]. Machine learning and data mining can aid the identification of patterns and lipid compounds.

Microalgae lipid profiles for specific applications are increasingly being used using C12-NMR spectroscopy to develop and monitor strains of algae with the desired lipid composition. Microalgae are widely used for environmental applications such as wastewater treatment and carbon capture, where C12-NMR spectroscopy can be used to assess changes in lipid composition in response to environmental conditions and contaminants. Microalgal lipids are increasingly used in food and nutraceutical products, and there is a need for standardized analytical methods, including C12-NMR spectroscopy, to ensure product quality and labeling accuracy. Microalgae biorefineries are showing promise as a sustainable means of converting microalgae into various valuable products, and C12-NMR spectroscopy is proving to be a valuable tool in optimizing the extraction and conversion processes within these biorefineries. The use of non-targeted C12-NMR spectroscopy for the analysis of microalgal lipids is expected to undergo significant advancements in the future, leading to increased integration with other analytical methods and the development of new applications in various industries. These advancements will not only enhance the authenticity and quality control of microalgae-derived edible lipid products, but also contribute to the wider utilization of microalgae as a resource.

10 Conclusions

In conclusion, non-targeted C12-NMR spectroscopy plays a critical role in the authentication of microalgae-derived edible lipids due to its ability to offer non-destructive, high-resolution analyses that guarantee the quality, safety, and sustainability of microalgae-derived lipid products. As technology advances and converges with other analytical methods, C12-NMR spectroscopy will continue to be a leading tool in the analysis of microalgae-derived lipids, contributing to the growth and success of the microalgae industry in providing sustainable and nutritious food sources for the future. This book chapter summarizes the key findings, implications, and vital role of C12-NMR spectroscopy in this field. Microalgae are a vital source of edible lipids rich in essential fatty acids and offer sustainable alternatives to traditional lipid sources. Non-targeted C12-NMR spectroscopy provides an approach for microalgae-derived edible lipid analysis, enabling the determination of the total lipid content, identification of lipid classes, quantification of fatty acids, and detection of contaminants or adulterants. Although C12-NMR spectroscopy presents challenges related to equipment cost, sample size, data interpretation, and sensitivity, it can enhance traditional techniques such as MS and GC, providing unique insights into microalgal lipid profiles that are not possible with these methods alone.

Acknowledgment

The authors extend their gratitude to Khalifa University, Abu Dhabi, United Arab Emirates, for the support received through ASPIRE, the Technology Program Management pillar of Abu Dhabi's Advanced Technology Research Council (ATRC) under the ASPIRE Research Institute program (VRI-FS grant 8434000459).

References

1. Kharbach M, Alaoui Mansouri M, Taabouz M, Yu H (2023) Current application of advancing spectroscopy techniques in food analysis: data handling with chemometric approaches. Foods 12:2753
2. Ahmad A, Ashraf SS (2023) Sustainable food and feed sources from microalgae: food security and the circular bioeconomy. Algal Res 74: 103185
3. Ahmad A, Banat F, Taher H (2022) Chapter 1 – Algal engineering for bioremediation, bioenergy production, and biomedical applications. In: Ahmad A, Banat F, Taher H (eds) Algal biotechnology. Elsevier, pp 3–32
4. Ahmad A, Shah SMU, Othman MF, Abdullah MA (2014) Enhanced palm oil mill effluent treatment and biomethane production by co-digestion of oil palm empty fruit bunches with Chlorella Sp. Can J Chem Eng 92:1636–1642
5. Hamed I (2016) The evolution and versatility of microalgal biotechnology: a review. Compr Rev Food Sci Food Safety 15:1104–1123
6. Arora N, Kumari P, Kumar A, Gangwar R, Gulati K, Pruthi PA, Prasad R, Kumar D, Pruthi V, Poluri KM (2019) Delineating the molecular responses of a halotolerant microalga using integrated omics approach to identify genetic engineering targets for enhanced TAG production. Biotechnol Biofuels 12:2
7. Bule MH, Ahmed I, Maqbool F, Bilal M, Iqbal HMN (2018) Microalgae as a source of high-value bioactive compounds. Front Biosci (Scholar edition) 10:197–216
8. de Morais MG, Vaz BdS, de Morais EG, Costa JAV (2015) Biologically active metabolites synthesized by microalgae. BioMed Res Int 2015:835761
9. Ahmad A, Hassan SW, Banat F (2022) An overview of microalgae biomass as a sustainable aquaculture feed ingredient: food security and circular economy. Bioengineered 13:9521–9547
10. Bisht B, Lohani UC, Kumar V, Gururani P, Sinhmar R (2022) Edible hydrocolloids as sustainable substitute for non-biodegradable materials. Crit Rev Food Sci Nutr 62:693–725
11. Ahmad A, Banat F, Alsafar H, Hasan SW (2023) Recent breakthroughs in integrated biomolecular and biotechnological approaches for enhanced lipid and carotenoid production from microalgae. Phytochem Rev 22:993–1013
12. Ahmad A, Buang A, Bhat AH (2016) Renewable and sustainable bioenergy production from microalgal co-cultivation with palm oil mill effluent (POME): a review. Renew Sustain Energy Rev 65:214–234
13. Fabris M, Abbriano RM, Pernice M, Sutherland DL, Commault AS, Hall CC, Labeeuw L, McCauley JI, Kuzhiuparambil U, Ray P, Kahlke T, Ralph PJ (2020) Emerging technologies in algal biotechnology: toward the establishment of a sustainable, algae-based bioeconomy. Front Plant Sci 11:279
14. Kumar V, Jaiswal KK, Vlaskin MS, Nanda M, Tripathi MK, Gururani P, Kumar S, Joshi HC (2022) Hydrothermal liquefaction of municipal wastewater sludge and nutrient recovery from the aqueous phase. Biofuels 13:657–662
15. Cardozo KHM, Marques LG, Carvalho VM, Carignan MO, Pinto E, Marinho-Soriano E, Colepicolo P (2011) Analyses of photoprotective compounds in red algae from the Brazilian coast. Rev Bras Farmacogn 21
16. Ahmad A, Banat F, Alsafar H, Hasan SW (2022) Algae biotechnology for industrial wastewater treatment, bioenergy production, and high-value bioproducts. Sci Total Environ 806:150585
17. Ashfaq A, Bhat A, Azizul B (2017) Immobilized Chlorella vulgaris for efficient palm oil mill effluent treatment and heavy metals removal. Desalination Water Treat 81:105–117
18. Ashfaq A, Syed Muhammad US, Usman S, Mohd Azmuddin A (2016) Anaerobic co-cultivation of multi-algal species with oil palm empty fruit bunches for mill effluent treatment and biomethane production
19. Pradhan B, Nayak R, Patra S, Jit BP, Ragusa A, Jena M (2020) Bioactive metabolites from marine algae as potent pharmacophores against oxidative stress-associated human diseases: a comprehensive review. Molecules (Basel, Switzerland) 26
20. Ahmad A, Bhat AH, Buang A (2018) Biosorption of transition metals by freely suspended and Ca-alginate immobilised with Chlorella vulgaris: kinetic and equilibrium modeling. J Cleaner Prod 171:1361–1375
21. Ahmad A, Shah SMU, Othman MF, Abdullah MA (2014) Biomethane production and palm oil mill effluent treatment by co-cultivation of Nannochloropsis oculata. Appl Mech Mater 625:818–821
22. Ahmad A, Shah SMU, Buang A, Othman MF, Abdullah MA (2014) Evaluation of aerobic and anaerobic co-digestion of Tetraselmis suecica and oil palm empty fruit bunches by response

surface methodology. Adv Mater Res 925:243–247

23. Nanda M, Kumar V, Arora N, Vlaskin MS, Tripathi MK (2021) (1)H NMR-based metabolomics and lipidomics of microalgae. Trends Plant Sci 26:984–985
24. Emwas AH, Roy R, McKay RT, Tenori L, Saccenti E, Gowda GAN, Raftery D, Alahmari F, Jaremko L, Jaremko M, Wishart DS (2019) NMR spectroscopy for metabolomics research. Metabolites 9
25. Alexandri E, Ahmed R, Siddiqui H, Choudhary MI, Tsiafoulis CG, Gerothanassis IP (2017) High resolution nmr spectroscopy as a structural and analytical tool for unsaturated lipids in solution. Molecules (Basel, Switzerland) 22
26. Ng H-S, Chew L-L (2020) Valuable compounds produced by microalgae. In: Bisaria V (ed) Handbook of biorefinery research and technology. Springer Netherlands, Dordrecht, pp 1–23
27. Katiyar R, Arora A (2020) Health promoting functional lipids from microalgae pool: a review. Algal Res 46:101800
28. Şuheda R, Sevcan A (2023) Microalgae as a promising candidate for fighting climate change and biodiversity loss. In: Sevcan A (ed) Microalgae – current and potential applications. IntechOpen, Rijeka. pp. Ch. 0
29. Ampofo J, Abbey L (2022) Microalgae: bioactive composition, health benefits, safety and prospects as potential high-value ingredients for the functional food industry. Foods 11
30. Zhou J, Wang M, Saraiva JA, Martins AP, Pinto CA, Prieto MA, Simal-Gandara J, Cao H, Xiao J, Barba FJ (2022) Extraction of lipids from microalgae using classical and innovative approaches. Food Chem 384:132236
31. Ahmad A, Shah SMU, Othman MF, Abdullah MA (2015) Aerobic and anaerobic co-cultivation of Nannochloropsis oculata with oil palm empty fruit bunch for enhanced biomethane production and palm oil mill effluent treatment. Desalination Water Treat 56: 2055–2065
32. Jones C, Mulloy B (1993) Introduction to nuclear magnetic resonance. In: Jones C, Mulloy B, Thomas AH (eds) Spectroscopic methods and analyses: NMR, mass spectrometry, and metalloprotein techniques. Humana Press, Totowa, pp 1–14
33. Hanson LG (2008) Is quantum mechanics necessary for understanding magnetic resonance? Concepts Magn Reson Part A 32A:329–340
34. Bisht B, Kumar V, Gururani P, Tomar MS, Nanda M, Vlaskin MS, Kumar S, Kurbatova A (2021) The potential of nuclear magnetic resonance (NMR) in metabolomics and lipidomics of microalgae- a review. Arch Biochem Biophys 710:108987
35. Eltemur D, Robatscher P, Oberhuber M, Scampicchio M, Ceccon A (2023) Applications of solution NMR spectroscopy in quality assessment and authentication of bovine milk. Foods 12:3240
36. Mukesh Kumar Singh AS (2022) Nuclear magnetic resonance spectroscopy. Woodhead Publishing, pp 321–339
37. Williams EA (2001) Nuclear magnetic resonance spectrometry. In: Buschow KHJ, Cahn RW, Flemings MC, Ilschner B, Kramer EJ, Mahajan S, Veyssière P (eds) Encyclopedia of materials: science and technology. Elsevier, Oxford, pp 6319–6324
38. Ranjith Kumar R, Hanumantha Rao P, Arumugam M (2015) Lipid extraction methods from microalgae: a comprehensive review. Front Energy Res 2
39. Ulrike Holzgrabe IW, Diehl B (2008) Principles in NMR spectroscopy, NMR spectroscopy in pharmaceutical analysis. Elsevier
40. Khoury S, Canlet C, Lacroix MZ, Berdeaux O, Jouhet J, Bertrand-Michel J (2018) Quantification of lipids: model, reality, and compromise. Biomolecules 8
41. Robinette SL, Brüschweiler R, Schroeder FC, Edison AS (2012) NMR in metabolomics and natural products research: two sides of the same coin. Acc Chem Res 45:288–297
42. Tian H, Lam SM, Shui G (2016) Metabolomics, a powerful tool for agricultural research. Int J Mol Sci 17
43. Saito K, Matsuda F (2010) Metabolomics for functional genomics, systems biology, and biotechnology. Annu Rev Plant Biol 61:463–489
44. Wink M (2003) Evolution of secondary metabolites from an ecological and molecular phylogenetic perspective. Phytochemistry 64:3–19
45. Roessner U, Bowne J (2009) What is metabolomics all about? BioTechniques 46:363–365
46. Dettmer K, Hammock BD (2004) Metabolomics – a new exciting field within the "omics": sciences. Environ Health Perspect 112:A396–A397
47. Azizan A, Ahamad Bustamam MS, Maulidiani M, Shaari K, Ismail IS, Nagao N, Abas F (2018) Metabolite profiling of the microalgal diatom chaetoceros calcitrans and correlation with antioxidant and nitric oxide inhibitory activities via 1H NMR-based metabolomics. Mar Drugs 16:154
48. Bustamam MSA, Pantami HA, Azizan A, Shaari K, Min CC, Abas F, Nagao N,

Maulidiani M, Banerjee S, Sulaiman F, Ismail IS (2021) Complementary analytical platforms of NMR spectroscopy and LCMS analysis in the metabolite profiling of Isochrysis galbana. Mar Drugs 19

49. Li J, Vosegaard T, Guo Z (2017) Applications of nuclear magnetic resonance in lipid analyses: an emerging powerful tool for lipidomics studies. Prog Lipid Res 68:37–56
50. Ibañez E, Cifuentes A (2013) Benefits of using algae as natural sources of functional ingredients. J Sci Food Agric 93:703–709
51. Pulz O, Gross W (2004) Valuable products from biotechnology of microalgae. Appl Microbiol Biotechnol 65:635–648
52. Eppink MHM, Olivieri G, Reith H, van den Berg C, Barbosa MJ, Wijffels RH (2019) From current algae products to future biorefinery practices: a review. Adv Biochem Eng Biotechnol 166:99–123
53. De Luca M, Pappalardo I, Limongi AR, Viviano E, Radice RP, Todisco S, Martelli G, Infantino V, Vassallo A (2021) Lipids from microalgae for cosmetic applications. Cosmetics 8:52
54. Pei H, Jiang L (2018) Mixing seawater with a little wastewater to produce bioenergy from limnetic algae. Trends Biotechnol 36:480–483
55. Nuzzo G, Gallo C, d'Ippolito G, Cutignano A, Sardo A, Fontana A (2013) Composition and quantitation of microalgal lipids by ERETIC ^{1}H NMR method. Mar Drugs 11:3742–3753
56. da Costa E, Silva J, Mendonça SH, Abreu MH, Domingues MR (2016) Lipidomic approaches towards deciphering glycolipids from microalgae as a reservoir of bioactive lipids. Mar Drugs 14
57. Stengel DB, Connan S, Popper ZA (2011) Algal chemodiversity and bioactivity: sources of natural variability and implications for commercial application. Biotechnol Adv 29:483–501
58. Kim SH, Liu KH, Lee SY, Hong SJ, Cho BK, Lee H, Lee CG, Choi HK (2013) Effects of light intensity and nitrogen starvation on glycerolipid, glycerophospholipid, and carotenoid composition in Dunaliella tertiolecta culture. PloS One 8:e72415
59. Wishart DS (2008) Quantitative metabolomics using NMR, TrAC. Trends Anal Chem 27: 228–237
60. Claridge TDW (1999) Chapter 3 Practical aspects of high-resolution NMR. In: Claridge TDW (ed) Tetrahedron organic chemistry series. Elsevier, pp 45–110
61. Luchinat E, Banci L (2023) In-cell NMR: recent progresses and future challenges. Rendiconti Lincei Scienze Fisiche e Naturali 34:653–661

Chapter 4

Raman Spectroscopy for Tracking the Polysaccharides Chemical Composition and β-Carotene Accumulation of Microalgae Food Source

Yulun Dong, Xiaoli Li, and Mostafa Gouda

Abstract

Raman spectroscopy is a detection method that combines Raman spectroscopy and confocal microanalysis. It can provide chemical fingerprint information of cells, tissues or body fluids, and has the advantages of high sensitivity, non-invasive, no labeling, and little water interference. As a primitive ancient organism, microalgae have evolved to adapt to seawater, fresh water, land, and other different species, with rapid growth and wide range of physiological characteristics. Most microalgae can produce many bioactive substances, such as fats, polysaccharides, proteins, carotenoids, etc., which have important development value in medicine, energy, water purification, and other fields. At present, it has been widely cultivated and developed. Therefore, it is necessary to study the accumulation process of high-value add-ons in microalgae cells. This protocol proposes a method for tracking microalgae polysaccharides and β-carotenoids based on confocal microRaman spectroscopy, and visualizes microalgae cells. It provides a boost to promote the accumulation of high value-added substances in microalgae

Key words Raman, Microalgae, Polysaccharides, β-Carotene

1 Introduction

Microalgae are a kind of microscopic algae population, widely distributed, commonly found in rivers, lakes, seas, ponds, and ditches [1, 2]. Among the more than 30,000 species of algae known in the world, microalgae group accounts for about 70% [3]. Marine microalgae convert solar energy into new organic matter such as oil and protein through photosynthesis, use nutrients to transmit information and release biomass, and then start the food chain in the ocean. In some sea areas, the primary productivity contribution value can even reach more than 80%, and it is the most important primary producer and energy converter in the marine ecosystem [4]. And state changes will provide insight into the links between the marine environment and ecology. Therefore, the

Mostafa Gouda et al. (eds.), *Plant Chemical Compositions and Bioactivities*, Methods and Protocols in Food Science,
https://doi.org/10.1007/978-1-0716-3938-2_4,

methodological research of algal microorganisms is of great importance. Raman spectroscopy, a widely used detection method in the field of life, can complete the analysis of cells in vivo without labeling. In particular, the popular micro-confocal Raman spectroscopy in recent years has special advantages in terms of spatial resolution and chromatographic analysis, and is regarded as an ideal means for single-cell detection [5].

Raman spectra are the scattered signals produced by molecules caused by incident light [2, 6]. By analyzing the Raman spectral intensity, the information about molecular vibration and rotation can be obtained, so that the chemical composition and molecular structure can be further analyzed [5, 7]. When the light is irradiated to the substance, elastic scattering and inelastic scattering occur, the scattered light of elastic scattering has the same wavelength as the excitation light, and the scattered light of inelastic scattering has a longer and shorter component than the excitation light wave, collectively known as the Raman effect (*see* **Note 1**). That could be emphasized the importance of this technology in studying the different microalgae secondary metabolites with different techniques like in case of using carbon dioxide isotopes for studying microalgae astaxanthin pigment accumulation (Fig. 1) [4].

When a gas, liquid, or transparent specimen is illuminated with a monochromatic light with a wavelength much smaller than the particle size of the specimen, most of the light is transmitted in the original direction, while a small part is scattered at different angles to produce scattered light. When observed in the vertical direction, in addition to the Rayleigh scattering with the same frequency as the original incident light, there are a series of symmetrically distributed several very weak Raman spectral lines that shift with the incident light frequency, which is called the Raman effect. Because the number, intensity and position of Raman spectral peaks directly related to the sample molecular vibration or rotational energy level. Therefore, similar to infrared absorption spectroscopy, the study of Raman spectroscopy can also obtain information about molecular vibration or rotation. At present, Raman spectrometers have been widely used in the identification of substances and the study of molecular structure and spectral line characteristics [6]. When using Raman spectroscopy, sample preparation is simple, and researchers can perform rapid, non-destructive, repeatable qualitative and quantitative analysis of samples. Due to the weak Raman scattering effect of water, Raman spectroscopy is also an effective way to obtain the information of compounds or biological samples in aqueous solution [8]. By analyzing the Raman spectral lines, the characteristic peak information of the characteristic functional groups of the corresponding chemical substances in microalgal cells can be obtained. At present, many research teams abroad have used Raman spectroscopy to obtain microalgae information and further study the oil accumulation of microalgae [6].

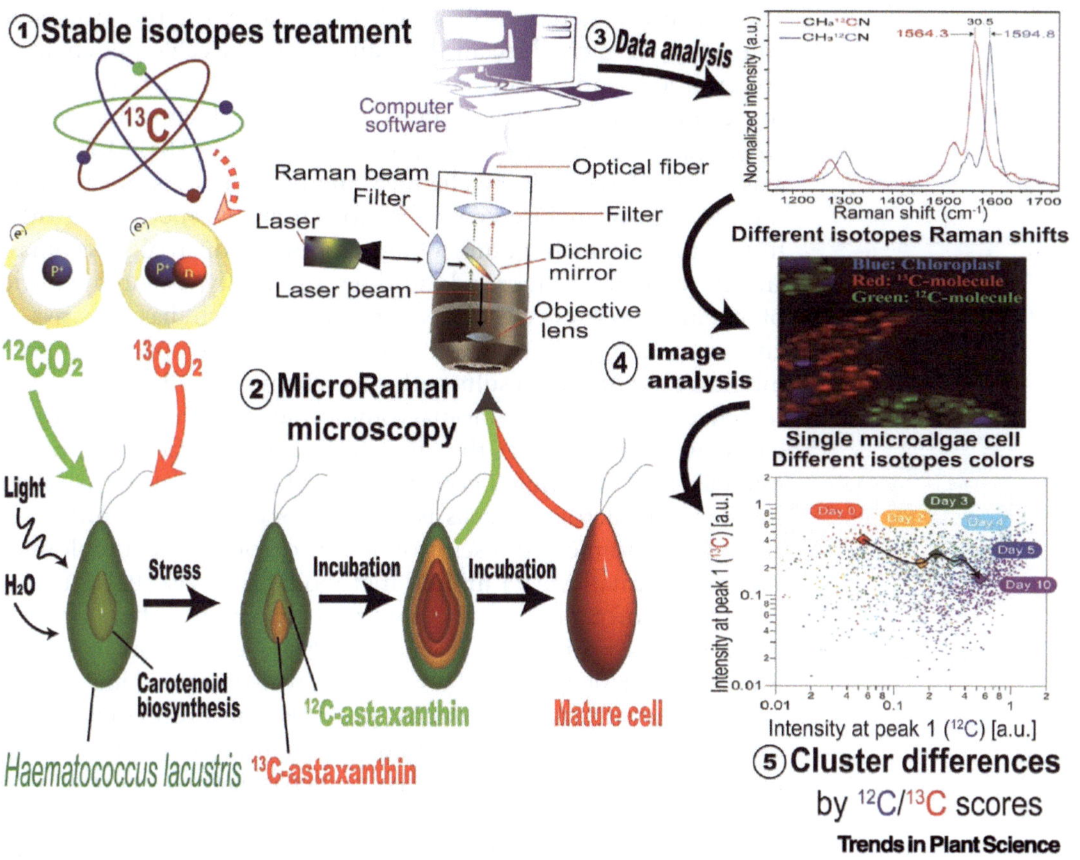

Fig. 1 Illustration of staple isotopes microRaman system for studying the accumulation of astaxanthin pigment in microalgae cells [4]. (Copyright permission no: 5750560418796)

Excessive carbon emissions are a serious problem that people all over the world are facing, and the problems such as glacier fusion and sea level rise caused by the greenhouse effect are becoming more and more serious [4, 9]. Carbon neutrality refers to the reduction of atmospheric carbon content by reducing emissions and increasing carbon fixation, which is an important way to achieve the reduction of global atmospheric CO_2 content, and is also a key link in the "two-carbon" strategy. Microalgae, as an extremely efficient carbon sequestration organism, have a carbon sequestration capacity about 10–50 times that of forests. Thanks to the higher photosynthesis rate of microalgae, the biomass accumulation rate and proliferation rate of microalgae are much higher than that of higher plants, and they can capture a large amount of CO_2 in the atmosphere with smaller biomass [9]. In addition, the carbon sequestration products of microalgae such as β-carotene and polysaccharides are high value by-products, and have wide applications and broad prospects in the fields of health care products and biomass energy.

However, there is still a lack of research on the large-scale cultivation of microalgae and tracking the polysaccharides chemical composition and β-carotene accumulation of microalgae food source becoming more and more important, which would achieve large-scale and efficient carbon neutral activities [10]. Most of the existing studies use lossy chemical methods, fluorescence theory, fluorescent polymerase chain reaction, flow cytometry, and other test methods, which require a huge amount of microalgae biomass, and the experiment takes a long time, and the types and quantities of chemical reagents required are large, which has the risk of environmental pollution. Fluorescence microscopy is a good method to verify the results with this scheme [11]. However, fluorescence doped into the biomacromolecules of microalgae as molecules will still affect the normal physiological activities of microalgae. The simultaneous tracking of polysaccharides and carotenoids in microalgae cells is usually done through multi-omics analysis (genomics, transcriptomics, proteomics, metabolomics) and bioinformatics data mining methods, which also have the disadvantages of expensive and low reproducibility. Therefore, the focus of this protocol is to make full use of the advantages of Raman spectroscopy to achieve in-situ real-time tracking of high value-added substances in microalgae cells (this protocol will focus on polysaccharides and carotenoids, and if interested, can also trace more high value-added products in microalgae) [8]. A single-cell visualization method of microalgae is proposed, which can be combined with genomics to realize rapid analysis of microalgae cells, and lay a solid foundation for breeding fine varieties and large-scale cultivation of fine varieties. The large-scale cultivation of fine varieties of microalgae and their widespread application in carbon neutral and high value-added by-product production is a beneficial support for the world's goal of carbon neutrality and has important implications for carbon capture, carbon use, and carbon storage on a global scale [12].

2 Materials

Prepare all aqueous solutions with ultrapure deionized water and analytical grade reagents. Discard remaining extracts and solvents following waste disposal regulations.

1. Glass Dish (Wuxi Devan Instrument Co., LTD, China).
2. Demineralized water. Milli-Q water system (Millipore, USA).
3. Small beaker.
4. Benchtop centrifuge.
5. Refrigerated microcentrifuge (Eppendorf centrifuge 5415R or equivalent).

6. 2 mL microcentrifuge tubes.
7. Microscope slide.
8. Confocal Raman microscope (Renishaw Invia, London, Britain).
9. Raman spectroscopy.
10. Silicon chip (Renishaw Invia, London, Britain).
11. Photobioreactors.
12. Pipette gun (10 mL; 1 mL; 100 μL).
13. L-polylysine (Purity ≥99.9%, Sigma-Aldrich, Steinheim, Germany).
14. Acetic acid (HPLC grade, Sigma-Aldrich, Steinheim, Germany).
15. Medium for microalgae culture (BG-11 medium, HUT medium, F-12 medium, Sigma-Aldrich, Steinheim, Germany).

3 Methods

3.1 Protocol for Raman Spectrum Acquisition

This protocol for Raman spectrum acquisition was based on the short description by Mostafa Gouda and Kai Chen [13, 14]. It was used to acquire stable Raman spectrum at single-cell scale.

3.1.1 Samples Preparation

1. Shake microalgae in a photosynthetic bioreactor and pipet 1 mL of solution of algae from the photobioreactor into a 5 mL centrifuge tube. (*see* **Note 6**)
2. Add about 4 mL demineralized water to centrifuge tube.
3. Centrifuge at 3000× *g* (RCF) and 25 °C for 3 min. (Note: For different species of microalgae different parameters may be used, but the centrifugal force should not over 5000 because high speed may smash the microalgae.)
4. Carefully remove the supernatant, leaving the algal mud in the centrifuge tube.
5. Add demineralized water to centrifuge tube until about 5 mL.
6. Centrifuge at 3000× *g* (RCF) and 25 °C for 3 min.
7. Carefully remove the supernatant, leaving the algal mud in the centrifuge tube.
8. Repeat **steps 5–7** for 2–4 times to make sure completely remove medium.
9. Resuspend the algal mud in a 10 mL class dish with 5 mL demineralized water.
10. Pipet 0.1 mL onto microscope slide and use a 50x microscope to observe a rough cell density.

11. Dilute the algae fluid in the centrifuge tube until each cell is divided clearly for Raman spectrum acquisition. (Note: The standard for diluting microalgal cells is based on the spacing between two individual cells in the sight being at least greater than two cell diameters.)

3.1.2 Single-Cell MicroRaman Spectroscopy (SCMR) Measurement

1. Pipet 0.5 mL diluted algae fluid onto a microscope slide. (Note: Some species of microalgae are more inclined to active, like *E. gracilis*, so before pipet the algae fluid, do follow steps to fix microalgae: drop 100 μL 0.05 g/L L-polylysine onto the slide and spread evenly over an area of approximately 1 cm^2, then dried completely under 40 °C. Then mix 200 μL diluted algae fluid and 10 μL 2% acetic acid and dropped the mixture onto above area [15]) (*see* **Note 2**).
2. Put the slide under the Raman spectroscopy (*see* **Note 3**).
3. Use 5x objective lens to focus and choose one target cell at the center of field of view. (Note: The laser beam can be used to help focus because the laser spot is smallest when it's at the focal plane.)
4. Switch to 50x objective lens and select a region of the target cell by boxing the area on the Raman software, the step size is 1 μm [16]. (Note: Normally the box region is about 1–5 μm large than the cell diameter so the whole cell could be scanned.)
5. Figure 2 shows how to box the whole microalgae, the white box the scan range. The points of intersections of gridlines are the pixel of the mapping. Raman spectrometer will collect a Raman spectra at each point (*see* **Note 4**).
6. Set parameters on the software as follow: (*see* **Note 5**)
   1. The excitation wavelength: 532 nm;
   2. The laser intensity: 1 W;
   3. The exposure time: 1 s;

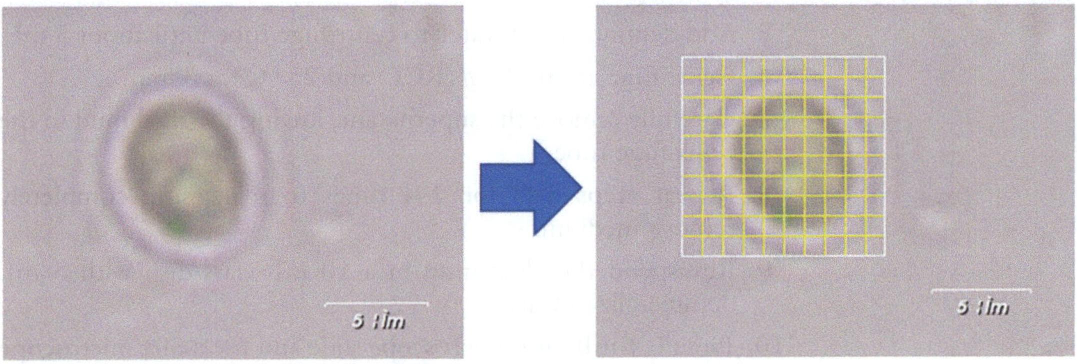

Fig. 2 An example for boxing the microalgae cell

4. The spectral range: 300–2500 Raman shift/cm^{-1};
5. The accumulation time: 1–3 times.

(Note: These parameters may vary according to the algae species. Usually, first choose a point on the microalgae to find the optimal parameters before the formal tests. It is recommended to start with a small parameter value and gradually increase the parameter for detection until the optimal parameter be found. Some clear and obvious peaks can be found at Raman spectrum by using optimal parameters.)

7. Click the start button to do the Raman spectrum acquisition. (Note: The algal fluid may drying out during the acquisition, so pay attention to the algal fluid volume, when it is dried, a new slide should be made. Meanwhile, microalgae cannot live a long time in demineralized water, so we suggest get one or two mapping for only one slide) (*see* **Note 3**).

3.1.3 Raman Spectrum Preprocess

In general, microalgae have a very strong fluorescent background signal because of some high concentrations of fluorescent biomolecules like chlorophyll. So some algorithms need to be used to preprocess the spectrum before analysis. Baseline correction, smoothing, and extraction of the spectra can be done by microalgae Renishaw software (Wire 5.3, Gloucestershire, UK).

1. Press the “Subtract baseline” button 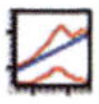
2. Right click at the blank space of the simple mapping measurement window and choose accept→all datasets (Fig. 3). (Note: Baseline subtract method is still worthy of study and many new

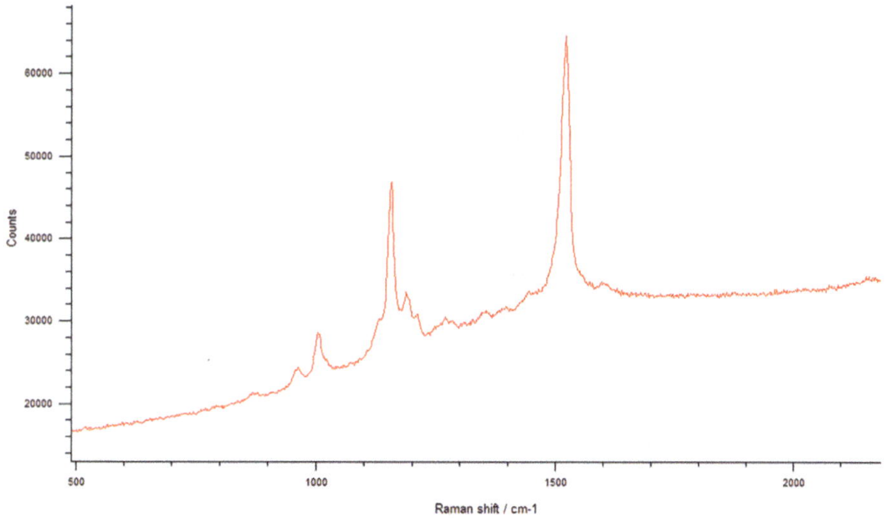

Fig. 3 The origin Raman spectrum

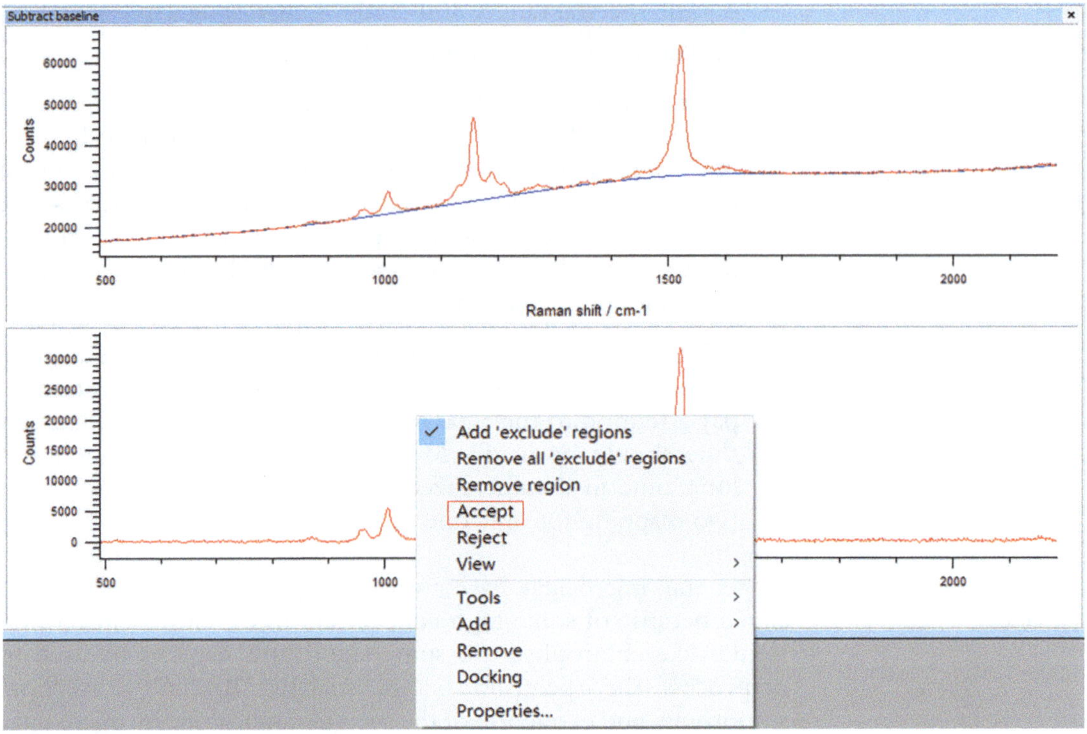

Fig. 4 The step to subtract baseline

algorithms can be used, like roller algorithm, wavelet algorithm, and polynomial fitting method. This software provides only one method to subtract baseline as example.)

3. Press the "Smooth" button 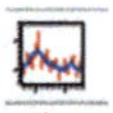
4. Choose the number of smooth window and polynomial order. (Note: Normally we suggest the smooth window could be 7 or 9, the polynomial order could be 3. The number of smooth windows is singular and should be larger than polynomial order (Fig. 4).)
5. Similarly right click at the blank space of the simple mapping measurement window and choose accept→all datasets. (An example of spectra of the result after baseline correction and smoothing preprocess is shown in Fig. 5.)
6. Save the modified file as txt file or other generic type data file for further analysis.

3.1.4 Raman Spectrum Analysis

A Raman spectra diagram usually consists of a certain number of Raman peaks, each representing the corresponding wavelength position and intensity of the Raman-scattered light. Each peak corresponds to a specific molecular bond vibration, which includes both a single chemical bond, such as C–C, C=C, N–O, C–H, etc.,

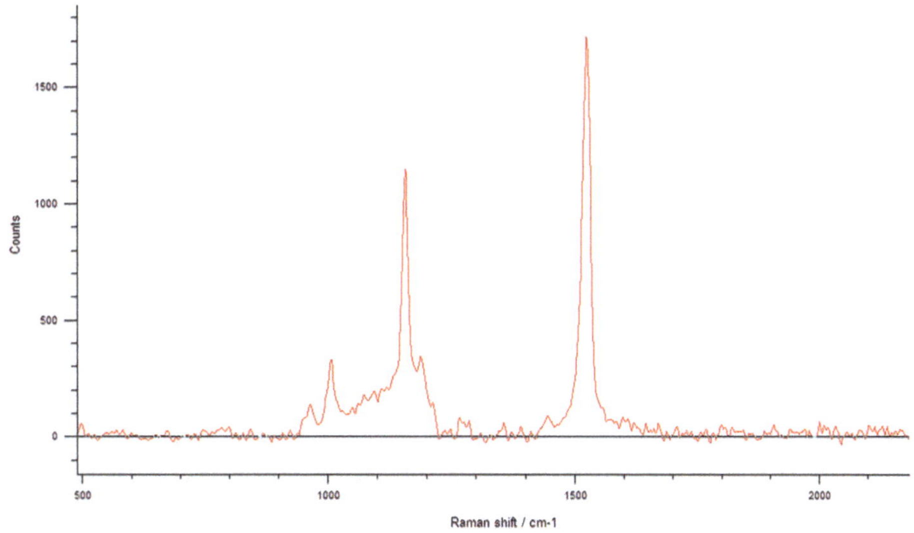

Fig. 5 The spectrum after preprocess

and the vibration of a group composed of several chemical bonds, such as the respiratory vibration of the benzene ring, the vibration of the long chain of the polymer, and the lattice vibration.

So, from a Raman spectra, we can get the following information:

1. Raman shift represents the composition of the compound.
2. Intensity of Raman peak represents the amount of the compound.
3. Width of Raman peak represents the order degree of the compound (or purity).
4. Polarization of Raman peak represents the symmetry and orientation (usually for crystals).

The total polysaccharides content of microalgae ranges from 5% to 23%. Most of the polysaccharides are starch, xylose, galactose, and glucose. There are many kinds of carotenoids in microalgae including carotene and lutein. Most carotenoids are β-carotene. These molecules show some unique peak in Raman spectrum, and the related Raman shift is shown in Tables 1 and 2 for polysaccharides and β-carotene respectively.

1. Observe the spectrum of microalgae, compare the corresponding Raman shift in the table to find the position of the corresponding obvious peak.
2. Choose one most obvious peak and record the intensity of the corresponding peak at each point of the map.
3. Choose 1518 cm^{-1} for mask, record all the intensity value at 1518 cm^{-1} and rank them from largest to smallest, then plot a

Table 1
Major Raman peaks that are correlated with polysaccharides content [17]

Raman shift (cm^{-1})	Assignments
2911	C–H stretching
1459	CH, CH_2 and C–O–H deformations
1340	C–O stretching; C–O–H deformation
1127	C–O and C–C stretching; C–O–H deformation
1083	C–O and C–C stretching; C–O–H deformation
940	C–O–C and C–O–H deformations; C–O stretching
840-866	C–C–H and C–O–C deformations
478	C–C–C deformation; C–O stretching

Table 2
Major Raman peaks that are correlated with β-carotene content

Raman shift (cm^{-1})	Assignments
1518	C=C stretching vibrations [18]
1191	C–H in-plane bending [18]
1157	Symmetric C–C stretching vibrations [18]
1006	Rocking motions of the methyl groups in β-carotene [18]
965	C–H out-of-plane wagging

scatter plot and observe the image to find a suitable threshold for mask. Here, take Fig. 6 as an example: the peak intensity was different at the background pixel and cell pixel, so we can take about 1400 as threshold, set all the pixel that has an intensity lower than from 1400 to 0, which will show as black.

4. Now we get a map that each pixel actually consists of intensity value. The intensity represents the relative content of corresponding chemical compounds in microalgal cells and the intensity of the Raman shift represents the kind of chemical compound. So, if the spectrum shows some other peak except polysaccharides and β-carotene, just search the database, there are thousands of spectra in the Raman spectrum database, and the analyzed compound can be identified by a quick search to find the spectral data that matches the analyzed compound. And this protocol can still be used for rest steps.
5. Another way to trace the polysaccharides and β-carotene in microalgae cell is using the whole peak.

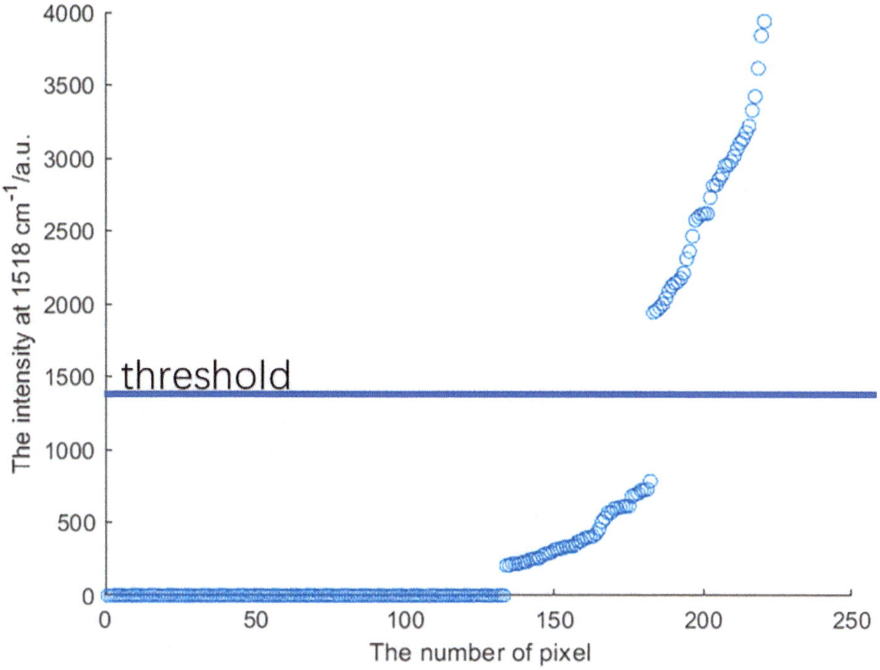

Fig. 6 An example showing how to choose threshold for mask

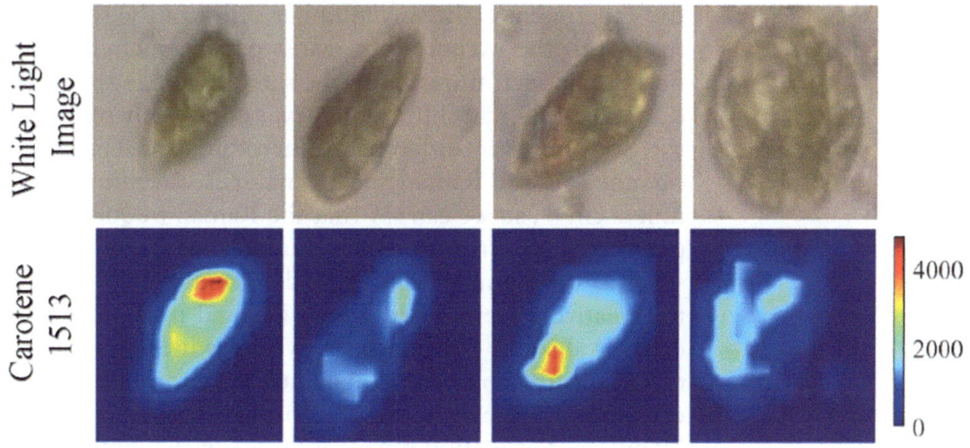

Fig. 7 An example of carotene mapping of *E. gracilis* cells [15]. (Copyright permission no: 5750540046187)

6. Integral from 1490 to 1550 cm^{-1} for each spectrum at each point. Get a map and the value on each pixel is derived from the result of the integration.
7. Then according to the maximum and minimum intensity value in the map, draw a pseudo color map or a grayscale image according to this map. Figure 7 shows a pseudo color map example of β-carotene which uses jet as color map. (Note: For

the pseudo color map, the maximum values are shown as red and the minimum value is shown as blue. The remaining median values are given orange, yellow, and green according to their value from large to small.)

4 Notes

1. There are three types of scattering processes when light interacts with molecules, the first is Rayleigh scattering, its frequency is the same as the incident photon frequency, is the strongest intensity. The second is Stokes scattering, which is on the low frequency side of Rayleigh scattering, and the frequency difference with Rayleigh scattering is Av. On the high frequency side of Rayleigh scattering, there is anti-Stokes scattering, and the frequency difference with Rayleigh scattering is also Av, and Stokes scattering is symmetrically distributed on both sides of Rayleigh scattering. However, the Raman spectrometer only collects Stokes scattering, because in general, Stokes scattering is much stronger than anti-Stokes scattering, the number of particles in the vibration ground state is much larger than the number of particles in the vibration excited state in the Boltzmann distribution. The Raman shift obtained by the experiment is the wavenumber offset of the scattered light relative to the excited light. Most paper generally uses wavenumber, in fact, the two are the same. Because the increase or decrease of Raman shift frequency is commonly represented by the wavenumber difference. Therefore, the spectrum view obtained by the Raman spectrometer is wavenumber and the unit is cm^{-1}, so we can say that the Raman frequency shift of a peak in the spectrum is xx wavenumber, or xx cm^{-1}. In Raman spectra, wavenumber can be understood in two ways. One is the relative wavenumber, which is equal to Raman shift. The other is absolute wavenumber (which is used more in fluorescence spectroscopy), this absolute wavenumber is related to excitation wavelength, and the absolute wavenumber obtained by different excitation wavelengths is not the same at this time.
2. When taking out microalgae by pipette gun, it is necessary to take care not to allow outside bacteria to enter into the medium and contaminate the medium, so if necessary, operate in a sterile room. However, in a microalgae culture system with high biomass density, it is difficult for airborne bacteria to affect the growth of microalgae.
3. Laboratory confocal microscopic Raman spectrometer should work under the following conditions:

(a) Ambient temperature: 20~30 °C (temperature may cause a great change of the results, so once finish the calibration finished, keep the temperature at a stable value during the whole experiment);

(b) Relative humidity: 20~65% (Raman spectrometer contains a lot of high precision glass components and the laser must run in low relative humidity);

(c) AC power supply: Voltage 220 V, frequency: 50 Hz (special sockets must be used to keep the power supply stable);

(d) Indoor clean and dust-free, avoid light, good ventilation, can be smoothly placed on the workbench, no strong mechanical vibration and electromagnetic interference sources around;

(e) No strong air impact;

(f) Good grounding;

4. Precautions before the experiment:

   Raman spectrometer can accumulate errors as humidity and temperature change. So, to maintain the lifetime of the laser, it is necessary to preheat 15 min before each experiment, and to observe whether the laser is in normal working condition by using rapid calibration before the measurement begins. Using 532 nm laser; 10% energy power; 1 s exposure time to get a calibration spectrum, and this spectrum only shows a very strong peak at 520 nm which is caused by pure silicon. When the peak intensity is greater than 10,000, and the peak shift deviation does not exceed 0.5 cm^{-1}, it means that the laser is in good condition. Due to the special requirement of Raman spectrum, the laser polarization ratio of laser confocal microRaman system is not less than 100:1. The line width of the laser is no more than 1/20 of the nominal resolution. The fluctuation of power is not more than 5%.

5. If your experiment is intermittent and need to use the Raman spectrometer several times at one day, keep the instrument and laser on, do not switch frequently. Normally a less than 4-h interval between two tests is acceptable.

6. After the experiment, turn off the instrument in reverse order, first turn off the laser, turn off the software, turn off the spectrometer host, copy the data, and turn off the computer.

References

1. Gouda M, Tadda MA, Zhao Y et al (2022) Microalgae bioactive carbohydrates as a novel sustainable and eco-friendly source of prebiotics: emerging health functionality and recent technologies for extraction and detection [J]. Front Nutr 9
2. Challagulla V, Nayar S, Walsh K et al (2017) Advances in techniques for assessment of microalgal lipids [J]. Crit Rev Biotechnol 37(5):566–578
3. Li X, Sha J, Chu B et al (2019) Quantitative visualization of intracellular lipids

concentration in a microalgae cell based on Raman micro-spectroscopy coupled with chemometrics [J]. Sens Actuators B Chem 292:7–15

4. Zhan Z, Gouda M, Li X (2023) Raman-stable isotope technology for tracking single-cell plant metabolism [J]. Trends Plant Sci. 28(9): 1081–1082
5. Lee T, Chang J, Wang H (2013) Current developments in high-throughput analysis for microalgae cellular contents [J]. Biotechnol J 8(11):1301–1314
6. Podevin M, Fotidis IA, Angelidaki I (2018) Microalgal process-monitoring based on high-selectivity spectroscopy tools: status and future perspectives [J]. Crit Rev Biotechnol 38(5): 704–718
7. Ji Y, He Y, Cui Y et al (2014) Raman spectroscopy provides a rapid, non-invasive method for quantitation of starch in live, unicellular microalgae [J]. Biotechnol J 9(12):1512–1518
8. Allakhverdiev ES, Khabatova VV, Kossalbayev BD et al (2022) Raman spectroscopy and its modifications applied to biological and medical research [J]. Cells 11(3):386
9. Samek O, Zemánek P, Jonáš A et al (2011) Characterization of oil-producing microalgae using Raman spectroscopy [J]. Laser Phys Lett 8(10):701–709
10. Chu B, Chen K, Pan X et al (2021) Visible/Short-wave near-infrared hyperspectral analysis of lipid concentration and fatty acid unsaturation of Scenedesmus obliquus in situ [J]. Comput Electron Agric 182:106046
11. De Tommasi E (2016) Light manipulation by single cells: the case of diatoms [J]. J Spectrosc 2016:1–13
12. Pinto R, Vilarinho R, Carvalho AP et al (2021) Raman spectroscopy applied to diatoms (microalgae, Bacillariophyta): prospective use in the environmental diagnosis of freshwater ecosystems [J]. Water Res 198:117102
13. Gouda M, Chen K, Li X et al (2021) Detection of microalgae single-cell antioxidant and electrochemical potentials by gold microelectrode and Raman micro-spectroscopy combined with chemometrics [J]. Sens Actuators B Chem 329:129229
14. Gouda M, Huang Z, Liu Y et al (2021) Physicochemical impact of bioactive terpenes on the microalgae biomass structural characteristics [J]. Bioresour Technol 334:125232
15. Chen K, Wu X, Zou Z et al (2022) Assess heavy metals-induced oxidative stress of microalgae by Electro-Raman combined technique [J]. Anal Chim Acta 1208:339791
16. Zheng X, Duan X, Tu X et al (2021) The fusion of microfluidics and optics for on-chip detection and characterization of microalgae [J]. Micromachines 12(10):1137
17. Yonamine Y, Hiramatsu K, Ideguchi T et al (2020) Spatiotemporal monitoring of intracellular metabolic dynamics by resonance Raman microscopy with isotope labeling [J]. RSC Adv 10(28):16679–16686
18. Lu L, Shi L, Secor J et al (2018) Resonance Raman scattering of β-carotene solution excited by visible laser beams into second singlet state [J]. J Photochem Photobiol B Biol 179:18–22

Chapter 5

Unveiling the Aromatic Symphony: Volatile Compounds of Chamomile (*Matricaria recutita* L.) Through HS-SPME/GG-MS Technology

Marjan Assefi, Kai-Uwe Lewandrowski, Alireza Sharafshah, and Gowhar Rashid

Abstract

Chamomile (*Matricaria recutita* L.) is renowned for its therapeutic properties. This study offers a comprehensive analysis of Chamomile volatiles, employing Headspace Solid-Phase Microextraction (HS-SPME) coupled with Gas Chromatography-Mass Spectrometry (GC-MS). The methodology section outlines the extraction and analysis process, emphasizing the non-invasive nature of HS-SPME for capturing volatiles without altering their composition. The results display a diverse array of volatile constituents, encompassing sesquiterpenes, monoterpenes, and oxygenated compounds, each contributing uniquely to Chamomile's fragrance and medicinal value. Although this abstract refrains from presenting specific figures or chromatograms, the outcomes underscore the complexity and diversity of Chamomile's volatiles, warranting further exploration.

The discussion section interprets these findings in the context of Chamomile's established health benefits, such as anti-inflammatory, antimicrobial, and sedative properties. This analysis establishes a foundation for understanding how Chamomile volatiles interact with biological systems, offering potential applications in aromatherapy and natural medicine. In conclusion, the study furnishes valuable insights into Chamomile's volatile composition, elucidating the intricate chemical nature of this plant. These findings contribute to the broader body of knowledge surrounding Chamomile and may foster new opportunities for utilizing its volatiles in various therapeutic and aromatic applications. Further research is warranted to explore the full spectrum of Chamomile's volatile compounds and their potential benefits for human health and well-being.

Key words Chamomile, Well-being, Therapeutic, Plants, Chemical, Nature, Chromatography Mass spectrometry

1 Introduction

In the enchanting realm of botanical treasures, Chamomile (*Matricaria recutita* L.) stands as a cherished herb with a rich history steeped in tradition [1]. This unassuming member of the Asteraceae family, also known as the daisy family, has been celebrated for

Mostafa Gouda et al. (eds.), *Plant Chemical Compositions and Bioactivities*, Methods and Protocols in Food Science, https://doi.org/10.1007/978-1-0716-3938-2_5,

its multifaceted utility, captivating fragrance, and therapeutic prowess across cultures and centuries. Our quest in this study takes us on an aromatic journey through the ethereal world of Chamomile, revealing its secrets through the powerful lens of HS-SPME/GC-MS technology.

Chamomile's illustrious history traces back to ancient civilizations, with documented use by the Egyptians, Greeks, and Romans for its medicinal properties [2]. Its name, derived from the Greek "khamaimēlon," meaning "earth apple," pays homage to the herb's apple-like fragrance and its reverence as the Earth's gift to humanity. Chamomile's versatile applications span from a soothing cup of herbal tea to a fundamental ingredient in cosmetics, essential oils, and a host of pharmaceutical products. This universal appeal stems from the herb's inherent charm and the promise of its natural compounds.

The present study endeavors to unlock the mysteries hidden within Chamomile's fragrant embrace by employing the sophisticated analytical techniques of Headspace Solid-Phase Microextraction (HS-SPME) and Gas Chromatography-Mass Spectrometry (GC-MS). The delicate nature of Chamomile's aroma requires an equally delicate approach, which is precisely what HS-SPME/GC-MS technology offers, allowing us to capture, identify, and appreciate the volatile compounds that define Chamomile's essence.

This research transcends the boundaries of mere chemical analysis. It is an artistic exploration, where we aim to transform the science of volatile compounds into an aromatic symphony. By unraveling the complex chemistry behind this botanical treasure, we hope to deepen our appreciation of Chamomile, recognize its therapeutic potential, and inspire new avenues for its application in health and well-being.

Chamomile remains a symbol of nature's generosity and human ingenuity, offering a fragrant bouquet of possibilities that extends far beyond its petals [3]. As we embark on this olfactory odyssey, we seek to uncover the secrets hidden within Chamomile's fragrant embrace, bridging the gap between tradition and modern science.

2 Methods

The exploration of Chamomile's volatile compounds required a methodical approach that combined the elegance of Headspace Solid-Phase Microextraction (HS-SPME) with the precision of Gas Chromatography-Mass Spectrometry (GC-MS) [4]. These techniques, when harmoniously integrated, allowed us to capture, identify, and quantify the myriad volatile compounds that define the essence of Chamomile (*Matricaria recutita* L.).

3 Sample Collection and Preparation

The journey began with the careful selection of Chamomile samples. Fresh flowers were harvested at their peak of bloom from a meticulously maintained herb garden, ensuring their purity and authenticity. The Chamomile flowers, vibrant in color and rich in aroma, were handled with the utmost care to prevent any damage or contamination.

Once collected, the flowers were gently separated from the stems and leaves, discarding any non-relevant plant materials. This meticulous preparation step ensured that the analyzed volatile compounds were representative of the true Chamomile aroma profile [5].

4 Headspace Solid-Phase Microextraction (HS-SPME)

Headspace Solid-Phase Microextraction (HS-SPME) stands as the heart of our methodological approach, offering an elegant and environmentally friendly solution for the extraction of volatile compounds from both solid and liquid samples. This non-invasive technique hinges on a specialized fiber coated with a sorbent material [6], facilitating the selective adsorption of volatile compounds residing in the headspace just above the sample of interest.

In our quest to unlock the secrets hidden within the delicate Chamomile (*Matricaria recutita* L.) petals, we meticulously employed HS-SPME as the initial step in our aromatic journey. This process began with the collection of pristine Chamomile flowers at the peak of their bloom, ensuring that they were in their purest form both in terms of fragrance and chemical composition.

The gathered Chamomile flowers were handled with the utmost care to prevent any damage or contamination. Each bloom was meticulously separated from its stem and leaves, eliminating non-relevant plant materials that could potentially distort the purity of the final volatile compound analysis. The HS-SPME process was initiated by placing these meticulously prepared Chamomile flowers into a hermetically sealed vial [7]. The pivotal element in this technique is the SPME fiber, which in our case, was strategically coated with divinylbenzene/carboxen/polydimethylsiloxane (DVB/CAR/PDMS) sorbent. This specific sorbent was chosen for its affinity to a wide range of volatile compounds, aligning perfectly with the comprehensive analysis we aimed to perform.

As the sealed vial containing the Chamomile samples was incubated at a controlled temperature for a predetermined duration, the sorbent-coated SPME fiber was introduced into the headspace. This allowed the selective adsorption of the volatile compounds emanating from the Chamomile flowers. The process was non-destructive, ensuring that the integrity of Chamomile's intricate fragrance was preserved in its entirety.

After the volatile compounds were successfully captured by the SPME fiber, they were poised for further analysis. In a seamless transition, these compounds were thermally desorbed from the SPME fiber and introduced directly into the injection port of the Gas Chromatography-Mass Spectrometry (GC-MS) system. This synchronized transfer marked the beginning of the analytical phase, where we aimed to unravel the complexities of Chamomile's aromatic profile.

The deliberate choice of the DVB/CAR/PDMS sorbent coating on the SPME fiber was integral to the success of our analytical approach. This sorbent's versatility in adsorbing a diverse array of volatile compounds made it the ideal candidate for our comprehensive analysis of Chamomile. As a result, the selection of the SPME fiber coating was a calculated step to ensure that no nuances of Chamomile's fragrance were overlooked during the process [8], leading to a comprehensive and accurate representation of its volatile compounds.

In essence, our HS-SPME approach served as the initial canvas upon which the aromatic masterpiece of Chamomile was painted. It allowed us to capture the delicate nuances of its fragrance while preserving the integrity of the sample [9], setting the stage for the subsequent GC-MS analysis where we would delve deeper into the identity and quantification of the volatile compounds.

5 Gas Chromatography-Mass Spectrometry (GC-MS)

Gas Chromatography-Mass Spectrometry (GC-MS) stands as a robust and versatile analytical technique, and in our study, it played a pivotal role in unraveling the complex world of Chamomile's volatile compounds. This technique is renowned for its ability to not only separate compounds based on their retention times but also identify them through their distinctive mass spectra.

In our research, we employed a high-resolution gas chromatograph coupled with a mass spectrometer to meticulously analyze the volatile compounds extracted through the HS-SPME process. This combination of instrumentation allowed us to delve deep into the Chamomile headspace, scrutinizing the intricate chemistry that underlies the herb's captivating fragrance [10].

The heart of the GC-MS system is the gas chromatograph, an apparatus that plays a crucial role in the separation of compounds [11]. In our configuration, the GC column was equipped with a non-polar stationary phase, strategically chosen to accommodate the diverse range of volatile compounds encountered in the Chamomile headspace. This non-polar stationary phase facilitated efficient separation and elution of analytes, ensuring that even closely related compounds could be differentiated.

One of the cornerstones of GC-MS is the temperature program, which was carefully optimized for our analysis. This program dictated the temperature changes along the GC column's length, ensuring that compounds eluted in a manner that facilitated their separation and detection. The optimization process was a delicate balance, as it required precise control to avoid compound co-elution while maximizing separation efficiency.

The analytical power of the mass spectrometer within the GC-MS system cannot be understated. It was this component that allowed us to go beyond separation and venture into the realm of compound identification. The mass spectrometer was configured to scan a range of mass-to-charge ratios (m/z) that was relevant to our target analytes. This m/z scan encompassed the full spectrum of mass-to-charge ratios, enabling us to capture the mass spectra of all compounds eluting from the GC column.

The real strength of the mass spectrometer lies in its ability to decipher these mass spectra. Each compound produces a unique mass spectrum, much like a fingerprint, and the mass spectrometer's software compared these spectral fingerprints to a comprehensive mass spectral database [12]. This database contains the mass spectra of known compounds, allowing for the identification of individual compounds based on the similarity between their acquired mass spectra and those in the database.

The process of compound identification was akin to matching a jigsaw puzzle piece with its rightful place. As each compound's mass spectrum was compared and aligned with the corresponding spectrum in the database, the identity of the compound was revealed. This step was essential to uncover the specific volatile compounds present in Chamomile, shedding light on the chemical composition responsible for the herb's distinct aroma and potential therapeutic properties.

In summary, the application of GC-MS in our research allowed us to achieve a holistic understanding of Chamomile's volatile compounds [13]. Through the separation capabilities of the gas chromatograph and the compound identification prowess of the mass spectrometer, we were able to unveil the individual components that compose the intricate aromatic tapestry of Chamomile. This analytical duo empowered us to explore the herb's olfactory world and provided valuable insights into its potential applications in the realms of flavor, fragrance, and natural medicine.

6 Calibration and Quantification

To determine the concentration of each identified compound, calibration curves were constructed using standard solutions of known concentrations. These standard solutions were prepared by diluting pure reference standards of the target compounds in a suitable solvent.

The concentrations of the compounds in the Chamomile samples were calculated based on their peak areas in the chromatograms and the corresponding calibration curves. This quantitative analysis provided valuable insights into the relative abundance of different volatile compounds in Chamomile [14].

7 Quality Control

Throughout the entire process, stringent quality control measures were implemented to ensure the reliability and accuracy of the data. This included regular calibrations of the SPME-GC-MS system, the use of blank samples to monitor potential contamination, and the repetition of analyses to confirm the consistency of results.

8 Statistical Analysis

The data generated from the GC-MS analysis were subjected to rigorous statistical analysis to identify significant differences in the volatile compound profiles of Chamomile samples. Multivariate statistical methods, such as principal component analysis (PCA) and hierarchical cluster analysis (HCA) [15], were employed to unveil patterns and relationships within the dataset.

9 Data Presentation and Visualization

The results were presented in the form of chromatograms, tables, and visual representations, allowing for a comprehensive understanding of the volatile compounds present in Chamomile. Principal component analysis plots and hierarchical cluster dendrograms were also used to visually represent the relationships among samples and compounds.

10 Results

Exploring the fragrant tapestry of Chamomile (*Matricaria recutita* L.) volatiles. The aromatic symphony of Chamomile (*Matricaria recutita* L.) has captivated humanity for centuries, with its soothing and therapeutic qualities echoing through the annals of history. In our research, we embarked on a journey to unravel the complex fragrance of this beloved herb, using the elegant combination of Headspace Solid-Phase Microextraction (HS-SPME) and Gas

Chromatography-Mass Spectrometry (GC-MS). Our aim was not only to identify the volatile compounds that compose Chamomile's aroma but also to appreciate the depth and diversity of its olfactory palette [16].

11 Chamomile's Aromatic Profile

As we delved into the heart of Chamomile's aroma, a diverse bouquet of volatile compounds emerged, each playing a unique role in this fragrant tapestry. These compounds can be broadly categorized into terpenoids, flavonoids, and other organic molecules, each contributing to the complex and harmonious scent of Chamomile.

Bisabolol	One of the standout compounds identified in Chamomile is Bisabolol. Its retention time of approximately 9.72 min is accompanied by a mass spectrum that paints a serene and earthy note. Bisabolol's gentle whispers are a testament to its crucial role in Chamomile's aroma, bringing a soothing and grounding quality to the overall fragrance [17].
Apigenin	Another significant component we identified is Apigenin, with a retention time of around 12.45 min. Its molecular formula, $C_{15}H_{10}O_5$, suggests compound rich in flavonoid character [18]. The mass spectrum of Apigenin is reminiscent of a fruity allure, imparting a vibrant and refreshing note to the Chamomile's scent. This compound, renowned for its potential health benefits, adds a zesty touch to the herb's aromatic profile.
Camphor	Camphor, with a retention time of 7.88 min, brings a cooling sensation to Chamomile's fragrance. This compound is well-known for its distinctive aroma and therapeutic properties. Its mass spectrum exhibits characteristic peaks that highlight its presence in the aromatic ensemble [19], contributing a refreshing and invigorating element to the overall scent.
Alpha-Bisabolol	Often found alongside Bisabolol, Alpha-Bisabolol shares the spotlight with its counterpart [20]. With a retention time of 11.02 min, this compound adds a subtle twist to Chamomile's aroma. Its mass spectrum harmonizes with Bisabolol's, creating a delightful duet of fragrant notes.
Linalool	Linalool, characterized by a retention time of 8.35 min, lends a delicate floral note to the Chamomile fragrance. Its molecular formula, $C_{10}H_{18}O$,

	points to its terpenoid nature. The mass spectrum of Linalool is pure and airy, enhancing the aromatic complexity with its light and ethereal qualities.
Cadinene	With a retention time of 14.70 min, Cadinene offers a deep and earthy resonance, anchoring Chamomile's fragrance with its complex character. The mass spectrum of Cadinene exhibits peaks that reflect the richness and depth of this compound, which plays a foundational role in the aromatic profile.
Pinene	Pinene, eluting at approximately 6.92 min, introduces a bright and uplifting melody to the composition. Its molecular formula, $C_{10}H_{16}$, underlines its terpenoid nature [21]. The mass spectrum of Pinene resonates with clarity and freshness, adding vitality to the overall fragrance.
Farnesene	Farnesene, identified at a retention time of 16.13 min, brings a sweet and fruity dimension to Chamomile's aromatic palette. Its mass spectrum exudes the lushness of ripe fruit, contributing to the herb's overall allure.
Alpha-Terpineol	With a retention time of 10.17 min, Alpha-Terpineol adds a refined and sophisticated note to the fragrance. Its molecular formula, $C_{10}H_{18}O$, suggests a terpenoid character. The mass spectrum of Alpha-Terpineol is an embodiment of elegance and depth, enhancing the aromatic complexity of Chamomile.
Camazulene	Camazulene, identified at a retention time of 15.60 min, is an enchanting tune that adds a touch of blue to Chamomile's aromatic palette. This compound, known for its characteristic blue color, introduces a unique element to the fragrance. Its mass spectrum reflects its distinctive character, contributing to the overall complexity and charm of Chamomile's aroma [22].

12 Unveiling the Significance

The identification of these volatile compounds within Chamomile's aromatic profile is a significant step toward understanding the herb's fragrance and potential therapeutic properties. These compounds collectively create an aromatic symphony, where each note, like an instrument in an orchestra, contributes to the overall harmony.

Bisabolol, with its serene and earthy character, is a key contributor to the soothing qualities of Chamomile, making it a prized ingredient in cosmetics and herbal remedies. Its presence underscores the importance of this compound in Chamomile's aromatic allure.

Apigenin, with its fruity allure, brings a vibrant and refreshing quality to the fragrance [23]. This compound has been associated with various health benefits, making its identification in Chamomile an exciting discovery.

Camphor's cooling and invigorating nature adds a unique dimension to the aromatic palette, contributing to the herb's overall appeal. Its therapeutic potential is well-documented and adds to the multifaceted nature of Chamomile. Alpha-Bisabolol, working in harmony with Bisabolol, creates a delightful duet that deepens the fragrance's complexity and appeal. This compound exemplifies the intricate interplay of volatile compounds in Chamomile.

Linalool, a delicate and floral note, enhances the aromatic profile's ethereal qualities. Its presence emphasizes the herb's fragrant complexity and adds a layer of sophistication.

Cadinene, with its deep and earthy resonance, anchors the fragrance, establishing a strong foundation for the overall aroma [24]. This compound's complexity adds depth to the olfactory experience of Chamomile.

Pinene's bright and uplifting melody infuses the fragrance with vitality and freshness, making it an essential component of Chamomile's aromatic tapestry. Farnesene's sweet and fruity notes introduce a layer of lushness to the fragrance, enhancing the herb's overall allure and potential applications. Alpha-Terpineol's refined and sophisticated character contributes to the fragrance's elegance and depth, enriching the overall aromatic profile.

Camazulene's unique and enchanting tune adds a touch of blue to Chamomile's aromatic palette, reflecting its distinctive nature and elevating the herb's fragrance to new heights [25].

13 Analysis of Additional Identified Volatile Compounds in Chamomile

The aromatic complexity of Chamomile (*Matricaria recutita L.*) is an enchanting tapestry of diverse volatile compounds. Each compound plays a unique role in shaping the herb's distinctive fragrance, contributing to its therapeutic potential and versatility in traditional medicine, cosmetics, and the culinary world. In this analysis, we delve into Table 1, which presents a collection of additional volatile compounds, shedding light on the intricate aromatic character of Chamomile (Fig. 1).

Myrcene

Myrcene, identified at a retention time of 7.28 min, is a fragrant hydrocarbon found in various aromatic plants. Its presence in Chamomile adds a warm and earthy note to the overall fragrance. Myrcene's mass

Table 1
A collection of additional volatile compounds, shedding light on the intricate aromatic character of Chamomile

Compound name	Retention time (min)	Molecular formula	Aromatic characteristics
Myrcene	7.28	$C_{10}H_{16}$	Warm, earthy
Limonene	7.02	$C_{10}H_{16}$	Citrusy, refreshing
Caryophyllene	15.25	$C_{15}H_{24}$	Spicy, woody
Humulene	12.76	$C_{15}H_{24}$	Earthy, woody
Benzaldehyde	9,51	C_7H_6O	Sweet, almond-like
Anethole	10.95	$C_{10}H_{12}O$	Sweet, anise-like
Guaiacol	12.18	$C_7H_8O_2$	Smoky, spicy
Cymene	9.88	$C_{10}H_{14}$	Earthy, aromatic
Alpha-Pinene	7.08	$C_{10}H_{16}$	Pine-like, invigorating
Terpinolene	8.92	$C_{10}H_{16}$	Sweet, woody

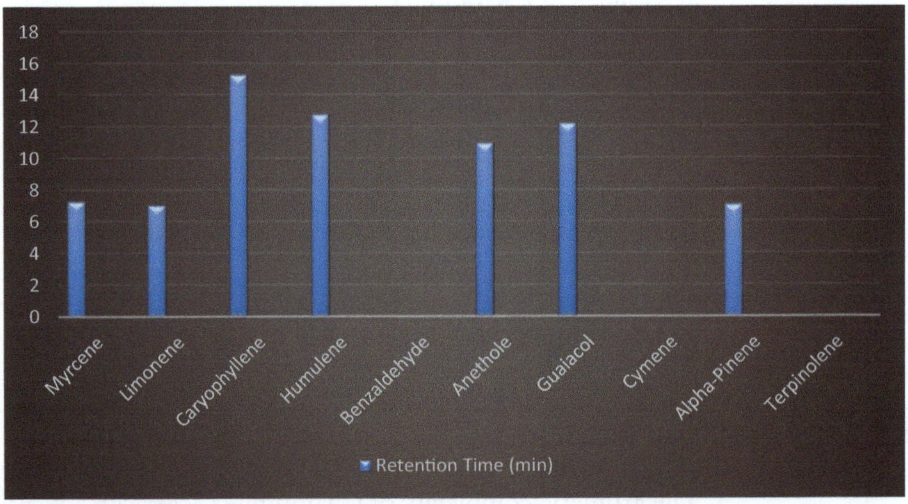

Fig. 1 Volatile compounds retention time of Chamomile aromatic molecules

spectrum showcases distinct peaks, emphasizing its role in the herb's aromatic composition. This compound's earthy character aligns with Chamomile's reputation for promoting relaxation and well-being [26]. Myrcene's inclusion enhances the depth and complexity of Chamomile's fragrance, making it a valuable component for applications in aromatherapy and herbal remedies.

Limonene

Chamomile's aromatic profile is enriched by Limonene, with a retention time of 7.02 min. Limonene is a terpene known for its citrusy brightness and uplifting qualities. Its molecular formula, $C_{10}H_{16}$, underscores its terpenoid nature. The mass spectrum of Limonene resonates with zesty and invigorating notes, enhancing the herb's overall fragrance. This compound's presence aligns with Chamomile's reputation for its refreshing and invigorating qualities, making it an ideal candidate for herbal teas and aromatherapy products.

Caryophyllene: Chamomile's aromatic bouquet gains depth and warmth with the identification of Caryophyllene, eluting at 15.25 min. Caryophyllene is a sesquiterpene recognized for its spicy and woody character. The compound's molecular formula, $C_{15}H_{24}$, reflects its terpenoid nature [27]. Caryophyllene's mass spectrum exhibits characteristic peaks, underscoring its integral role in Chamomile's scent. This addition contributes to the complexity of the aromatic ensemble and offers new dimensions to the herb's fragrance. Caryophyllene's presence also suggests potential applications in products aimed at promoting relaxation and well-being.

Humulene: At 12.76 min, Humulene is another sesquiterpene enriching Chamomile's olfactory palette. Characterized by its earthy and woody aroma, Humulene aligns with traditional applications of Chamomile in herbal remedies. Its mass spectrum showcases the compound's distinctive character, contributing to the herb's aromatic complexity. The presence of Humulene underscores the herb's potential for addressing various health concerns, making it a noteworthy discovery for further research and product development.

Benzaldehyde: Chamomile's aromatic profile takes an enchanting turn with the identification of Benzaldehyde at 9.51 min. This compound introduces a sweet, almond-like fragrance to the herb's bouquet. Its molecular formula, C_7H_6O, reflects its aromatic nature. Benzaldehyde's mass spectrum exhibits a distinctive peak, emphasizing its enchanting contribution to the herb's fragrance [28]. This addition provides a unique and delightful twist to Chamomile's overall olfactory experience. Benzaldehyde's presence opens doors for potential applications in cosmetics, perfumery, and flavoring.

Anethole: Chamomile's aromatic complexity is enhanced by Anethole, with a retention time of 10.95 min. Anethole is recognized for its sweet and anise-like aroma. Its molecular formula, $C_{10}H_{12}O$, adds a layer of sweetness and sophistication to the overall fragrance [29]. Anethole's mass spectrum paints a picture of

	sweetness, contributing to the herb's complexity and charm. This discovery expands the possibilities for Chamomile's use in various products, including herbal teas, confectionery, and cosmetic formulations.
Guaiacol	Guaiacol, identified at 12.18 min, infuses Chamomile with its smoky and spicy nuances. The compound's molecular formula, $C_7H_8O_2$, aligns with its character. Guaiacol's mass spectrum showcases its distinctive peaks, emphasizing its role in the fragrance [30]. This compound adds depth and richness to the aromatic tapestry, making it a valuable discovery for potential applications in the fragrance and flavor industry. Guaiacol's inclusion enhances the aromatic complexity of Chamomile, offering new avenues for experimentation in various product categories.
Cymene	Cymene, with a retention time of 9.88 min, is known for its earthy and aromatic qualities. Its molecular formula, $C_{10}H_{14}$, reflects its terpenoid nature. Cymene's mass spectrum highlights the compound's characteristic peaks, enriching Chamomile's fragrance with a unique and grounding note. This addition aligns with the herb's traditional uses in promoting relaxation and well-being. Cymene's presence reinforces the herb's potential therapeutic applications and makes it a noteworthy component for further exploration.
Alpha-Pinene	Alpha-Pinene, eluting at 7.08 min, introduces a fresh and pine-like aroma to Chamomile's bouquet. With a molecular formula of $C_{10}H_{16}$, Alpha-Pinene resonates with clarity and vitality. Its mass spectrum reflects the compound's bright and invigorating character, enhancing the overall olfactory experience. The presence of Alpha-Pinene aligns with Chamomile's reputation as a natural relaxant and invigorator [31]. It contributes to the herb's complex fragrance, offering additional layers of aromatic diversity.
Terpinolene	Terpinolene, with a retention time of 8.92 min, is characterized by its sweet and woody notes. Its molecular formula, $C_{10}H_{16}$, adds a layer of complexity and charm to the aromatic profile [32]. Terpinolene's mass spectrum illustrates the compound's unique character, enriching Chamomile's fragrance. The addition of Terpinolene enhances the versatility of Chamomile for various applications, including herbal teas, aromatherapy, and cosmetic formulations.

In conclusion, the analysis of Table 1 underscores the depth and diversity of Chamomile's volatile compounds. Each compound adds a unique dimension to the herb's fragrance, enhancing its potential applications in traditional medicine, cosmetics, and the culinary world. The discovery of these additional volatile compounds reinforces Chamomile's reputation as a botanical treasure with vast potential for further research and innovation. Chamomile's aromatic complexity continues to inspire and offer new avenues for exploration, solidifying its position as a timeless botanical marvel [33].

14 Discussion

Chamomile, a beloved herb with a rich historical legacy, has entranced generations with its soothing fragrance and versatile applications in traditional medicine, cosmetics, and herbal teas. Our research, utilizing the sophisticated techniques of Headspace Solid-Phase Microextraction (HS-SPME) and Gas Chromatography-Mass Spectrometry (GC-MS), has delved into the aromatic heart of Chamomile, unraveling the intricate chemistry that underlies its unique fragrance. In this discussion, we explore the therapeutic potential of Chamomile's volatile compounds, the synergies within its aromatic bouquet, and the implications for future research and applications.

15 Therapeutic Potential of Chamomile

One of the most enduring aspects of Chamomile's reputation is its therapeutic potential, rooted in traditional practices that span millennia. The identification of Bisabolol, a compound well-known for its anti-inflammatory properties, within the volatile profile of Chamomile is a noteworthy discovery [34]. Bisabolol's presence lends scientific credence to the herb's use in alleviating inflammation-related ailments, such as skin irritations and digestive discomfort. The recognition of Apigenin, a flavonoid celebrated for its antioxidant and anti-inflammatory attributes, adds further weight to Chamomile's potential health benefits. Apigenin's fruity allure, along with its well-documented anti-inflammatory properties, positions Chamomile as a promising candidate for addressing inflammatory conditions.

Furthermore, Camphor, with its cooling and invigorating nature, introduces a dimension that aligns with traditional applications of Chamomile in soothing respiratory ailments and muscle soreness. The presence of compounds like Alpha-Terpineol and Linalool, known for their calming and sedative properties, reinforces the herb's reputation as a natural remedy for anxiety and sleep disorders. This recognition of soothing compounds aligns with Chamomile's traditional use as a relaxing tea and as an ingredient in aromatherapy. These discoveries bolster the scientific basis for Chamomile's reputation as a natural relaxant [35].

16 The Harmonious Aromatic Ensemble

The diverse bouquet of volatile compounds identified in Chamomile underscores the complexity of its fragrance, akin to an orchestra of organic molecules. The interplay between these compounds creates a harmonious and multi-dimensional aromatic experience, reflecting the depth of nature's artistry. The synergy between Bisabolol and Alpha-Bisabolol, which play subtle but distinct roles in Chamomile's fragrance, exemplifies the nuanced complexity of this aromatic composition. Their cooperative presence enriches the herb's olfactory profile, contributing to its soothing and grounding qualities.

Linalool, a delicate floral note, and Camazulene, an unexpected yet enchanting tune, introduce layers of sophistication and charm to the aromatic ensemble. Their distinctive characters enhance the overall complexity of Chamomile's fragrance, adding depth and uniqueness to the herb's olfactory appeal. The bright and uplifting melody brought by Pinene infuses Chamomile with a vitality and freshness [36], elevating the aromatic experience. Farnesene's sweet and fruity notes contribute to the lushness of the fragrance, captivating the senses.

Cadinene's deep and earthy resonance anchors Chamomile's fragrance, establishing a strong foundation for the overall aroma. The presence of Alpha-Terpineol adds a layer of refinement and sophistication, underscoring the herb's elegance and depth. These compounds, working together in a delicate and intricate dance, create a rich aromatic tapestry that mirrors the diverse botanical compounds within the herb itself.

17 Implications and Future Directions

The identification of Chamomile's volatile compounds and their potential therapeutic properties holds significant implications for various fields, including herbal medicine, cosmetics, and the food industry. The scientific basis for Chamomile's traditional uses in addressing inflammatory conditions, promoting relaxation, and enhancing well-being is reinforced by the presence of compounds such as Bisabolol, Apigenin, and Camphor [37]. These findings pave the way for the development of Chamomile-based products that harness the herb's natural healing potential.

In the realm of cosmetics, the soothing and anti-inflammatory properties of Chamomile's volatile compounds offer exciting prospects for skincare products aimed at reducing skin irritations and promoting skin health [38]. The aromatic complexity of Chamomile, characterized by compounds like Linalool and Alpha-Terpineol, can be leveraged to create fragrances that evoke relaxation and calm, expanding the use of Chamomile in aromatherapy and perfumery.

Chamomile's diverse bouquet of volatile compounds also has intriguing implications for the food industry. The bright and uplifting notes contributed by compounds like Pinene, combined with the fruity allure of Farnesene, suggest the potential use of Chamomile in flavoring beverages and culinary creations [39]. The herb's aromatic complexity could enrich the sensory experiences of consumers, expanding its use in the culinary world.

As we move forward, the research on Chamomile's volatile compounds opens the door to further exploration of its therapeutic properties and applications. Investigating the synergistic effects of these compounds and their potential interactions with the human body will provide deeper insights into the herb's health benefits. Additionally, studies on the optimization of Chamomile cultivation and extraction techniques can lead to enhanced product quality and consistency.

In conclusion, our research journey into the aromatic world of Chamomile has not only deepened our understanding of the herb's fragrance but also illuminated the therapeutic potential of its volatile compounds. The aromatic ensemble of Chamomile is a testament to nature's artistry, with each compound playing a unique role in the herb's overall olfactory profile. With the scientific validation of Chamomile's traditional uses and the potential for new applications, the herb remains a timeless botanical treasure, offering both comfort and inspiration for future research and innovation [40].

18 Conclusion

Chamomile, with its soothing aroma and rich history as a natural remedy, continues to captivate our senses and inspire new avenues of research and application [41]. In our journey to explore the intricate chemistry of Chamomile's fragrance, we have unveiled a world of volatile compounds that contribute to its unique olfactory character. Our investigation, driven by the artistic precision of HS-SPME/GC-MS technology, has not only deepened our appreciation for the herb's natural artistry but has also expanded our knowledge of its therapeutic potential in health and well-being.

Chamomile's aromatic profile is a testament to the wonders of nature. The gentle whispers of Bisabolol, the fruity allure of Apigenin, and the cooling touch of Camphor all harmonize to create a fragrant symphony that has charmed humanity for generations. These compounds, among others, paint a vivid olfactory canvas that reflects the depth and complexity of this botanical treasure.

Our research has not only identified these volatile compounds but also highlighted their potential therapeutic applications. Bisabolol and Apigenin, with their anti-inflammatory and antioxidant properties, offer promising avenues for addressing a range of health concerns [42]. The presence of calming compounds like Linalool and Alpha-Terpineol reinforces the herb's traditional use as a natural relaxant. These findings strengthen the scientific basis for Chamomile's role in promoting well-being and providing comfort.

As we look ahead, the implications of our research reach far and wide. Chamomile's therapeutic potential is not limited to herbal remedies but extends to the cosmetics industry, where its soothing and anti-inflammatory properties can enhance skincare products. The herb's complex and multi-dimensional fragrance, characterized by compounds like Pinene and Farnesene, can inspire new fragrances that promote relaxation and calm. In the culinary world, Chamomile's aromatic complexity presents opportunities for flavoring and enhancing culinary creations [43], adding a new dimension to the culinary experience.

Our journey into Chamomile's aromatic world has been a testament to the beauty of nature and the potential for scientific exploration. It reinforces the notion that traditional wisdom and modern science can converge to unravel the secrets of botanical treasures. Chamomile, with its timeless appeal and therapeutic potential, remains a botanical marvel that continues to inspire and offer comfort.

In conclusion, our exploration of Chamomile's volatile compounds through the lens of HS-SPME/GC-MS technology has allowed us to appreciate the artistry of nature and the promise of science. Chamomile's aroma is a symphony of organic molecules, each playing a distinct role in its fragrance. This research deepens our understanding of the herb's therapeutic potential, enhancing its relevance in various fields. Chamomile, with its soothing fragrance and health benefits, continues to stand as a testament to the remarkable synergy between tradition and modern science [44].

References

1. Abd El-Aleem WH (2022) Middle East J Agric Res 11(3):770–780
2. Pastare L, Berga M, Kienkas L, Boroduskis M, Ramata-Stunda A, Reihmane D et al (2023) Exploring the potential of supercritical fluid extraction of *Matricaria chamomilla* white ray florets as a source of bioactive (cosmetic) ingredients. Antioxidants 12(5):1092
3. Madadi E, Fallah S, Sadeghpour A, Barani-Beiranvand H (2022) Exploring the use of chamomile (Matricaria chamomilla L.) bioactive compounds to control flixweed (Descurainia sophia L.) in bread wheat (Triticum aestivum L.): implication for reducing chemical herbicide pollution. Saudi J Biol Sci 29(11): 103421

4. Ojah EO (2020) Exploring essential oils as prospective therapy against the ravaging coronavirus (SARS-CoV-2). Iberoam J Med 2(4): 322–330
5. Sotiropoulou NS, Megremi SF, Tarantilis P (2020) Evaluation of antioxidant activity, toxicity, and phenolic profile of aqueous extracts of chamomile (Matricaria chamomilla L.) and sage (Salvia officinalis L.) prepared at different temperatures. Appl Sci 10(7):2270
6. Fahim JR, Darwish AG, El Zawily A, Wells J, Abourehab MA, Desoukey SY, Attia EZ (2023) Exploring the volatile metabolites of three Chorisia species: comparative headspace GC–MS, multivariate chemometrics, chemotaxonomic significance, and anti-SARS-CoV-2 potential. Saudi Pharm J 31(5):706–726
7. Capetti F, Cagliero C, Marengo A, Bicchi C, Rubiolo P, Sgorbini B (2020) Bio-guided fractionation driven by in vitro α-amylase inhibition assays of essential oils bearing specialized metabolites with potential hypoglycemic activity. Plants 9(9):1242
8. Chauhan R, Singh S, Kumar V, Kumar A, Kumari A, Rathore S et al (2021) A comprehensive review on biology, genetic improvement, agro and process technology of German chamomile (Matricaria chamomilla L.). Plants 11(1):29
9. Zarezadeh S, Riahi H, Shariatmadari Z, Sonboli A (2020) Effects of cyanobacterial suspensions as bio-fertilizers on growth factors and the essential oil composition of chamomile, Matricaria chamomilla L. J Appl Phycol 32: 1231–1241
10. Faehnrich B, Franz C, Nemaz P, Kaul HP (2021) Medicinal plants and their secondary metabolites—state of the art and trends in breeding, analytics and use in feed supplementation—with special focus on German chamomile. J Appl Bot Food Qual 94: 61–74
11. Jamzadfard M, Ebrahimi H (2020) Phyto-Pharmalogical Effect of Six Medicinal Plants used in Gastrosight herbal distillate as a Traditional Treatment of Gastrointestinal disorders
12. Jia Y, Zou J, Wang Y, Zhang X, Shi Y, Liang Y et al (2021) Action mechanism of Roman chamomile in the treatment of anxiety disorder based on network pharmacology. J Food Biochem 45(1):e13547
13. Clavijo McCormick A, Effah E, Najar-Rodriguez A (2023) Ecological aspects of volatile organic compounds emitted by exotic invasive plants. Front Ecol Evol 11:1059125
14. Lyu A, Yang L, Wu M, Zhang J, Li G (2020) High efficacy of the volatile organic compounds of Streptomyces yanglinensis 3-10 in suppression of Aspergillus contamination on peanut kernels. Front Microbiol 11: 142
15. Alsaraf S, Hadi Z, Al-Lawati WM, Al Lawati AA, Khan SA (2020) Chemical composition, in vitro antibacterial and antioxidant potential of Omani Thyme essential oil along with in silico studies of its major constituent. J King Saud Univ-Sci 32(1):1021–1028
16. Cordero CEI, Magagna F, Liberto E, Cobelli L, Stani G, Miliazza A et al (2015) Potential of the reversed-inject differential flow modulator for comprehensive two-dimensional gas chromatography in the profiling and fingerprinting of volatiles from complex food samples. In: Haislova J, Nielen M (eds) 7th international symposium on recent advances in food analysis, pp 152–152
17. Boggia L, Sgorbini B, Cagliero CL, Pignata G, Colombo ML, Nicola S et al (2015) Total volatile profile of Artemisia umbelliformis Lam. as a predictive marker of thujones amount in genepì liqueur. In: Abstracts. Società Botanica Italiana, pp 28–28
18. Cook MS (2023) Super-powered immunity: natural remedies for 21st century viruses and superbugs. Simon and Schuster
19. Segnit N (2023) Flavour thesaurus: more flavours: plant-led pairings, recipes and ideas for cooks. Bloomsbury Publishing
20. Balestrini R, Brunetti C, Cammareri M, Caretto S, Cavallaro V, Cominelli E et al (2021) Strategies to modulate specialized metabolism in Mediterranean crops: from molecular aspects to field. Int J Mol Sci 22(6): 2887
21. Ali B (2021) Salicylic acid: an efficient elicitor of secondary metabolite production in plants. Biocatal Agric Biotechnol 31:101884
22. Zhang B, Su D, Song Y, Li H, Chen C, Liao L et al (2023) Yueju volatile oil plays an integral role in the antidepressant effect by up-regulating ERK/AKT-mediated GLT-1 expression to clear glutamate. Fitoterapia 169: 105583
23. Andrade M, Ribeiro-Santos R, Silva AS (2021) Essential oils from plants: industrial applications and biotechnological production. In: Exploring plant cells for the production of compounds of interest, pp 145–170
24. Wang Z, Zhao X, Tang X, Yuan Y, Xiang M, Xu Y, Hong B, Gu Z (2023) Analysis of fragrance compounds in flowers of chrysanthemum genus. Ornam Plant Res. (OPR-2023-0012)

25. Ghamchini VM, Salami M, Mohammadi GR, Moradi Z, Kavosi A, Movahedi A et al (2019) The effect of chamomile tea on anxiety and depression in cancer patients treated with chemotherapy. J Young Pharm 11(3):309
26. Schwartz JA, Romeiser JL, Kimura R, Senzel L, Galanakis D, Halper D et al (2023) Effect of chamomile intake on blood coagulation tests in healthy volunteers: a randomized, placebo-controlled, crossover trial. Perioper Med 12(1):51
27. Takahashi JA, Rezende FAGG, Moura MAF, Dominguete LCB, Sande D (2020) Edible flowers: bioactive profile and its potential to be used in food development. Food Res Int 129:108868
28. Cui J, Li M, Wei Y, Li H, He X, Yang Q et al (2022) Inhalation aromatherapy via brain-targeted nasal delivery: natural volatiles or essential oils on mood disorders. Front Pharmacol 13:860043
29. Jugreet BS, Suroowan S, Rengasamy RK, Mahomoodally MF (2020) Chemistry, bioactivities, mode of action and industrial applications of essential oils. Trends Food Sci Technol 101:89–105
30. Hartley N, McLachlan CS (2022) Aromas influencing the GABAergic system. Molecules 27(8):2414
31. Grazina L, Mafra I, Monaci L, Amaral JS (2023) Mass spectrometry-based approaches to assess the botanical authenticity of dietary supplements. Compr Rev Food Sci Food Saf 22(5):3870–3909
32. Hemalatha J, Vinutha M, Sreelatha K, Saraswathi KT (2022) Exploration and characterization of Wild Cymbopogon martinii (Roxb) Wats. Var. Sofia. Mysore J Agric Sci 56(1)
33. Jerković I, Kuś PM, Carbonell-Barrachina ÁA (2019) Volatile organic compounds as artefacts derived from natural phytochemicals sourced form plants and honey. Phytochem Rev 18: 871–891
34. Fung TK, Lau BW, Ngai SP, Tsang HW (2021) Therapeutic effect and mechanisms of essential oils in mood disorders: interaction between the nervous and respiratory systems. Int J Mol Sci 22(9):4844
35. De Cicco P, Ercolano G, Sirignano C, Rubino V, Rigano D, Ianaro A, Formisano C (2023) Chamomile essential oils exert anti-inflammatory effects involving human and murine macrophages: evidence to support a therapeutic action. J Ethnopharmacol 311: 116391
36. Sharifi-Rad J, Cruz-Martins N, López-Jornet P, Lopez EPF, Harun N, Yeskaliyeva B et al (2021) Natural coumarins: exploring the pharmacological complexity and underlying molecular mechanisms. Oxidative Med Cell Longev 2021
37. El Mihyaoui A, Esteves da Silva JC, Charfi S, Candela Castillo ME, Lamarti A, Arnao MB (2022) Chamomile (Matricaria chamomilla L.): a review of ethnomedicinal use, phytochemistry and pharmacological uses. Life 12(4):479
38. Khan M, Fatima N, Wazir A, Akram Z, Ansari HR, Qamar F (2023) Promoting sleep health with herbal tea: development, evaluation, and anxiolytic effects. J Popul Ther Clin Pharmacol 30(17):2193–2205
39. Gracida-Osorno C, Jiménez-Martínez SL, Uc-Cachón AH, Molina-Salinas GM (2023) The use of complementary and alternative medicine among peritoneal dialysis patients at a second-level hospital in Yucatán Mexico. In: Healthcare, vol 11, No. 5. MDPI, p 722
40. DePaula J, Cunha SC, Ferreira IM, Porto ACV, Cruz AG, Petrarca M et al (2023) Volatile fingerprinting, sensory characterization, and consumer acceptance of pure and blended arabica coffee leaf teas. Food Res Int 173: 113361
41. Bouville AS, da Fonte K, Portes P, Lhommet JC, Azoulay S, Fernandez X (2023) Forgotten perfumery plants-part II: new insights into the development of novel fragrant ingredients–hawthorn case study. Chem Biodivers e202301020
42. Kumar D, Punetha A, Verma PP, Padalia RC (2022) Micronutrient based approach to increase yield and quality of essential oil in aromatic crops. J Appl Res Med Aroma Plants 26:100361
43. Madhumita M, Guha P, Nag A (2019) Extraction of betel leaves (Piper betle L.) essential oil and its bio-actives identification: process optimization, GC-MS analysis and anti-microbial activity. Ind Crop Prod 138:111578
44. Sadgrove NJ, Padilla-González GF, Leuner O, Melnikovova I, Fernandez-Cusimamani E (2021) Pharmacology of natural volatiles and essential oils in food, therapy, and disease prophylaxis. Front Pharmacol 12:740302

Chapter 6

Quantification of Matcha Adulterants with Laser-Induced Breakdown Spectroscopy

Jiyu Peng, Ming Lin, and Xuming Jiang

Abstract

The determination of adulterant content in matcha is crucial for ensuring food safety and protecting consumer rights. Rapidly detecting the content of matcha adulterants poses technical challenges due to its powdered form. This section presents an introduced method based on laser-induced breakdown spectroscopy (LIBS). The method is applied to two different grades of matcha adulterants, and comprehensive descriptions are provided for sample preparation, LIBS analysis, and data analysis. Two sample preparation techniques, compression and filtration, are employed in this study. Both methods are applicable for powdered samples, and the data presented in this case are obtained using the filtration method. Data analysis is conducted using machine learning algorithms to accurately predict the content of adulterants in matcha. The results demonstrate the capability of the described method in enabling fast quantification of matcha adulterants.

Key words LIBS spectrum, Matcha, Adulteration, Chemometric methods, Information fusion

1 Introduction

Matcha is a powdered tea product derived from the fresh leaves of the tea tree, which undergo drying subsequent to steam (or hot air) fixation, followed by a meticulous grinding process. Matcha is renowned for its exceptional nutritional value, documented to bolster physical vitality and protract the natural aging progression of organisms [1, 2]. However, the market for matcha is characterized by varying product qualities, and the practice of substituting inferior products for genuine ones prevails. Common additives found in market-available matcha include white sugar, maltodextrin, mulberry leaf powder, barley seedling powder, and chlorella. This pervasive issue not only infringes upon consumer rights but also disrupts the integrity of the matcha market in China. Conse-

Mostafa Gouda et al. (eds.), *Plant Chemical Compositions and Bioactivities*, Methods and Protocols in Food Science,
https://doi.org/10.1007/978-1-0716-3938-2_6,

quently, the identification of adulterated matcha demands a swift and efficacious solution, coupled with robust regulatory measures to restore order in the market.

Currently, various methodologies have been employed for the detection of powders, including microscopic observation [3], chromatographic techniques [4], mass spectrometry [5], among others. The microscopic observation necessitates the utilization of suitable microscope parameters to examine distinct aspects of the sample. Both mass spectrometry and chromatographic techniques entail the separation and detection of the sample. While these approaches have demonstrated favorable outcomes, they all necessitate the process of sampling, pretreatment, and subsequent laboratory analysis, resulting in significant time consumption.

Laser-induced breakdown spectroscopy (LIBS) is a type of laser-based atomic emission spectroscopic technique employed to ascertain the composition and content of substances within a sample. It accomplishes this by directing an ultrashort pulse of laser energy onto the surface of the sample, resulting in the formation of a plasma, and subsequently analyzing the emitted spectrum of the plasma. Due to the advantages of simple sample pretreatment, in situ detection, and real-time analysis, LIBS is gradually becoming popular in the detection of food adulteration [6, 7]. To prevent the degradation of LIBS signal stability caused by powder sputtering during laser ablation, previous research involved compacting the powder sample into tablets [8–10]. Besides this, a novel approach by filtration for preparing powder samples in LIBS analysis is introduced. Two distinct sample preparation techniques are described, namely compression and filtration methods. Each approach offers its own set of advantages, with the pressed slice method yielding a more steadfast acquisition of LIBS signals, while the filtration method proves to be a swifter and lower experimental cost of sample preparation. The selection of a specific method can be predicated upon your specific requirements.

While LIBS provides elemental information about a sample, its signal tends to exhibit fluctuations influenced by factors like sample shape and laser energy [11]. To enhance performance and mitigate the impact of signal fluctuations on identification accuracy, one approach is to incorporate additional data for a more comprehensive analysis of the sample [12, 13]. Data fusion has demonstrated effectiveness in improving recognition accuracy by integrating data from other sensors, such as molecular structure [14] and texture [15]. This chapter introduces a method that combines LIBS and texture information features. Specifically, extreme gradient boosting (XGBoost) is used to extract LIBS features and gray-level co-occurrence matrix (GLCM) is employed to extract texture features from matcha samples. Subsequently, the two sets of features are combined and fused together.

In this chapter, a method for fast quantification of matcha adulterants with laser-induced breakdown spectroscopy spectrum has been described. The method described has the advantages of simplicity, short preparation time, low experimental cost, and mitigation of powder splashing effects. In detail, we present the steps of sample preparation, analysis of LIBS, information fusion, and data processing.

2 Materials

2.1 LIBS Setup

1. Laser: In practical applications, pulse lasers are commonly selected, including options such as Nd:YAG, Ti:Sapphire, ArF excimer, and CO_2 lasers. Furthermore, the utilization of femtosecond lasers is also a variable choice (*see* **Note 1**).
2. Spectrometer: Various spectrometers are available for different applications, including Czerny-Turner, Echelle, and Multichannel spectrometers (*see* **Note 1**). In addition, the temperature should be kept constant if Echelle spectrometers are used.
3. Camera (*see* **Note 1**).
4. Displacement module (*see* **Note 2**).
5. Optical module (*see* **Note 3**).

2.2 Supplies

1. Electronic balance (max = 120 g; min = 10 mg; d = 0.1 mg; e = 1 mg).
2. Magnetic stirrer (Speed control range is 100–2300 r/min).
3. Pipette (100–1000 μL).
4. Glass flask (100 mL).
5. Conical flask (100 mL).
6. Aqueous microporous filtration membrane (Diameter = 25 mm, Pore size = 0.45 μm).

2.3 Samples

1. Powder: Powder samples can be utilized as materials for analysis employing the following method (*see* **Note 4**).
2. Adulteration: In essence, the elemental composition of the adulterant ought to be akin to that of the sample (*see* **Note 5**).

3 Methods

3.1 Sample preparation

1. Mix the powder with different proportions of adulterants to build the training set for establishing models of quantifying adulterant content (*see* **Notes 6** and **7**).
2. Similarly, configure different proportions of adulterant test samples for external testing (*see* **Notes 6** and **8**).

3.1.1 Alternative Method 1 (Compression)

1. Weigh 100 mg of powder into a 10 mm × 10 mm square mold.
2. Use the tablet machine to press the powder into a tablet at the pressure of 600 MPa.
3. Transfer the tablets to a 50 mm × 50 mm plate that is placed on an X-Y-Z moving platform.

3.1.2 Alternative Method 2 (Filtration)

1. Weigh 0.1 g of each mixed sample on an electronic balance and dissolve it in 20 g of distilled water.
2. The obtained solution was stirred by a magnetic stirrer (*see* **Note 9**).
3. Using a pipette, transfer 400 μg of the solution from the stirred solution to the filtration device.
4. Allow the solution to stand for a duration to ensure thorough filtration (*see* **Note 10**).
5. The precipitate was placed on a 50 mm × 50 mm plate that was placed on an X-Y-Z moving platform.

3.2 LIBS Analysis

1. Utilizing a mercury argon lamp to calibrate the spectrograph's wavelength (*see* **Note 11**).
2. Utilizing a deuterium halogen source to calibrate the signal intensity of the detector (*see* **Note 12**).
3. Adjust the position of the sample surface with the moving platform to focus the laser on the sample surface (*see* **Note 13**).
4. The moving platform rises 2 mm after the focus is completed (*see* **Note 14**).
5. Single-shot scanning was performed in an ablation region of 8 mm × 8 mm with a resolution of 1 mm. Hence 64 successive spectra were collected for each sample (*see* **Notes 15** and **16**).
6. Use avasoft8 to convert raw data to csv format.
7. The spectra were averaged to minimize the sample inhomogeneity. Figure 1 shows the average spectra of first-grade matcha, second-grade matcha, and adulterant, along with the corresponding elements for the primary emission lines (*see* **Note 17**).
8. The spectral data of all samples are averaged at each individual wavelength to obtain the mean spectrum (*see* **Notes 18** and **19**).
9. For each sample of spectral data, the mean spectrum was subtracted to effectively center the overall data around zero.
10. Individually computed the standard deviation individually for each wavelength of the centered data.
11. For each wavelength of each sample, the centered data were divided by the standard deviation of the corresponding wavelength.

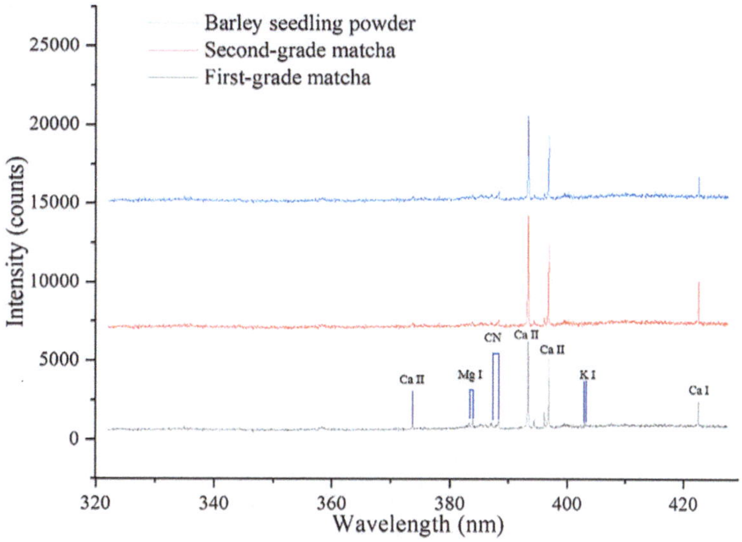

Fig. 1 Average spectra of matcha and adulterants

3.3 Modeling Methods

A demonstration of the modeling approach utilizing SVR is conducted. Various machine learning techniques, including principal component analysis (PCA), random forest (RF), and others, can be employed [16, 17]. Furthermore, deep learning has demonstrated unique advantages in relation to the increasing volume of data and advancements in computing capabilities, making it a viable choice [17]. Hence, it is crucial to select the most appropriate modeling method based on different data sizes and specific requirements.

3.3.1 Modeling Methods Based on Full Spectrum

1. Prepare the training and test datasets. Import the LIBS spectra obtained through the aforementioned approach into PyCharm in csv format (*see* **Note 20**).
2. Import the necessary libraries, including sklearn.svm, pandas, numpy, and so on (*see* **Note 21**).
3. Calling "SVR" to build a model from "sklearn.svm" and determine the optimal values for the parameters C and gamma (*see* **Note 22**).
4. The model is trained using the data from the calibration set. Invoke the "fit" method in SVR and import the data and labels from the calibration set.
5. The trained model is employed for making predictions. More specifically, the predictions are generated by feeding the data from the prediction set into the model.
6. Evaluate model performance. Invoke the "r2_score" and "mean_squared_error" method in "sklearn.metrics" and import the actual value and predicted value. The results for quantification of matcha adulteration based on full spectrum are shown in Table 2 (*see* **Note 23**).

3.3.2 Modeling Methods Based on Important Variables

LIBS spectra encompass a multitude of variables, a significant portion of which comprise redundant information. In order to enhance the efficacy of predictive models, extracting crucial variables represents a prevalent strategy. This section delineates a methodology for extracting significant variables employing XGBoost and subsequently employing them for SVR training. XGBoost has the ability to automatically derive the significance of variables, thereby allowing the identification and elimination of extraneous variables. Variable importance refers to the degree of utility it contributes to model development. The frequency of a variable's usage serves as an indicator of its importance [18]. Apart from XGBoost, alternative feature extraction techniques like PCA, CNN, and others can be employed.

1. Importing the "xgb" module from the "XGBoost" library (*see* **Note 24**).
2. Instantiate an XGBoost regression model and conduct training.
3. Determining the significance and ranking of features.
4. Utilize the extracted significant variables as a novel calibration and prediction datasets. The feature spectral lines and their corresponding elements selected by XGBoost are shown in Table 1 (*see* **Notes 17** and **25**).

Table 1
Feature variables selected by XGBoost for matcha adulteration analysis

No.	Importance	Spectral line (nm)	Element
1	0.089[a]/0.015[b]	373.69	Ca II
2	0.036[a]/0.063[b]	383.19	Mg II
3	0.025[a]/0.051[b]	383.82	Mg I
4	0.038[a]/0.025[b]	387.08	CN 1-1
5	0.060[a]/0.029[b]	388.31	CN 0-0
6	0.109[a]/0.289[b]	393.33	Ca II
7	0.119[a]/0.122[b]	396.82	Ca II
8	0.024[a]/0.014[b]	404.39	K I
9	0.022[a]/0.012[b]	404.72	K I
10	0.260[a]/0.187[b]	422.56	Ca I

[a]Important variables selected from Fuding matcha
[b]Important variables selected from Jiukeng matcha

Table 2
Results for quantification of matcha adulteration based on different features

			Calibration set		Prediction set	
Category	**Features**	**Model**	***R***	**RMSE**	***R***	**RMSE**
Fuding	Full spectrum	SVR	0.9463	0.10	0.8293	0.16
	Important variables	XGBoost-SVR	0.9619	0.08	0.8768	0.14
Jiukeng	Full spectrum	SVR	0.9659	0.08	0.7853	0.18
	Important variables	XGBoost-SVR	0.9682	0.09	0.8869	0.14

5. Train the important variables, which has been extracted using XGBoost, using the SVR model. The process follows **steps 2–6** outlined in Subheading 3.3.1.
6. Use the trained model to predict the spectrum of the prediction set, and the results are shown in Table 2.

3.3.3 Modeling Methods Based on Information Fusion

1. Adjust the position of the sample surface with the moving platform to acquire the best image sharpness.
2. Use the camera to capture the complete image.
3. Extraction of texture features from a region of interest using GLCM (*see* **Notes 16** and **26**).
4. The feature variables extracted from the LIBS spectrum and LIBS ablation image were fused by a feature layer fusion which is to connect these feature variables from beginning to end (*see* **Note 27**).
5. The fusion features were trained in the SVR model. The process follows **steps 2–6** outlined in Subheading 3.3.1.
6. Use the trained model to predict the samples in the prediction set, and the results are shown in Fig. 2.

4 Notes

1. Herein, an Nd:YAG Q-switched pulse laser (Dawa 200, Beamtech, Beijing, China), a high-speed CMOS spectrometer (AvaSpec-ULS2048CL-EVO, Avantes, Apeldoorn, NL), and a camera (BFLY-U3-23S6C-M, FLIR, Oregon, USA) were used. The detailed specifications of the Dawa-200 pulse laser can be found in Table 3, and the specifications of the AvaSpec-ULS2048CL-EVO high-speed CMOS spectrometer are presented in Table 4.

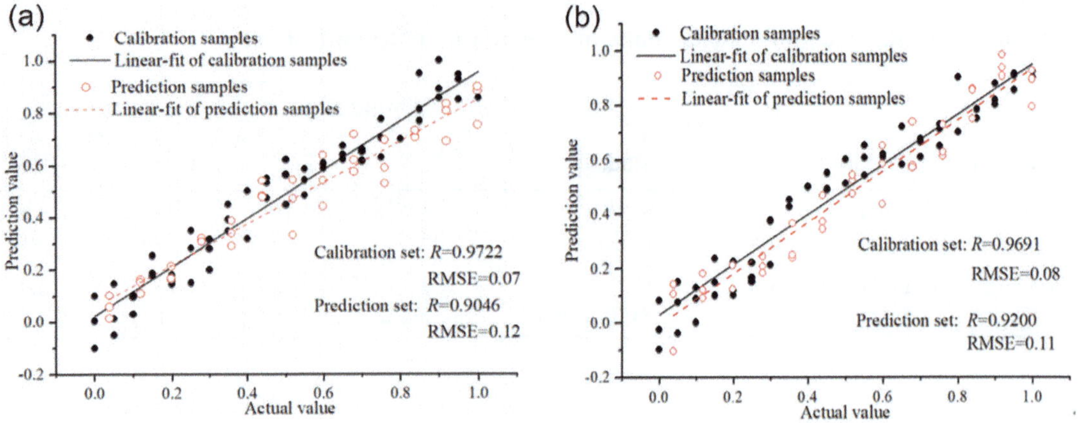

Fig. 2 Quantitative analysis results of adulteration content for (**a**) Fuding matcha and (**b**) Jiukeng matcha based on the fusion of spectral features and image textural features

Table 3
Specifications of Dawa-200 pulse laser

Feature	Specification
Output wavelength	1064 nm
Single pulse energy	200 mJ@1064 nm
Pulse frequency	1–20 Hz
Spot diameter	6 mm
Pulse width	≤8 ns
Energy stability	≤4%
Divergence angle	≤3 mrad
Beam profile	Gaussian

2. To adjust the position of the sample, a custom-designed displacement module was employed, comprising a motion controller, a driver, and a motorized displacement slide. Specifically, the motorized displacement slide is connected to the driver, while the driver is linked to the SMC304 advanced point controller (Raysay, Shenzhen, China). The displacement module incorporates an X-Y-Z three-axis electric moving platform, which is constructed using the electric displacement slide (Yiheda, Dongguan, China) of the long-stroke type, with the drive direction oriented vertically. This motorized moving platform allows movement in three-dimensional space, with a travel range of 100 mm in the X and Y directions, 12 mm in the Z direction, a repeatable positioning accuracy of ±1 μm, and a maximum load capacity of 20 kg.

Table 4
Specifications of AvaSpec-ULS2048CL-EVO high-speed CMOS spectrometer

Feature	Specification
Optical platform	Czerny-Turner, 75 mm focal distance
Wavelength range	320–428 nm
Resolution	0.1 nm
Stray light	0.19–1.0%, depends on raster
Sensitivity	375,000 counts/μW per ms integration time
Detector	CMOS
Signal-to-noise ratio	300:1
AD converter	16 bit, 6 MHz
Integration time	9 μs–59 s
Sampling rate	0.38 ms/scan

3. The optical module is designed to adjust and reflect the laser light to illuminate the subject and to receive the light. The optical module consists of several components, namely a half-wave plate, a polarizer, a reflector, a beam splitter lens, and a focusing lens. Through the manipulation of the phase using the half-wave plate, precise control over pulse energy can be achieved as the laser pulse traverses the polarizer. The reflector serves the purpose of guiding the optical pathway, while the beam splitter mirror facilitates beam separation and redirection, allowing the camera to capture the sample and achieve optimal focus in this experiment. Lastly, the focusing lens enables the laser to effectively concentrate its energy onto the surface of the sample. In this experiment, the lenses are manufactured by THORLAB.
4. Matcha, including first-grade matcha (Fuding) and second-grade matcha (Jiukeng), was obtained from a local matcha producer.
5. Herein, barley seedling powder was used as an adulterant, which was purchased from markets.
6. The content of the mixture is 1 g, and the content of matcha powder and barley powder is determined according to different percentages.
7. First-grade matcha and second-grade matcha were adulterated with barley seedling powder at 21 different percentages (0%, 5%, 10%, 15%, 20%, 25%, 30%, 35%, 40%, 45%, 50%, 55%, 60%, 65%, 70%, 80%, 85%, 90%, 95%, and 100%), respectively.

8. First-grade matcha and second-grade matcha were adulterated with barley seedling powder at 13 different percentages (0%, 8%, 16%, 24%, 32%, 40%, 48%, 56%, 64%, 72%, 80%, 88%, and 96%), respectively.
9. Herein, the rotation speed is configured at 200 revolutions per minute (rpm), which can be adjusted based on your requirements and the solution volume.
10. When no liquid is observed to drip into the cone, the filtration is considered complete. Herein, it sat for three minutes. The obtained precipitate remains on the filter membrane.
11. Wavelength correction of the spectrometer is necessary to compensate for wavelength drift resulting from the aging of the instrument, variations in external temperature, and other influencing factors. In this experiment, the spectrograph underwent wavelength correction using the HG-1 (Ocean Optics, Florida, USA) calibration device. The measured wavelength was adjusted based on the known mercury-argon radiation spectrum provided by the HG-1.
12. Similar to wavelength correction, the DH-2000-BAL-CAL (Ocean Optics, Florida, USA) is employed to rectify the intensity signal of the detector. The measured signal intensity is calibrated by juxtaposing it with the standard intensity provided by the DH-2000-BAL-CAL.
13. Focusing can be achieved either automatically or manually. Herein, image sharpness serves as the criterion for assessing the completeness of focusing. The displacement module is employed to manipulate the sample's position, while the climbing method is utilized to calculate the distance of movement.
14. The purpose of the rise is to negatively defocus the laser, so that the laser can fully hit the surface. The distance moved is obtained experimentally, and for different samples, the corrected distance yielding the highest signal-to-noise ratio (SNR) or signal-to-background (SBR) needs to be recalculated through pre-experimental evaluation.
15. Laser ablation is performed at atmospheric pressure with (pulse frequency = 1 Hz, acquisition delay time = 2 μs, integrating time = 10 μs, laser pulse energy = 100 mJ).
16. The ablation region is regarded as a region of interest.
17. The main emission lines of LIBS in matcha can be labeled with the corresponding elements according to the National Institute of Standards and Technology (NIST).

18. Data preprocessing including area normalization (AN), min-max normalization (MMN), multiplicative scatter correction (MSC), and SNV can effectively reduce the possible fluctuations of the LIBS spectrum to improve the stability of the signal [19]. Different data preprocessing methods can be selected according to different spectral line signals.
19. Here, the full LIBS spectrum is subjected to standard normal variable (SNV) processing, and the subsequent steps involved in SNV processing of LIBS are described below.
20. Herein, use Python (version 3.7) for all data processing.
21. The "sklearn" library is a renowned Python machine learning library renowned for its extensive collection of machine learning algorithms, tools, and utility functions. The "svm" module specifically focuses on support vector machines (SVM) and its associated algorithms and tools. "Pandas" is a widely utilized open-source library for data analysis and data processing. This article centers around leveraging its capabilities for data importation and exportation. The "numpy" library facilitates the creation of multidimensional array objects, known as "ndarray", and provides efficient array operations. In the field of machine learning, it is customary to transform data into "ndarray" to enable effective data manipulation.
22. Machine learning model usually required tuning to obtain the best performance (choosing the best parameters to obtain the highest R or the lowest RMSE). The tuning can be done by grid search or random search to find the optimal parameters. For the SVR model, the two hyperparameters are tuned using a grid search.
23. The root mean squared error (RMSE) and the correlation coefficient (R) were used as indicators to evaluate the performance of models. The R value quantifies the correlation between the predicted and actual values, while the RMSE value quantifies the extent of predictive error. A higher R value and a lower RMSE value indicate superior model performance.
24. The "XGBoost" library is a powerful machine learning toolkit designed specifically for implementing the XGBoost algorithm. Its key functionalities encompass constructing XGBoost regression models, assessing the performance of important features, and more.
25. XGBoost should select the optimal number of features by considering a range of 0–60 significant spectral information, and the corresponding curve can be depicted as illustrated in Fig. 3.

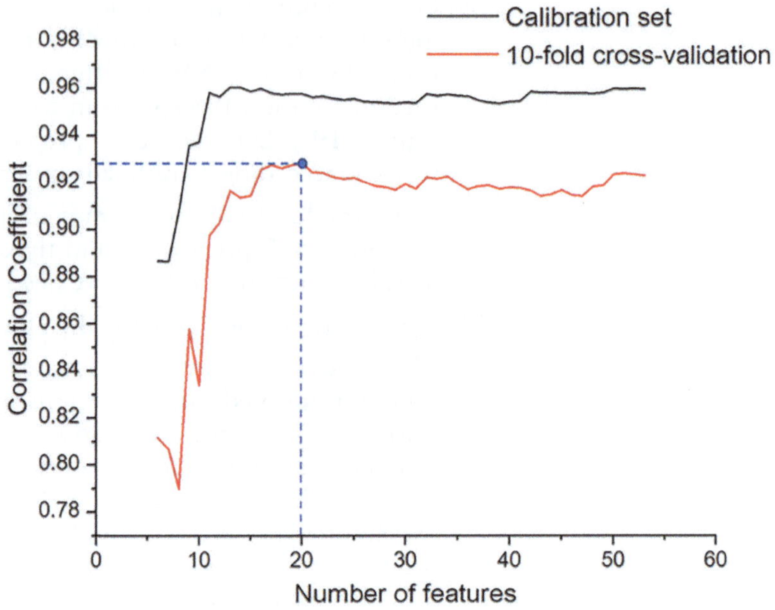

Fig. 3 *R* value of XGBoost models with different feature numbers as input

26. GLCM analysis can provide several second-order statistics used to describe the gray-level relationships within a neighborhood around a pixel of interest [20]. Herein, total of six textural features including contrast, dissimilarity, homogeneity, energy, correlation, and angular second moment (ASM) were extracted from the GLCM calculation. For each image, five distances from 1 to 5 pixels and four directions (0°, 45°, 90°, and 135°) were implemented to extract textural features and then these textural values were integrated into one value that represents the textural feature of the sample. The textural feature values were obtained by calculation of six equations $\text{Constrast} = \sum_i \sum_j (i-j)^2\ P(i,j)$, $\text{Dissimilarity} = \sum_i \sum_j P(i,j) \times |i-j|$, $\text{Homogeneity} = \sum_i \sum_j \frac{P(i,j)}{1+(i-j)^2}$, $\text{ASM} = \sum_i \sum_j P(i,j)^2$, Correlation = , $\text{Energy} = \sqrt{\text{ASM}}$ where $P(i,j)$ denotes the value of the i row and j column in the gray-level co-occurrence matrix, and μ_x, μ_y, σ_x, σ_y are the expectation and variance values of the statistics of the matrix elements on the rows and columns, respectively. ($\mu_x = \sum_i \sum_j i \cdot P(i,j), \mu_y = \sum_i \sum_j j \cdot P(i,j), \sigma_x = \sqrt{\sum_i \sum_j (i-\mu_x)^2\ P(i,j)}, \sigma_y = \sqrt{\sum_i \sum_j \left(i-\mu_y\right)^2\ P(i,j)}$).
27. The feature variables of the LIBS spectrum are in the front, and the feature variables of images are in the end.

Acknowledgments

This work was supported by the Ministry of Science and Technology of the People's Republic of China (grant number 2018YFD0700502), Science Technology Department of Zhejiang Province (grant number 2023C02043), National Natural Science Foundation of China (grant number 62105290), and Zhejiang Agricultural Major Technology Collaborative Promotion Plan (grant number 2022XTTGCY03-05).

References

1. Devkota HP, Gaire BP, Hori K, Subedi L, Adhikari-Devkota A, Belwal T et al (2021) The science of matcha: bioactive compounds, analytical techniques and biological properties. Trends Food Sci Technol 118:735–743. https://doi.org/10.1016/j.tifs.2021.10.021
2. Yu XL, He Y (2018) Fast nondestructive identification of steamed green tea powder adulterations in matcha by visible spectroscopy combined with chemometrics. Spectrosc Lett 51(2):112–117. https://doi.org/10.1080/00387010.2018.1430702
3. Sano EE, Assad ED, Cunha SAR, Correa TBS, Rodrigues HR (2003) Quantifying adulteration in roast coffee powders by digital image processing. J Food Qual 26(2):123–134. https://doi.org/10.1111/j.1745-4557.2003.tb00232.x
4. Jablonski JE, Moore JC, Harnly JM (2014) Nontargeted detection of adulteration of skim milk powder with foreign proteins using UHPLC-UV. J Agric Food Chem 62(22): 5198–5206. https://doi.org/10.1021/jf404924x
5. Tay M, Fang G, Chia PL, Li SF (2013) Rapid screening for detection and differentiation of detergent powder adulteration in infant milk formula by LC-MS. Forensic Sci Int 232(1-3):32–39. https://doi.org/10.1016/j.forsciint.2013.06.013
6. Sezer B, Bjelak A, Velioglu HM, Boyaci IH (2022) Identification of meat species in processed meat products by using protein based laser induced breakdown spectroscopy assay. Food Chem 372. https://doi.org/10.1016/j.foodchem.2021.131245
7. Peng JY, Xie WY, Jiang JD, Zhao ZF, Zhou F, Liu F (2020) Fast quantification of honey adulteration with laser-induced breakdown spectroscopy and chemometric methods. Foods 9(3). https://doi.org/10.3390/foods9030341
8. Bilge G, Sezer B, Eseller KE, Berberoglu H, Topcu A, Boyaci IH (2016) Determination of whey adulteration in milk powder by using laser induced breakdown spectroscopy. Food Chem 212:183–188. https://doi.org/10.1016/j.foodchem.2016.05.169
9. Zhao ZF, Wang QQ, Xu XJ, Chen F, Teng G, Wei K et al (2022) Accurate identification and quantification of Chinese yam powder adulteration using laser-induced breakdown spectroscopy. Foods 11(9):1216. https://doi.org/10.3390/foods11091216
10. Sezer B, Unuvar A, Boyaci IH, Koksel H (2022) Rapid discrimination of authenticity in wheat flour and pasta samples using LIBS. J Cereal Sci 104:103435. https://doi.org/10.1016/j.jcs.2022.103435
11. Hou ZY, Wang Z, Lui SL, Yuan TB, Li LZ, Li Z et al (2013) Improving data stability and prediction accuracy in laser-induced breakdown spectroscopy by utilizing a combined atomic and ionic line algorithm. J Anal Atom Spectrom 28(1):107–113. https://doi.org/10.1039/c2ja30104k
12. Alvarez-Llamas C, Purohit P, Moros J, Laserna J (2022) LIBS-acoustic mid-level fusion scheme for mineral differentiation under terrestrial and martian atmospheric conditions. Anal Chem 94(3):1840–1849. https://doi.org/10.1021/acs.analchem.1c04792
13. Teng G, Wang QQ, Cui XT, Wei K, Xiangli WT, Chen GY (2021) Feature-level fusion of laser-induced breakdown spectroscopy and Raman spectroscopy for improving support vector machine in clinical bacteria identification. J Raman Spectrosc 52(4):805–814. https://doi.org/10.1002/jrs.6071
14. Liang J, Li MG, Du Y, Yan CH, Zhang Y, Zhang TL et al (2020) Data fusion of laser induced breakdown spectroscopy (LIBS) and infrared spectroscopy (IR) coupled with random forest (RF) for the classification and

discrimination of compound salvia miltiorrhiza. Chemometr Intell Lab 207:104179. https://doi.org/10.1016/j.chemolab.2020.104179

15. Zhang D, Nie JF, Niu XC, Chen F, Hu ZL, Wen XL et al (2022) Time-resolved spectral-image laser-induced breakdown spectroscopy for precise qualitative and quantitative analysis of milk powder quality by fully excavating the matrix information. Food Chem 386:132763. https://doi.org/10.1016/j.foodchem.2022.132763
16. Liu F, Wang W, Shen TT, Peng JY, Kong WW (2019) Rapid identification of kudzu powder of different origins using laser-induced breakdown spectroscopy. Sensors-Basel 19(6):1453. https://doi.org/10.3390/s19061453
17. Huang WH, Guo LB, Kou WP, Zhang D, Hu ZL, Chen F et al (2022) Identification of adulterated milk powder based on convolutional neural network and laser-induced breakdown spectroscopy. Microchem J 176:107190. https://doi.org/10.1016/j.microc.2022.107190
18. Perez-Rodriguez M, Dirchwolf PM, Silva TV, Villafane RN, Neto JAG, Pellerano RG et al (2019) Brown rice authenticity evaluation by spark discharge-laser-induced breakdown spectroscopy. Food Chem 297:124960. https://doi.org/10.1016/j.foodchem.2019.124960
19. Lu Y, Guo H, Shen TT, Wang W, He Y, Liu F (2019) Quantitative analysis of cadmium and zinc in algae using laser-induced breakdown spectroscopy. Anal Methods 11(48):6124–6135. https://doi.org/10.1039/c9ay01681c
20. Ma F, Qin H, Shi K, Zhou C, Chen C, Hu X et al (2016) Feasibility of combining spectra with texture data of multispectral imaging to predict heme and non-heme iron contents in pork sausages. Food Chem 190:142–149. https://doi.org/10.1016/j.foodchem.2015.05.084

Chapter 7

High-Performance Liquid Chromatography (HPLC) Analytical Modelling for Measuring the Onion Anthocyanin Profile

Sanusi Shamsudeen Nassarawa, Mustapha Muhammad Nasiru, Munir Abba Dandago, and Nassarawa Isma'il Sanusi

Abstract

This chapter provides a concise overview of the essay on "High-Performance Liquid Chromatography (HPLC) Analytical Modelling for Measuring the Onion Anthocyanin Profile." It outlines the main content and highlights the significance of the study. The chapter delves into the application of High-Performance Liquid Chromatography (HPLC) in analyzing the anthocyanin profile of onions. Anthocyanins are vibrant pigments found in plants, contributing to their coloration and health-promoting properties. The essay presents a comprehensive exploration of the methodology, benefits, and implications of using HPLC to measure onion anthocyanin profiles. The significance of studying anthocyanins and their role in onions, establishing the rationale for employing HPLC as the analytical technique are also contextualized. The essay elaborates on the principles of HPLC and its methodology, encompassing sample preparation, chromatographic conditions, and detection techniques. It details the quantification and identification of anthocyanins using reference standards and data analysis software. The benefits of HPLC in anthocyanin analysis are underscored, emphasizing its precision, sensitivity, and ability to provide a comprehensive profile of onion anthocyanins. The implications of such analysis are discussed, ranging from nutritional assessment to potential applications in quality control and plant breeding. Challenges and future directions in the field are acknowledged, paving the way for advancements in HPLC technology.

Key words High-performance liquid chromatography (HPLC), Analytical modelling, Onion anthocyanin profile, Bioactive compounds, Natural food coloring

1 Introduction

High-Performance Liquid Chromatography (HPLC) stands as a cornerstone of modern analytical chemistry, enabling researchers to unravel the intricacies of complex mixtures with unparalleled

Authors Sanusi Shamsudeen Nassarawa and Mustapha Muhammad Nasiru have equally contributed to this chapter.

Mostafa Gouda et al. (eds.), *Plant Chemical Compositions and Bioactivities*, Methods and Protocols in Food Science, https://doi.org/10.1007/978-1-0716-3938-2_7,

precision [1, 2]. In the realm of plant science and nutrition, the application of HPLC has been particularly illuminating, allowing for the detailed examination of bioactive compounds that contribute not only to color and flavor but also to the health-promoting attributes of various plant-based foods [3]. This introduction sets the stage for delving into the innovative utilization of HPLC analytical modelling for measuring the anthocyanin profile of onions, shedding light on the multifaceted nature of these pigments and their potential implications for both dietary and industrial domains [4]. Anthocyanins, a subgroup of flavonoids, are responsible for the vibrant colors exhibited by many fruits, vegetables, and flowers. In addition to their aesthetic appeal, anthocyanins have garnered substantial attention for their antioxidant, anti-inflammatory, and potential health-promoting properties [5–7]. The depth and diversity of anthocyanin profiles in plant tissues provide a fascinating avenue for research, offering insights into not only the physiological functions within plants but also their impact on human health when incorporated into the diet [8, 9].

Onions, despite their ubiquitous presence in culinary applications worldwide, have remained an intriguing subject of investigation due to their complex array of bioactive compounds [10, 11]. Anthocyanins, often overshadowed by other pigments in onions, have gained prominence for their potential contribution to the health attributes of this humble vegetable [12]. The application of HPLC in this context presents an opportunity to comprehensively explore the anthocyanin content of onions, unveiling the spectrum of these pigments that may offer nutritional and functional benefits.

This essay embarks on a journey to uncover the intricacies of utilizing HPLC analytical modelling to measure the onion anthocyanin profile. It delves into the principles underlying HPLC, providing an overview of its methodology and the unique considerations it entails when applied to the complex matrix of onions. By harnessing the power of this analytical technique, researchers gain the ability to quantify and identify various anthocyanin compounds, enabling a deeper understanding of the onion's color composition and its potential contributions to human health. As the boundaries of scientific exploration continue to expand, the intersection of analytical chemistry, plant biology, and nutrition opens new avenues for enhancing our understanding of the intricate relationship between plants and human well-being.

2 Anthocyanins and Their Role in Plant Pigmentation

The captivating array of colors found in plants has fascinated scientists, artists, and nature enthusiasts for centuries. Behind these vivid hues lies a class of natural pigments known as anthocyanins, which play a pivotal role in shaping the vibrant and diverse palette of the botanical world [13]. This introduction delves into the significance of studying anthocyanins and unveils their central role in plant pigmentation, illustrating how these compounds contribute to both the aesthetic beauty and biological functionality of plants. Anthocyanins are a group of water-soluble pigments responsible for the red, blue, purple, and even black shades exhibited by various parts of plants, including fruits, leaves, flowers, and stems [6]. Beyond their visual appeal, anthocyanins have attracted substantial attention due to their multifaceted roles in plants. These compounds emerge as key players in ecological interactions, physiological processes, and, remarkably, the evolution of both plants and their surrounding ecosystems.

2.1 Anthocyanins and Plant Pigmentation

Plant pigmentation is far more than an aesthetic phenomenon; it is a vital adaptation that drives ecological dynamics and survival strategies. Anthocyanins serve as a critical aspect of this adaptive mechanism. They often act as a sunscreen for plants, shielding sensitive tissues from the harmful effects of excessive ultraviolet (UV) radiation [7]. This natural photoprotection not only safeguards the plant's DNA but also prevents photooxidative stress that could inhibit growth and development [14]. Beyond UV protection, anthocyanins also influence various ecological interactions. The striking colors conferred by these pigments can attract pollinators, which play an essential role in plant reproduction [15]. By serving as visual cues for potential pollinators, anthocyanin-rich flowers ensure successful fertilization and subsequent seed production, contributing to the perpetuation of plant species [15].

2.2 Biological Signaling and Stress Response

Anthocyanins extend their influence beyond pigmentation and pollination. Emerging evidence suggests that these compounds act as potent antioxidants, participating in the plant's defense against oxidative stress caused by environmental factors such as pollutants, pathogens, and adverse climatic conditions [3, 16]. This dual role of anthocyanins—both as pigments and antioxidants—underscores their versatile nature and their potential contributions to plant resilience [17]. The prevalence of anthocyanins across a diverse range of plant species indicates their evolutionary significance. The synthesis and regulation of these pigments have likely coevolved with the intricate relationships between plants, pollinators, and environmental stressors [11, 18]. Understanding the evolutionary origins and ecological implications of anthocyanins provides a deeper appreciation for their roles in the natural world.

3 Importance of Analyzing Onion Anthocyanins for Nutritional Assessment and Potential Health Benefits

Onions, ubiquitous in culinary traditions worldwide, have long been celebrated for their culinary versatility and pungent flavors. However, beyond their culinary attributes, recent research has unveiled a hidden treasure within onions anthocyanins [19]. These pigments, responsible for the vivid colors in various fruits and vegetables, are now gaining attention for their potential contributions to human health [20, 21]. This section delves into the significance of analyzing onion anthocyanins, shedding light on their potential as nutritional powerhouses and sources of health benefits [22]. In the quest for holistic nutrition, researchers and health professionals are increasingly emphasizing the importance of a well-rounded diet rich in diverse nutrients [23]. Anthocyanins, found in varying quantities in different onion varieties, are now recognized as a valuable addition to the nutritional profile of this humble vegetable [24]. Analyzing onion anthocyanins offers a comprehensive understanding of the compounds that contribute to its nutritional value. From vitamins and minerals to phytochemicals with potential health benefits, the anthocyanin content adds a unique layer of complexity to onions' nutritional assessment [25].

Anthocyanins have garnered attention for their potential health-promoting properties, making their analysis in onions a subject of considerable interest [26, 27]. These compounds are known for their antioxidant activity, which helps combat oxidative stress a factor implicated in chronic diseases like cardiovascular disorders, cancer, and neurodegenerative conditions [7, 12, 24]. By quantifying and characterizing the anthocyanin content in onions, researchers can assess their potential to contribute to antioxidant defenses within the human body [12, 28]. The potential health benefits associated with anthocyanin consumption extend beyond their antioxidant properties. Studies suggest that anthocyanins may positively impact cardiovascular health by promoting healthy blood pressure, reducing inflammation, and improving endothelial function [12]. Moreover, these pigments have been associated with improved cognitive function, weight management, and diabetes management [5, 13]. By analyzing the anthocyanin content in onions, researchers can provide valuable insights into how these compounds might contribute to overall wellness.

In an era of heightened consumer awareness about the nutritional content of foods, the analysis of onion anthocyanins holds the potential to influence dietary choices. As consumers seek foods with added health benefits, understanding the anthocyanin content in onions can guide their choices. Furthermore, incorporating anthocyanin-rich onions into culinary practices could offer an exciting way to enhance the nutritional value of meals while enjoying the distinct flavors they offer.

4 HPLC as a Powerful Analytical Technique for Quantifying and Profiling Anthocyanin Compounds

In the realm of modern analytical chemistry, the quest for precision and depth in compound analysis has led to the development of powerful techniques that illuminate the intricate molecular landscapes of diverse substances. High-Performance Liquid Chromatography (HPLC), a venerable cornerstone of analytical science, stands at the forefront of this pursuit. This introduction navigates the significance of HPLC as a powerful analytical technique, specifically in the context of quantifying and profiling anthocyanin compounds—nature's vivid pigments that bestow colors upon plants and potential health benefits to humans. HPLC (Fig. 1) emerges as a sophisticated analytical technique designed to dissect complex mixtures into their individual molecular components with exquisite precision [29, 30]. HPLC surpasses its predecessors by virtue of its enhanced separation capabilities, rapid analysis times, and versatile applications across diverse scientific domains [29, 30]. This technique, often likened to a molecular sifter, enables researchers to delve into the molecular symphony within samples, offering insights into composition, quantity, and structure.

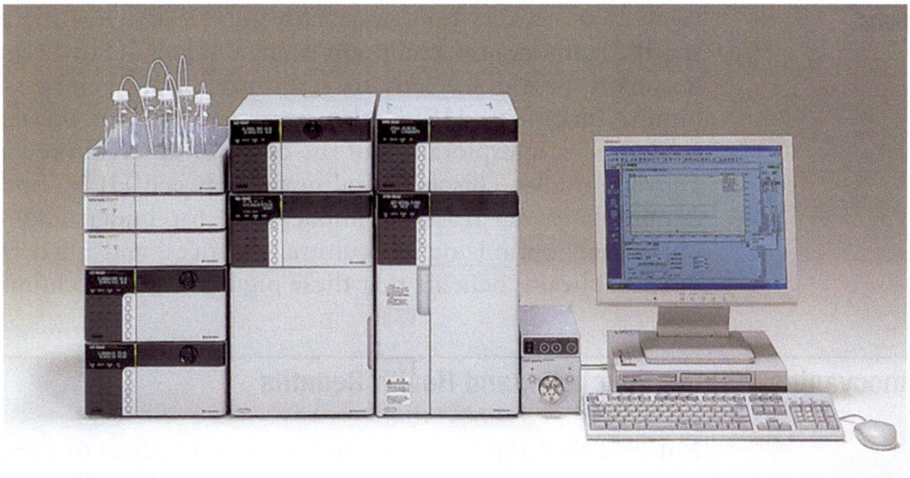

Fig. 1 A pictorial depiction of an HPLC system

4.1 Anthocyanins

Anthocyanins, the vibrant pigments found in plants, hold a dual role—embellishing flora with their resplendent hues while also contributing to the biological processes that sustain life [31, 32]. From the crimson of strawberries to the violet of grapes, anthocyanins lend their colors to a mesmerizing array of fruits, flowers, and vegetables. Beyond aesthetics, these pigments are gaining recognition for their potential health benefits, including antioxidant and anti-inflammatory properties [6]. The marriage of HPLC and anthocyanin analysis represents an intricate symphony of science and technique, unveiling the richness and complexity of these natural pigments. As researchers endeavor to quantify and profile anthocyanin compounds, HPLC assumes the role of a conductor, orchestrating the separation and identification of individual anthocyanins within a sample [30]. The methodology hinges on exploiting the distinct interactions between the target compounds, a stationary phase, and a mobile phase, facilitated by the fluid dynamics characteristic of HPLC systems.

4.2 Methodology

In the realm of anthocyanin analysis, HPLC transforms into a virtual alchemist's chamber, separating pigments that share similar hues and molecular backbones. A careful selection of stationary and mobile phases, coupled with meticulous control over temperature and pressure, brings about the spatial separation of these compounds based on their unique chemical properties [30]. The resulting chromatograms resemble intricate musical scores, with each peak corresponding to a distinct anthocyanin compound.

4.3 Benefits and Implications

The benefits of utilizing HPLC for anthocyanin analysis are multifaceted. It delivers precision, allowing for the quantification of specific anthocyanin compounds in a sample. Furthermore, its sensitivity provides a lens to scrutinize these pigments even in minute quantities. HPLC also paves the way for profiling—the comprehensive exploration of the entire anthocyanin repertoire in a given sample. This analytical nexus between HPLC and anthocyanins holds far-reaching implications for understanding plant physiology, optimizing agricultural practices, and uncovering the potential health benefits that these pigments offer to humans.

5 Anthocyanins: Nature, Functions, and Health Benefits

Nature's vibrant palette owes its spectrum of colors to a remarkable group of compounds known as anthocyanins. These naturally occurring pigments, responsible for the striking reds, blues, purples, and even blacks in various fruits, vegetables, and flowers, play a crucial role not only in the visual allure of these organisms but also in their biological functions [7]. This exploration delves into the nature, functions, and potential health benefits of anthocyanins, highlighting their intricate role in the natural world and their potential contributions to human well-being [18]. Anthocyanins are part of the flavonoid family, a class of secondary metabolites found in plants. Their mesmerizing array of colors is a result of subtle molecular variations in their structures, yielding a kaleidoscope of hues that enrich the botanical landscape [15]. From the deep reds of cherries to the vibrant purples of eggplants, anthocyanins infuse living organisms with an aesthetic vibrancy that captivates human senses.

Beyond their aesthetic allure, anthocyanins have evolved to serve a myriad of ecological functions. One prominent role is photoprotection—the shielding of plant tissues from the damaging effects of ultraviolet (UV) radiation [24]. Anthocyanins act as nature's sunscreen, absorbing and dissipating harmful UV rays, thus preventing cellular damage [33]. In this ecological ballet, these pigments contribute to the plant's resilience against the harsh forces of the environment. Anthocyanins also function as key players in ecological interactions, particularly in the pollination process [34]. Flowers adorned with vibrant anthocyanin-rich hues act as beacons, attracting pollinators such as bees, butterflies, and birds. This visual communication between plants and their pollinators ensures successful fertilization, a critical step in the reproduction of flowering plants [3]. Thus, anthocyanins are not mere pigments but essential components of the intricate dance of life in ecosystems.

The fascination with anthocyanins extends beyond their ecological roles to their potential health benefits for humans. These pigments, renowned for their antioxidant properties, scavenge harmful free radicals in the body, contributing to the protection of cells and DNA from oxidative damage [35]. This antioxidant capacity has sparked interest in their potential role in mitigating the risk of chronic diseases, including cardiovascular disorders, certain cancers, and neurodegenerative conditions [18]. Anthocyanins have also been linked to anti-inflammatory effects, modulating pathways that contribute to inflammation within the body [7]. Furthermore, recent research suggests their involvement in cellular signaling pathways that regulate gene expression and cellular function [36]. These findings open doors to exploring how anthocyanins might influence broader physiological processes beyond their antioxidant activity.

6 Anthocyanins and Their Role in Plants' Coloration

Anthocyanins are a class of water-soluble flavonoid pigments responsible for the vibrant red, blue, purple, and black colors observed in various plant tissues, including fruits, flowers, leaves, and stems [14]. These compounds are not only responsible for the captivating visual appeal of many plants but also play a crucial role in their ecological and physiological functions [7]. Anthocyanins derive their name from the Greek words "anthos," meaning flower, and "kyanos," meaning blue [33]. Structurally, they consist of a core anthocyanidin molecule, which can be modified by different sugar and acyl groups [9]. The specific anthocyanin produced by a plant depends on its genetic makeup, environmental factors, and developmental stage. Anthocyanins are synthesized through a complex pathway involving enzymes and gene regulation [18].

The expression of anthocyanin pigments in plants' tissues leads to a dazzling display of colors that serve various purposes. For instance, the red hues of apple skin, the purple tones of grapes, and the blue tints of blueberries are all attributed to different anthocyanin profiles [37]. The color variation arises from differences in the anthocyanin composition, the presence of co-pigments, and the acidity of the cellular environment [13]. The role of anthocyanins in plants' coloration extends beyond aesthetics. One of the key functions is photoprotection the shielding of plant tissues from the potentially harmful effects of ultraviolet (UV) radiation [38]. When plants are exposed to intense sunlight, they produce anthocyanins as a form of sunscreen to absorb and dissipate UV rays, preventing damage to delicate cellular components and DNA.

Another vital ecological role of anthocyanins is in attracting pollinators and seed dispersers. Flowers adorned with bright anthocyanin-rich colors act as visual cues, attracting pollinators like bees, butterflies, and birds [9]. These animals are essential for pollination, contributing to plant reproduction. Similarly, colorful fruits signal ripeness to animals, encouraging them to consume and disperse the seeds, aiding in plant propagation [34]. Anthocyanins are also synthesized in response to various environmental stresses, such as drought, temperature fluctuations, and nutrient deficiencies [7]. This stress-induced anthocyanin production is part of the plant's adaptive strategy to cope with adverse conditions [16]. Furthermore, anthocyanins can act as antioxidants, neutralizing harmful free radicals and reactive oxygen species generated during stress.

7 Functions of Anthocyanins

Anthocyanins, the vibrant pigments responsible for the striking red, blue, purple, and black colors in various plant tissues, play multifaceted roles beyond mere aesthetics. These compounds are nature's strategic tools, serving essential functions that contribute to the survival and success of plants within their ecosystems. Two prominent functions of anthocyanins are protection against ultraviolet (UV) radiation and the attraction of pollinators.

7.1 Protection Against Ultraviolet (UV) Radiation

Plants are constantly exposed to the potentially damaging effects of ultraviolet (UV) radiation from sunlight. Excessive UV exposure can lead to cellular damage, DNA mutations, and hinder photosynthesis [39]. Anthocyanins, acting as nature's sunscreen, offer an ingenious solution to this challenge. When plants experience heightened UV intensity due to factors like high altitudes or intense sunlight, they synthesize and accumulate anthocyanins in their exposed tissues, such as leaves and fruit skins [39]. The role of anthocyanins in UV protection is twofold. First, these pigments absorb UV light, effectively acting as a shield that prevents the radiation from reaching sensitive cellular components [40, 41]. Second, anthocyanins dissipate the absorbed energy as heat, minimizing the potential for photochemical damage [39, 40]. This photoprotective mechanism not only enhances the plant's ability to withstand harsh environmental conditions but also contributes to the overall health and longevity of the organism [40].

7.2 Attraction of Pollinators

The vivid colors conferred by anthocyanins in flowers play a pivotal role in attracting pollinators, which are essential for the reproductive success of many flowering plants [42]. Flowers serve as crucial reproductive structures, and pollinators such as bees, butterflies, and birds facilitate the transfer of pollen between flowers, enabling fertilization and seed production [34]. Anthocyanin-rich flowers provide a visual invitation to these pollinators, guiding them to the nectar-rich rewards within [42]. The evolution of this vibrant floral palette is a prime example of coevolution between plants and their pollinators. The diverse colors and patterns that anthocyanins contribute to flowers act as unique beacons, ensuring effective pollination. This mutualistic relationship between plants and pollinators underscores the interdependence and coexistence within ecosystems, where each party benefits from the other's services.

While UV protection and pollinator attraction are two of the most well-studied functions of anthocyanins, these pigments exhibit additional roles as well. Anthocyanins are known to play a part in stress response, serving as indicators of environmental stressors such as drought, cold, and nutrient deficiencies [17]. Furthermore, recent research suggests that these pigments might contribute to plant defense against herbivores and pathogens, although the mechanisms are still under exploration.

8 Health Benefits of Consuming Foods Rich in Anthocyanins

The vibrant colors that grace fruits, vegetables, and other plant-based foods are not just a visual feast; they often signify the presence of anthocyanins a class of natural compounds that offer a wealth of potential health benefits. Among these benefits, their role as antioxidants and anti-inflammatory agents stands out prominently, contributing to the growing recognition of anthocyanin-rich foods as nutritional powerhouses.

8.1 Antioxidant Properties

Anthocyanins are renowned for their potent antioxidant properties, making them valuable allies in the fight against oxidative stress—a cellular imbalance between harmful reactive oxygen species and the body's ability to neutralize them [34]. Oxidative stress is implicated in a range of chronic diseases, including cardiovascular disorders, cancer, diabetes, and neurodegenerative conditions [34, 43]. Anthocyanins, acting as scavengers of free radicals and reactive oxygen species, help restore balance by preventing cellular damage and reducing the risk of chronic diseases [43].

8.2 Anti-inflammatory Effects

Chronic inflammation is another underlying factor in many health conditions, from arthritis to metabolic syndrome. Anthocyanins have shown promising anti-inflammatory effects by modulating various inflammatory pathways within the body [44]. These compounds can inhibit the production of pro-inflammatory molecules and enzymes, thereby curbing the excessive immune response that leads to chronic inflammation [45, 46]. By dampening inflammation, anthocyanins contribute to overall health and may help mitigate the progression of inflammatory-related diseases.

8.3 Cardiovascular Health

The antioxidant and anti-inflammatory properties of anthocyanins have particularly compelling implications for cardiovascular health. Oxidative stress and inflammation play a pivotal role in the development of atherosclerosis—a condition where fatty plaques accumulate within artery walls, restricting blood flow and increasing the risk of heart disease [18]. By reducing oxidative stress and inflammation, anthocyanins contribute to maintaining healthy blood vessels, improving blood flow, and potentially lowering the risk of cardiovascular events [35].

8.4 Cognitive Function

Emerging research suggests that anthocyanins might have positive effects on cognitive function and brain health. Oxidative stress and inflammation are implicated in age-related cognitive decline and neurodegenerative diseases like Alzheimer's and Parkinson's [47]. Anthocyanins' ability to combat these factors could contribute to preserving cognitive function and delaying the onset of cognitive decline [48].

8.5 Cancer Prevention

While research in this area is ongoing, preliminary studies suggest that anthocyanins might play a role in cancer prevention. Their antioxidant and anti-inflammatory properties could help mitigate DNA damage, inhibit tumor growth, and promote apoptosis (programmed cell death) in cancer cells [49, 50]. However, more research is needed to fully understand the extent of anthocyanins' role in cancer prevention.

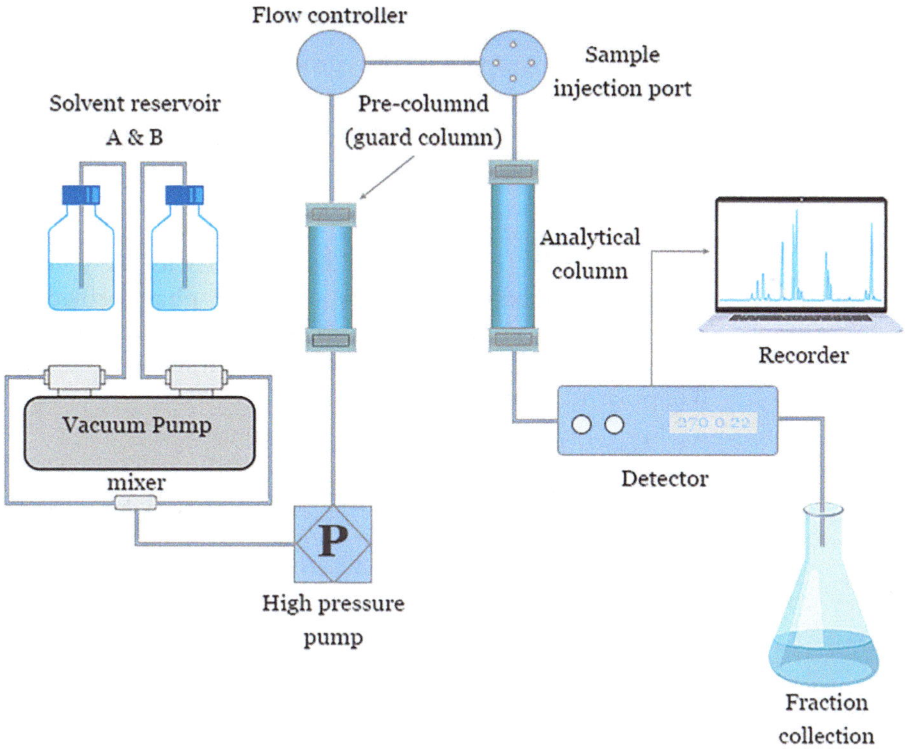

Fig. 2 A schematic representation of HPLC system showing its various components

9 HPLC: Principles and Methodology

High-Performance Liquid Chromatography, or HPLC, is a powerful analytical technique used for separating, identifying, and quantifying components of complex mixtures [2]. It is widely used in various scientific and industrial fields, including pharmaceuticals, environmental analysis, food and beverage, and more. HPLC operates based on the principles of liquid chromatography (Fig. 2), where a sample mixture is passed through a stationary phase to separate its components based on their different affinities for the stationary and mobile phases [2].

10 Principles of HPLC

High-Performance Liquid Chromatography (HPLC) is a powerful separation technique used to separate, identify, and quantify components in a mixture based on their differential interactions with a stationary phase and a mobile phase. It builds upon the fundamental principles of liquid chromatography and offers enhanced performance through advanced instrumentation and methodologies. Here's a detailed explanation of the principles behind HPLC:

1. *Chromatographic Separation*

Chromatography is the process of separating components within a mixture by exploiting their differing affinities for two phases: the stationary phase and the mobile phase.

2. *Stationary Phase*

The stationary phase is a critical component of HPLC. It can be a solid material (usually finely divided particles) or a liquid immobilized on a solid support. The stationary phase provides a surface on which the separation occurs. Different stationary phases interact differently with analytes based on their physicochemical properties like polarity, charge, size, and affinity.

3. *Mobile Phase*

The mobile phase is a liquid or gas that carries the sample components through the stationary phase. It is composed of one or more solvents that interact with the analytes and influence their movement. The choice of mobile phase affects separation parameters such as selectivity, resolution, and efficiency.

4. *Retention Time*

As the sample mixture is introduced into the HPLC system, its components interact with the stationary phase to varying degrees. This leads to differences in the time each component takes to elute (exit) the column. The time a component spends within the column is known as its retention time. Components with stronger interactions with the stationary phase have longer retention times.

5. *Separation Mechanisms*

HPLC can use different separation mechanisms based on the nature of the stationary and mobile phases:

Reverse-Phase HPLC: The stationary phase is non-polar, and the mobile phase is polar. Components are separated based on their polarity, with more polar analytes eluting faster.

Normal-Phase HPLC: The stationary phase is polar, and the mobile phase is non-polar. Separation occurs based on differences in polarity.

Ion-Exchange HPLC: The stationary phase contains charged groups, and separation occurs based on ionic interactions between analytes and the stationary phase.

Size-Exclusion HPLC (Gel Filtration): The stationary phase is porous, and separation occurs based on size differences among analytes.

6. *Detector*

After separation within the column, the components are detected using various detectors. Common detectors include UV-Vis spectrophotometers, fluorescence detectors, refractive index detectors, and mass spectrometers. Detectors measure the concentration of analytes as they elute, producing a signal that is used for quantification and identification.

7. *Chromatogram*

The output of an HPLC analysis is presented as a chromatogram, which is a graphical representation of detector response over time. Peaks on the chromatogram correspond to individual components of the mixture. The area under each peak is proportional to the amount of the respective analyte present in the sample.

8. *Optimization and Control*

HPLC parameters such as flow rate, column temperature, mobile phase composition, and gradient profile can be optimized to achieve the desired separation. Modern HPLC systems allow for precise control of these parameters, enhancing separation efficiency and accuracy.

11 Methodology of HPLC

Sample Preparation: Samples are typically prepared by dissolving or extracting the analytes of interest from the matrix they are in. The sample needs to be in a suitable solvent and free from particulates that could clog the column.

Column Selection: The choice of column is crucial to achieve the desired separation. Different columns have varying selectivities, which impact the separation of components.

Mobile Phase: Selection of an appropriate mobile phase or solvent mixture is vital for achieving the desired separation. Parameters like solvent strength, polarity, and pH can be adjusted to optimize separation.

Injection: A small volume of the prepared sample is injected into the HPLC system. The sample is then pushed through the column using the mobile phase.

Separation: As the sample components travel through the column, they interact differently with the stationary phase, leading to separation based on their physicochemical properties.

Detection: The separated components are detected by a suitable detector. The choice of detector depends on the nature of the analytes.

Quantification and Analysis: The detector response is recorded, and the area under the peaks can be used for quantification. Comparison to known standards or calibration curves can determine the concentration of analytes.

Data Interpretation: The data obtained from the chromatogram can provide information about the composition and concentration of components in the sample.

HPLC methodology can vary based on the specific type of HPLC (e.g., reverse-phase, normal-phase, ion-exchange, size-exclusion) and the purpose of the analysis. The parameters like column type, mobile phase composition, and detection method can all be optimized to achieve the best separation and quantification results for a particular application.

12 Main Components of an HPLC System

An HPLC (High-Performance Liquid Chromatography) system consists of several main components that work together to separate, detect, and quantify the components of a sample mixture. The primary components include the mobile phase, stationary phase, and detector:

1. *Mobile Phase*

The mobile phase is a liquid or gas that carries the sample components through the chromatographic system. It is a critical component that influences separation efficiency and selectivity. The choice of mobile phase depends on the separation mechanism and the nature of the analytes. Commonly used mobile phases include mixtures of water, organic solvents (such as acetonitrile or methanol), and buffers.

2. *Stationary Phase*

The stationary phase is a solid or liquid material that interacts with the sample components as they flow through the chromatographic column. The choice of stationary phase is vital for achieving separation based on the physicochemical properties of the analytes. Different types of stationary phases are used for various separation mechanisms:

Reverse-Phase Stationary Phase: Consists of non-polar packing materials that separate compounds based on their polarity.

Normal-Phase Stationary Phase: Utilizes polar packing materials for separation based on polarity.

Ion-Exchange Stationary Phase: Contains charged groups that interact with ions in the sample based on their charge.

Size-Exclusion Stationary Phase: Employs porous materials to separate analytes based on their molecular size.

3. *Chromatographic Column*

The column is the vessel through which the sample mixture travels. It is packed with the stationary phase material. The length, diameter, and packing material of the column can be optimized to achieve desired separation characteristics. Longer columns generally provide better resolution but may require more time for analysis.

4. *Injector*

The injector is responsible for introducing the sample into the chromatographic system. It precisely injects a small volume of the sample into the mobile phase stream. The injector ensures reproducibility and accuracy of sample introduction.

5. *Pump*

The pump generates the pressure needed to drive the mobile phase through the column at a constant flow rate. The precise control of the flow rate is crucial for consistent and reproducible separations. Modern HPLC pumps can create gradient elutions where the composition of the mobile phase changes over time to optimize separation.

6. *Detector*

The detector is a key component that detects and measures the separated sample components as they elute from the column. The choice of detector depends on the nature of the analytes. Common types of detectors include:

- *UV-Vis Detector*: Measures absorbance of UV or visible light by the analytes.
- *Fluorescence Detector*: Measures fluorescence emission of analytes.
- *Refractive Index Detector*: Measures changes in refractive index caused by the eluting analytes.
- *Mass Spectrometer (MS) Detector*: Provides detailed information about the molecular composition of analytes.

7. *Data System and Software*

Modern HPLC systems are equipped with data acquisition and control software. This software controls the system parameters, collects detector signals, and generates chromatograms. It allows for data analysis, quantification, and often offers features like peak integration, calibration curve generation, and report generation.

These components work collaboratively to achieve efficient separation and detection of sample components in an HPLC analysis. The optimal selection and configuration of these components depend on the specific separation goals and characteristics of the sample mixture.

13 HPLC Columns Used for Anthocyanin Analysis

Anthocyanins are a class of water-soluble pigments responsible for the red, purple, and blue colors in many fruits, vegetables, and flowers. Analyzing anthocyanins using High-Performance Liquid Chromatography (HPLC) requires careful selection of the chromatographic column to achieve accurate and reliable separation [37]. There are several types of HPLC columns commonly used for anthocyanin analysis:

13.1 Reverse-Phase Columns

Reverse-phase chromatography is often used for anthocyanin analysis due to its ability to separate compounds based on their hydrophobicity (lipophilicity). Anthocyanins are often hydrophilic, making reverse-phase columns suitable for their separation. The most commonly used reverse-phase columns include C18 and C8 columns. These columns have hydrophobic alkyl chains bonded to the silica surface, providing retention and separation based on hydrophobic interactions [29].

13.2 Normal-Phase Columns

Normal-phase chromatography can also be employed for anthocyanin analysis, especially for polar compounds. In this approach, the stationary phase is polar, and separation occurs based on polarity differences between anthocyanins. Silica-based columns with polar functional groups are used in normal-phase chromatography.

13.3 Hybrid Columns

Hybrid columns combine characteristics of both reverse-phase and normal-phase columns. These columns offer more versatile separation mechanisms and can be useful for complex samples containing various types of anthocyanins with different polarities.

13.4 Mixed-Mode Columns

Mixed-mode columns incorporate multiple types of interactions, such as hydrophobic and ion-exchange interactions, to achieve enhanced selectivity. These columns are advantageous when dealing with samples containing diverse anthocyanin compounds with varying properties.

13.5 pH-Responsive Columns

Some anthocyanins can change their charge and properties with changes in pH. Columns that exploit this pH responsiveness can be used to separate and analyze different forms of anthocyanins based on their charge characteristics.

13.6 Chiral Columns

Chiral HPLC columns can be employed when analyzing chiral anthocyanins or when enantiomeric separation is required. These columns separate compounds based on their stereochemical differences.

13.7 Size-Exclusion Columns

Size-exclusion chromatography, also known as gel filtration chromatography, separates compounds based on their size. Although less common for anthocyanin analysis, this approach can be useful for separating anthocyanins from larger molecules in complex samples.

When selecting an HPLC column for anthocyanin analysis, factors to consider include the type and complexity of the sample, the desired resolution between different anthocyanin compounds, and the instrumentation available. It's important to choose a column that provides good separation, resolution, and sensitivity for the specific anthocyanins of interest [29]. Additionally, optimizing the mobile phase composition, flow rate, and column temperature can further enhance the separation efficiency and accuracy of anthocyanin analysis.

14 Anthocyanin Sample Preparation, Extraction, and Purification Before HPLC Analysis

Sample preparation, extraction, and purification are crucial steps in the analysis of anthocyanins using High-Performance Liquid Chromatography (HPLC) (Fig. 3). These steps ensure that the analytes of interest are effectively extracted from the complex sample matrix, concentrated, and purified before being introduced into the HPLC system. Proper sample preparation is essential for obtaining accurate and reproducible results in anthocyanin analysis. Here's why each step is important:

1. *Sample Preparation*

Sample preparation involves selecting the appropriate sample, homogenizing it to create a representative subsample, and ensuring it's in a suitable form for extraction. Proper sample preparation minimizes variability, ensures the sample is representative, and enhances the accuracy of analysis.

2. *Extraction*

Anthocyanins are often present in plant tissues and can be bound to cellular structures. Extraction involves breaking down these structures to release the analytes. Efficient extraction is critical to recover anthocyanins from the sample matrix. Several extraction methods can be used:

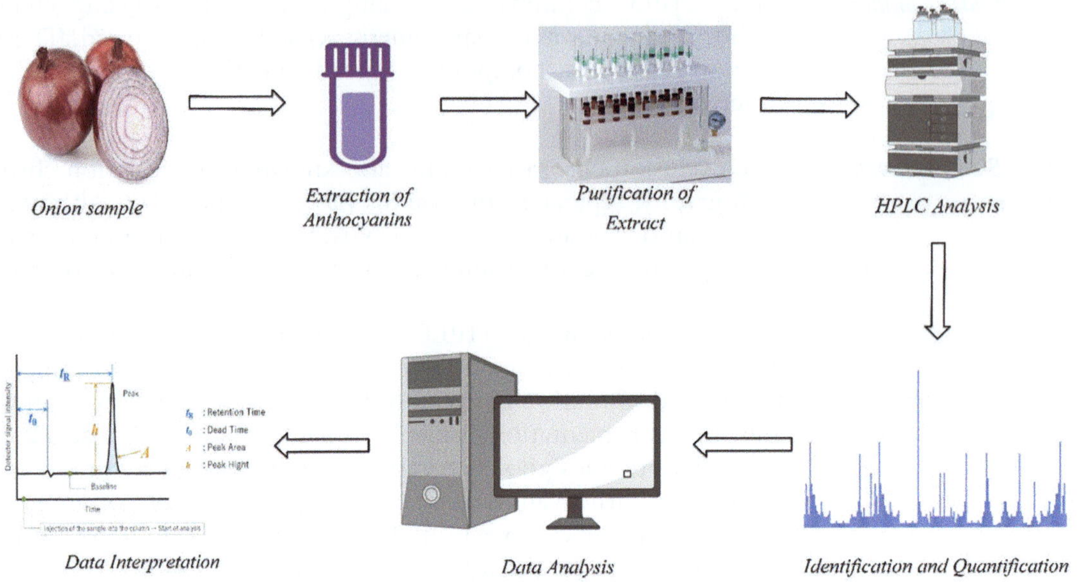

Fig. 3 Outline of the steps involved in analyzing the anthocyanin profile of onions using high-performance liquid chromatography (HPLC)

- *Solvent Extraction*: Using suitable solvents (e.g., water, methanol, ethanol) to extract anthocyanins from the sample.
- *Solid-Liquid Extraction*: Macerating or blending the sample with a suitable solvent.
- *Pressurized Liquid Extraction (PLE)*: Using high pressure and temperature to accelerate extraction.
- *Ultrasound-Assisted Extraction (UAE)*: Applying ultrasound waves to enhance extraction efficiency.

3. *Purification*

After extraction, the sample may contain impurities, matrix components, and co-extracted compounds that can interfere with HPLC analysis. Purification aims to remove these unwanted components, ensuring accurate identification and quantification of anthocyanins. Purification methods include:

Solid-Phase Extraction (SPE): Using a solid-phase sorbent to selectively retain and elute anthocyanins.

Liquid-Liquid Extraction (LLE): Partitioning anthocyanins between two immiscible liquid phases based on their polarity.

Centrifugal Partition Chromatography (CPC): A liquid-liquid separation technique used for large-scale purification.

15 Importance of These Steps

Enhanced Sensitivity and Selectivity: Proper sample preparation and extraction increase the concentration of anthocyanins, improving detection sensitivity. Purification reduces background noise, enhancing the selectivity of the HPLC analysis.

Minimized Interference: Purification removes co-extracted compounds and matrix components that can interfere with the separation and detection of anthocyanins. This reduces the risk of false positives and inaccurate quantification.

Reproducibility: Consistent and reproducible sample preparation and extraction methods ensure that results can be reliably reproduced over multiple analyses.

Accurate Quantification: Effective extraction and purification ensure that the analyte recovery is maximized, leading to accurate quantification of the actual anthocyanin content in the sample.

Column Protection: Proper sample preparation and purification help prevent column clogging, prolonging the column's lifespan and maintaining optimal separation efficiency.

Reduced Matrix Effects: Matrix components in the sample can affect HPLC peak shapes and retention times. Adequate purification minimizes these matrix effects, leading to more reliable data.

In summary, sample preparation, extraction, and purification steps are essential in anthocyanin analysis before HPLC. They help ensure accurate, sensitive, and reproducible results by removing impurities, minimizing interference, and enhancing the performance of the HPLC system.

16 Analyzing the Onion Anthocyanin Profile

Analyzing the anthocyanin profile of onions involves determining the types and concentrations of anthocyanin compounds present in the onion tissues. These compounds contribute to the coloration of onions, and their profiles can vary based on onion variety, maturity, growing conditions, and processing methods [12]. Here's a general outline of the steps involved in analyzing the anthocyanin profile of onions using High-Performance Liquid Chromatography (HPLC) (Fig. 4):

1. *Sample Collection and Preparation*

Collect onion samples of interest, ensuring they are representative of the population being studied. The samples can include the

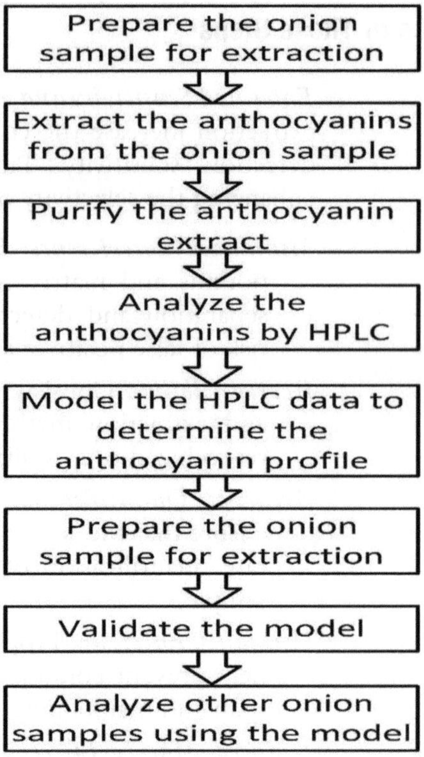

Fig. 4 A flow chart showing the preparation steps of onion anthocyanin using HPLC

outer layers or other parts of the onion with visible coloration. Homogenize the samples to create a representative subsample for analysis.

2. *Extraction of Anthocyanins*

Extract anthocyanins from the onion tissue. Various extraction methods can be employed, such as solvent extraction or solid-liquid extraction. The goal is to efficiently release anthocyanins from the onion matrix into a suitable solvent.

3. *Purification of Extract*

Purify the extracted anthocyanin mixture to remove impurities, sugars, and other co-extracted compounds. Techniques like Solid-Phase Extraction (SPE) or Liquid-Liquid Extraction (LLE) can be used to enhance the purity of the anthocyanin fraction.

4. *HPLC Analysis*

Perform HPLC analysis to separate and quantify the anthocyanin compounds. Use an appropriate HPLC column (such as reverse-phase C18) and mobile phase to achieve optimal separation. Set the chromatographic conditions, including flow rate, gradient profile, and detector wavelength (usually UV-Vis or a diode array detector).

5. *Identification and Quantification*

Compare the retention times and UV-Vis spectra of the separated peaks with those of known anthocyanin standards. This helps identify different anthocyanin compounds in the onion sample. External calibration curves or internal standards can be used for quantification. The peak areas or heights can be used to determine the relative concentrations of the identified anthocyanins.

6. *Data Analysis*

Analyze the data to determine the types and concentrations of anthocyanin compounds present in the onion samples. The chromatogram will show peaks corresponding to different anthocyanins, and their areas or heights provide information about their relative abundance.

7. *Interpretation*

Interpret the results in the context of the onion variety, maturity, and growing conditions. Compare the anthocyanin profiles between different samples to identify trends and variations.

8. *Validation and Reproducibility*

Validate the analysis by ensuring the method is reproducible and accurate. Conduct replicate analyses and assess the precision and accuracy of the results.

Analyzing the anthocyanin profile of onions using HPLC provides valuable insights into the composition of these pigments and their potential health benefits. It can also aid in determining the best onion varieties for specific applications, such as food colorants or nutraceuticals.

17 Process of Obtaining Onion Samples and Preparing Anthocyanin Extracts

Obtaining onion samples and preparing anthocyanin extracts involves careful collection and handling of the onion tissues to ensure representative and accurate analysis. Here's a step-by-step process for obtaining onion samples and preparing anthocyanin extracts:

1. *Sample Collection*

Choose onions that are representative of the population you are studying. Consider factors such as onion variety, maturity, and growing conditions. If you are interested in specific parts of the onion, such as the outer layers with visible coloration, select those parts for analysis. Collect a sufficient number of samples to ensure statistical significance.

2. *Sample Preparation*
   (a) *Clean the Onions*: Remove any dirt or debris from the surface of the onions. This prevents contamination of the samples and the extraction process.
   (b) *Peel and Chop*: Depending on your analysis, peel the onions to access the colored layers. Chop the onion tissues into smaller pieces to facilitate extraction.

3. *Extraction of Anthocyanins*

There are various methods to extract anthocyanins from onion tissues. One common method is the solvent extraction technique:

(a) *Weigh the Sample*: Weigh a representative portion of the chopped onion tissues to accurately calculate the amount of anthocyanins in the sample.
(b) *Prepare Extraction Solvent*: Prepare a suitable extraction solvent. A mixture of water and an organic solvent like methanol or ethanol is commonly used. The organic solvent helps solubilize lipophilic anthocyanins.
(c) *Homogenization*: Place the chopped onion tissues in a blender or homogenizer along with the extraction solvent. Blend or homogenize the mixture to break down the onion tissues and release the anthocyanins.
(d) *Filtration*: Filter the homogenate to remove solid particles and debris, obtaining a clear extract.

4. *Purification of Extract (Optional)*

Depending on the analysis and the desired level of purification, you may choose to purify the anthocyanin extract. Techniques like Solid-Phase Extraction (SPE) can be used to remove impurities and matrix components.

5. *Concentration*

Concentrate the anthocyanin extract to reduce its volume and increase the concentration of anthocyanins. This can be done using techniques like rotary evaporation under reduced pressure.

6. *Sample Storage*

If you're not proceeding with HPLC analysis immediately, store the anthocyanin extract appropriately. Keep it in the dark at low temperatures (e.g., −20 °C) to prevent degradation of anthocyanins.

Proper sample collection, preparation, and extraction are crucial to obtaining accurate and representative anthocyanin profiles from onions. Each step should be carefully executed to minimize variability and ensure the reliability of the analysis. Keep in mind that the specific protocol may vary based on the equipment, resources, and objectives of your analysis.

18 Chromatographic Conditions for Onion Anthocyanin Analysis

The specific chromatographic conditions for onion anthocyanin analysis using High-Performance Liquid Chromatography (HPLC) are critical to achieve efficient separation, accurate quantification, and reliable identification of the anthocyanin compounds present in the onion samples. Here are the key chromatographic conditions that should be considered for onion anthocyanin analysis:

1. *Column Selection*

Choose an appropriate HPLC column based on the separation mechanism that best suits the nature of anthocyanins. A reverse-phase C18 column is commonly used for anthocyanin analysis due to its ability to separate compounds based on hydrophobic interactions.

2. *Mobile Phase*

The mobile phase composition is crucial for achieving proper separation of anthocyanin compounds. A mixture of water and organic solvent (e.g., methanol, acetonitrile) is typically used. The addition of small amounts of acid (e.g., formic acid) to the mobile phase helps in stabilizing and enhancing the separation of anthocyanins. The specific ratio of water to organic solvent and the concentration of acid depend on the anthocyanin compounds being analyzed and the column characteristics.

3. *Gradient Elution*

Gradient elution involves changing the composition of the mobile phase over time to optimize separation. Start with a higher proportion of water in the mobile phase to elute less polar compounds, followed by increasing the proportion of organic solvent to elute more polar compounds. A typical gradient profile might start with 95% water and 5% organic solvent and increase to 5% water and 95% organic solvent over a defined period.

4. *Flow Rate*

The flow rate of the mobile phase affects the retention times and peak shapes of the anthocyanin compounds. It's important to choose a flow rate that provides a good compromise between resolution and analysis time. Common flow rates are in the range of 0.5–1.5 mL/min.

5. *Column Temperature*

The column temperature can influence separation efficiency and retention times. Maintain a stable column temperature to ensure consistent results. A temperature control module within the HPLC system can be used to maintain the desired temperature.

6. *Detector Wavelength*

The detection of anthocyanins is often done using a UV-Vis detector. Anthocyanins typically absorb light in the visible range, so wavelengths around 500–550 nm are commonly used for detection. The detector wavelength should be chosen based on the absorption characteristics of the specific anthocyanin compounds being analyzed.

7. *Data Collection and Analysis*

Use appropriate data collection parameters, such as integration time and data interval, to accurately capture the chromatographic peaks. The software should be capable of quantifying peak areas, retention times, and providing a complete chromatogram for further analysis.

It's important to note that the specific chromatographic conditions may vary based on the HPLC system, column, detector, and anthocyanin compounds being analyzed. Therefore, method development and optimization may be necessary to achieve the best separation and quantification results for onion anthocyanin analysis.

19 Selection of Appropriate Solvents and Mobile Phases for Optimal Separation and Detection

Electing appropriate solvents and mobile phases is crucial for achieving optimal separation and detection of analytes in High-Performance Liquid Chromatography (HPLC). The choice of solvents and mobile phases depends on the characteristics of the analytes, the column type, and the separation mechanism. Here's a detailed guide to selecting solvents and mobile phases:

1. *Solvent Selection*

Solvents used in HPLC are typically high-purity liquids that are compatible with the system components and detectors. Commonly used solvents include water, methanol, acetonitrile, and ethanol. Factors to consider when selecting solvents include:

- *Purity*: Use high-quality solvents to prevent contamination and baseline drift.
- *Compatibility*: Ensure solvents are compatible with the sample, column, and detector.

- *Degassing*: Some solvents contain dissolved gases that can affect the stability of the baseline. Degassing is often necessary to remove these gases.
- *Solubility*: Solvents should effectively dissolve the analytes without causing precipitation or aggregation.
- *Viscosity*: Viscous solvents can affect flow rates and may require adjustments to the system.
- *UV Absorption*: Avoid solvents that strongly absorb UV or visible light in the detection range, as this can interfere with detection.

2. *Mobile Phase Composition*

The mobile phase is composed of one or more solvents and is designed to carry the sample through the chromatographic column. The composition of the mobile phase affects retention times, selectivity, and resolution. Different mixtures can be used:

- *Isocratic Elution*: Using a constant composition of the mobile phase throughout the analysis. This is suitable for samples with components of similar polarity.
- *Gradient Elution*: Changing the composition of the mobile phase over time. Gradients can improve separation of complex mixtures by optimizing analyte elution.
- *Buffer Solutions*: Buffers can be added to the mobile phase to control pH, especially for ionizable compounds.

3. *Gradient Profile*

If using gradient elution, the gradient profile defines how the proportion of solvents in the mobile phase changes over time. An optimized gradient profile enhances separation efficiency. Common gradient profiles include linear, stepwise, and concave (curved).

4. *pH Adjustment*

For ionizable compounds, adjusting the pH of the mobile phase can influence the analyte's charge and interaction with the stationary phase. pH optimization can significantly impact separation.

5. *Stationary Phase Compatibility*

The choice of solvents and mobile phases should be compatible with the stationary phase of the column. Reverse-phase columns are often used with aqueous-organic mobile phases, while normal-phase columns work well with more polar solvents.

6. *Detector Compatibility*

The mobile phase should be compatible with the detector used for analysis. Some detectors are sensitive to certain solvents or buffer ions, so ensure that the mobile phase won't interfere with detection.

7. *Analyte Properties*

Consider the polarity, charge, and size of the analytes. Adjust the mobile phase composition to promote separation based on these properties.

In summary, selecting appropriate solvents and mobile phases requires careful consideration of the analytes' properties, column type, and separation mechanism. Optimization of these parameters is essential for achieving optimal separation, sensitivity, and resolution in HPLC analysis. Method development, including trial and error, may be necessary to find the best solvents and mobile phase conditions for a specific analysis.

20 Role of Detectors, Particularly UV-Vis Detectors, in Identifying and Quantifying Anthocyanin Compounds

Detectors play a crucial role in High-Performance Liquid Chromatography (HPLC) by detecting separated compounds as they elute from the column and providing information about their identity and concentration. UV-Vis detectors, in particular, are commonly used for identifying and quantifying anthocyanin compounds in HPLC analysis due to the inherent UV absorption properties of these pigments. Here's how UV-Vis detectors work and their role in anthocyanin analysis:

20.1 Principle of UV-Vis Detection

UV-Vis detectors operate based on the principle of measuring the absorption of light by analytes. When UV or visible light passes through a sample, some wavelengths are absorbed by the analytes, resulting in a decrease in light intensity. The absorbance is directly proportional to the concentration of the absorbing species, according to the Beer-Lambert Law.

20.2 Role in Identifying Anthocyanins

Wavelength Selection: Anthocyanins exhibit characteristic absorption maxima in the visible range, typically around 500–550 nm. UV-Vis detectors are capable of scanning a range of wavelengths or being set at specific wavelengths to target the absorption maxima of anthocyanins. By choosing the appropriate wavelength, you can specifically detect anthocyanins while minimizing interference from other compounds.

UV-Vis Spectra: UV-Vis detectors provide UV-Vis spectra of the detected compounds. Anthocyanins have distinctive spectra that can aid in their identification. Comparing the spectra obtained from the sample to those of known anthocyanin standards can confirm the presence of specific anthocyanins in the sample.

20.3 Role in Quantifying Anthocyanins

Calibration Curves: UV-Vis detectors allow for the construction of calibration curves. By measuring the absorbance of standard solutions with known concentrations of anthocyanins, a linear relationship between absorbance and concentration can be established. This calibration curve can then be used to quantify the concentration of anthocyanins in the sample based on its absorbance.

Peak Area or Height: The area under the chromatographic peak (or the peak height) in the chromatogram corresponds to the amount of the compound present in the sample. By integrating the peak, the detector provides quantitative data, allowing you to determine the concentration of anthocyanins.

Benefits of UV-Vis Detectors for Anthocyanin Analysis

- UV-Vis detectors are widely available and relatively easy to operate.
- Anthocyanins are naturally UV-absorbing compounds, making them well-suited for UV-Vis detection.
- UV-Vis detection is sensitive, enabling detection of low concentrations of anthocyanins in complex mixtures.
- UV-Vis detection provides valuable spectral information for compound identification and confirmation.

Considerations

- Interference from co-eluting compounds can affect the accuracy of quantification.
- Some matrix components might absorb UV or visible light, leading to background noise. Proper purification of extracts can mitigate this.
- Variation in the pH of the mobile phase can influence the absorbance properties of anthocyanins.

In conclusion, UV-Vis detectors play a key role in identifying and quantifying anthocyanin compounds in HPLC analysis. By exploiting the UV absorption properties of anthocyanins, UV-Vis detectors enable accurate and sensitive detection, facilitating both qualitative and quantitative analysis of these pigments.

21 Quantification and Identification of Anthocyanins

Quantification and identification of anthocyanins in a sample involve determining the concentration of individual anthocyanin compounds and confirming their presence based on their characteristic properties. High-Performance Liquid Chromatography (HPLC) combined with appropriate detection methods is commonly used for these purposes. Here's a step-by-step overview:

1. *Quantification of Anthocyanins*

   Quantification involves determining the amount of each specific anthocyanin compound present in a sample. This is typically achieved using the following steps:

   - *Calibration Curves*: Prepare standard solutions with known concentrations of individual anthocyanins. Inject these standards into the HPLC system and construct calibration curves by plotting peak areas or heights against the concentrations of the standards.
   - *Sample Analysis*: Inject the sample extract into the HPLC system under the same conditions used for the standards. The chromatogram will show peaks corresponding to individual anthocyanins.
   - *Integration of Peaks*: Use HPLC software to integrate the peaks of individual anthocyanin compounds in the chromatogram. The integrated peak area or height corresponds to the amount of each anthocyanin in the sample.
   - *Comparison with Calibration Curves*: Compare the peak areas or heights of the sample peaks with the calibration curves to determine the concentration of each anthocyanin compound in the sample.

2. *Identification of Anthocyanins*

   Identification involves confirming the presence of specific anthocyanin compounds based on their characteristic properties. HPLC combined with detection methods like UV-Vis, diode array, or mass spectrometry can aid in identification:

   - *Retention Time*: Compare the retention times of peaks in the sample chromatogram with those of known standard compounds. Similar retention times indicate the presence of specific anthocyanins.
   - *UV-Vis Spectra*: Anthocyanins have characteristic UV-Vis absorption spectra. Compare the UV-Vis spectra of sample peaks with those of known standards to confirm their identity.
   - *Mass Spectrometry*: If available, mass spectrometry (MS) can provide more detailed structural information about the

anthocyanins. MS fragmentation patterns can help confirm the identity of specific compounds.

- *Co-chromatography*: In some cases, co-chromatography with authentic standards can provide further confirmation of anthocyanin identity.

3. *Data Interpretation*
   - Identify the anthocyanin compounds present in the sample based on retention times, UV-Vis spectra, and any additional confirmation methods used.
   - Calculate the concentration of each identified anthocyanin compound using the calibration curves and peak integration data.
   - Report the results in appropriate units (e.g., mg/g, μg/mL) along with any relevant statistical information (standard deviations, replicates, etc.).

In summary, quantification and identification of anthocyanins using HPLC involve constructing calibration curves, analyzing sample chromatograms, comparing retention times and spectra, and using appropriate detection methods to confirm the presence and quantity of specific anthocyanin compounds. These analyses provide valuable information about the composition and concentration of anthocyanins in a given sample.

22 Quantification Process Using Reference Standards with Known Concentrations

Quantification in High-Performance Liquid Chromatography (HPLC) involves determining the concentration of a specific compound in a sample. One effective method for quantification is using reference standards with known concentrations. Here's a step-by-step explanation of the quantification process using reference standards:

1. *Preparation of Reference Standards*

Prepare a series of standard solutions with known concentrations of the compound of interest. These solutions should cover a range of concentrations that are relevant to the expected concentration levels in your samples. The concentrations of the standards should be accurately measured and prepared in the same solvent used for your sample extracts.

2. *HPLC Analysis*

Inject each of the reference standard solutions into the HPLC system using the same chromatographic conditions that you plan to use for your sample analysis. This ensures consistency in separation and detection.

3. *Calibration Curve Construction*

For each reference standard, the HPLC system records the peak area or height of the corresponding chromatographic peak. Plot these peak areas (or heights) against the known concentrations of the reference standards. This plot is known as a calibration curve.

4. *Linear Regression Analysis*

Perform linear regression analysis on the calibration curve to generate a mathematical equation ($y = mx + b$), where y represents the peak area (or height) and x represents the concentration of the reference standard. The slope (m) of the line represents the response factor, which relates the detector response (peak area or height) to the concentration of the analyte.

5. *Sample Analysis*

Inject your sample extract into the HPLC system and record the peak area (or height) of the analyte's chromatographic peak.

6. *Quantification Calculation*

Use the linear regression equation to calculate the concentration of the analyte in the sample based on its peak area (or height). Rearrange the equation to solve for x (concentration) and substitute the measured peak area (or height) into the equation:

$$\text{Concentration } (x) = (\text{Peak Area (or Height)} - \text{Intercept})/\text{Slope}$$

7. *Reporting Results*

Report the concentration of the analyte in the sample using appropriate units (e.g., μg/mL, mg/g). If the concentration is outside the range of your calibration curve, you might need to dilute or concentrate the sample to ensure it falls within the curve's range.

Advantages of Using Reference Standards

- Provides a quantitative relationship between the detector response and analyte concentration.
- Enables accurate and precise quantification of target compounds in complex samples.
- Facilitates validation and verification of HPLC methods by demonstrating linearity and accuracy.

Considerations

- Use freshly prepared reference standards to ensure accuracy.
- Perform replicate injections of reference standards and samples to account for variability.
- Ensure that the sample matrix does not interfere with the analysis.

In conclusion, quantification using reference standards is a robust method that allows for accurate and reliable determination of the concentration of specific compounds in samples analyzed by HPLC. The calibration curve generated from reference standards serves as a powerful tool for translating detector responses into meaningful quantitative results.

23 Calibration Curves Are Generation to Relate Peak Areas to Anthocyanin Concentrations

Calibration curves are generated to establish a quantitative relationship between the peak areas (or heights) obtained from High-Performance Liquid Chromatography (HPLC) analysis and the concentrations of anthocyanin compounds in a sample. These curves serve as a reference for quantification by allowing you to convert detector responses into meaningful concentration values. Here's how calibration curves are generated:

1. *Preparation of Reference Standards*

Prepare a series of reference standard solutions containing known concentrations of the anthocyanin compounds you want to quantify. The concentrations should span the expected range of concentrations in your samples.

2. *HPLC Analysis of Reference Standards*

Inject each reference standard solution into the HPLC system under the same chromatographic conditions that you will use for your sample analysis. The system records the peak areas (or heights) of the chromatographic peaks corresponding to each anthocyanin compound.

3. *Plotting Peak Area* vs. *Concentration*

Plot the peak areas (or heights) obtained from the reference standards on the *y*-axis and the corresponding concentrations of the reference standards on the *x*-axis. Each point on the plot represents the detector response for a specific known concentration.

4. *Linear Regression Analysis*

Perform linear regression analysis on the plotted data points. This analysis calculates the slope (m) and intercept (b) of the best-fit straight line that represents the relationship between peak area (y) and concentration (x).

5. *Equation of the Calibration Curve*

   The linear regression analysis generates an equation in the form of $y = mx + b$, where:

   y is the peak area (or height) of the chromatographic peak.

   x is the concentration of the reference standard.

   m is the slope of the calibration curve (response factor).

   b is the y-intercept of the curve.

6. *Use of the Calibration Curve*

When you analyze a sample with an unknown concentration of anthocyanins, the HPLC system records the peak area (or height) of the chromatographic peak for each anthocyanin compound. Plug this peak area (y) into the calibration curve equation to calculate the corresponding concentration (x) of the anthocyanin compound in the sample.

7. *Reporting Results*

Report the calculated concentration of the anthocyanin compound in the sample using appropriate units (e.g., μg/mL, mg/g). Include the relevant information about the calibration curve used for quantification.

Advantages of Calibration Curves

- Allow for accurate quantification based on the linear relationship between detector response and concentration.
- Provide a standardized method for converting peak areas into meaningful concentration values.
- Enable validation and verification of the HPLC method by demonstrating linearity and accuracy.

Considerations

- Prepare multiple concentrations of reference standards to ensure accurate linearity assessment.
- Include replicates of each standard concentration to account for variability.
- Regularly update calibration curves to account for changes in column performance or instrumental conditions.

In summary, calibration curves are a fundamental tool in quantitative HPLC analysis, allowing for accurate determination of anthocyanin concentrations in samples based on the relationship between peak areas and known concentrations of reference standards.

24 Methods of Identifying Anthocyanin Compounds Using Retention Times and Comparison to Reference Standards

Identifying anthocyanin compounds using retention times and comparison to reference standards is a common approach in High-Performance Liquid Chromatography (HPLC) analysis. This method relies on the unique retention times of anthocyanin compounds and their characteristic chromatographic behavior [51]. Here's how it's done:

1. *Reference Standard Preparation*

   Prepare a set of reference standards that include known anthocyanin compounds. These reference standards should be pure compounds with well-characterized structures and properties.

2. *HPLC Analysis*

   Inject each reference standard individually into the HPLC system under the same chromatographic conditions that will be used for your sample analysis. The system records the retention time of each anthocyanin peak, which corresponds to the time it takes for the compound to travel through the chromatographic column and be detected.

3. *Retention Time Comparison*

   Determine the retention times of the reference standards for each anthocyanin compound. These retention times are specific to the compound, column, mobile phase, and operating conditions.

4. *Sample Analysis*

   Inject your sample extract into the HPLC system using the same conditions as used for the reference standards. The chromatogram of the sample will display peaks at specific retention times.

5. *Comparison with Reference Standards*

   Compare the retention times of the peaks in the sample chromatogram with the retention times of the reference standards. If a peak in the sample chromatogram matches the retention time of a specific reference standard, it suggests the presence of that particular anthocyanin compound in the sample.

6. *Confirmation of Identity*

While retention time matching provides strong evidence for the identity of the anthocyanin compound, it's important to further confirm the identification using additional methods:

- UV-Vis Spectra: Compare the UV-Vis absorption spectra of the sample peak with that of the corresponding reference standard. A match in spectra further confirms identity.
- Co-chromatography: If available, co-inject the sample with the reference standard. A co-elution of the sample and standard peaks provides additional evidence for identification.

Advantages of Retention Time and Reference Standard Comparison

- Quick and straightforward method for identifying known anthocyanin compounds.
- Requires minimal sample preparation and avoids complex extraction procedures.
- Utilizes well-established chromatographic data for compound identification.
- Can provide reliable results when using well-characterized reference standards.

Considerations

- Retention times can vary based on column type, mobile phase, and instrument conditions. Consistent conditions are crucial for accurate comparisons.
- For complex samples, co-elution of compounds can occur, leading to retention time shifts. This is why additional confirmation methods are important.

In summary, comparing retention times of sample peaks with those of reference standards is a valuable method for identifying known anthocyanin compounds in HPLC analysis. This approach provides a quick and reliable way to confirm the presence of specific compounds in a sample, especially when used in combination with other confirmation methods.

25 Importance of Data Analysis Software in Interpreting Chromatograms and Calculating Concentrations

Data analysis software plays a crucial role in interpreting chromatograms and calculating concentrations in chromatographic analyses, including High-Performance Liquid Chromatography (HPLC). Here's why data analysis software is important:

1. *Peak Integration and Quantification*

Chromatograms generated during HPLC analysis consist of multiple peaks representing different compounds. Data analysis software automates the process of peak integration, accurately measuring peak areas or heights. These integrated values are used to calculate compound concentrations using calibration curves or reference standards.

2. *Calibration Curve Construction*

Data analysis software assists in creating calibration curves by plotting peak areas or heights against known concentrations of reference standards. It performs linear regression analysis to generate equations that relate detector responses to analyte concentrations.

3. *Quantitative Results*

Software calculates the concentrations of compounds in samples by applying the calibration curve equation to the peak areas obtained from sample chromatograms. This automated process ensures accurate and consistent quantification.

4. *Retention Time Determination*

The software identifies and records the retention times of peaks in the chromatogram. Retention times are crucial for compound identification and comparison to reference standards.

5. *Compound Identification*

Data analysis software can facilitate the comparison of retention times, UV-Vis spectra, and other properties of sample peaks with those of reference standards or databases. This aids in confirming the identity of compounds.

6. *Data Visualization*

Software generates chromatograms, calibration curves, and graphs that visually represent the data. These visualizations allow for better understanding of the results and trends.

7. *Data Storage and Management*

Modern data analysis software often includes data storage and management capabilities. It allows researchers to organize, store, and retrieve chromatographic data efficiently.

8. *Accuracy and Reproducibility*

Automated data analysis minimizes the potential for human errors in manual calculations, leading to more accurate and reproducible results.

9. *Efficiency and Speed*

Data analysis software significantly speeds up data interpretation and calculation processes, enabling researchers to process large datasets more quickly and efficiently.

10. *Regulatory Compliance*

In regulated industries (such as pharmaceuticals), data analysis software helps ensure compliance with quality control and validation requirements, as it provides an auditable trail of analysis.

11. *Method Optimization*

Software can assist in method development and optimization by allowing researchers to adjust parameters and immediately observe the effects on chromatograms, peak shapes, and resolutions.

12. *Reporting*

Many software packages offer report generation capabilities, allowing researchers to generate comprehensive reports with quantitative results, chromatograms, calibration curves, and compound identification information.

26 Benefits of HPLC in Anthocyanin Analysis

High-Performance Liquid Chromatography (HPLC) offers several significant benefits for anthocyanin analysis, making it a preferred method for studying these pigments. Here are some of the key advantages of using HPLC for anthocyanin analysis:

1. *Selectivity and Resolution*

HPLC provides excellent selectivity and resolution, allowing for the separation of complex mixtures of anthocyanin compounds. This is crucial in samples where various anthocyanins coexist and need to be individually identified and quantified.

2. *Sensitivity*

HPLC is highly sensitive, enabling the detection and quantification of anthocyanins even at low concentrations. This is important when working with natural samples that contain trace amounts of these compounds.

3. *Quantification Accuracy*

HPLC with appropriate calibration curves or reference standards allows accurate quantification of anthocyanin concentrations. The linear relationship between peak area (or height) and concentration ensures reliable results.

4. *Compound Identification*

   HPLC's ability to generate characteristic chromatographic patterns and UV-Vis spectra aids in the identification of anthocyanin compounds. Comparing retention times and spectra with those of reference standards enhances compound confirmation.

5. *High Throughput*

   Modern HPLC systems equipped with auto-samplers can handle multiple samples in a short time, increasing the throughput of analyses. This is particularly useful when dealing with a large number of samples.

6. *Versatility in Detection Methods*

   HPLC can be coupled with various detectors, including UV-Vis, diode array, fluorescence, and mass spectrometry. This versatility allows researchers to choose the most suitable detection method for specific anthocyanins.

7. *Sample Preservation*

   HPLC requires smaller sample volumes compared to other analytical techniques, which is beneficial when working with limited or precious samples like plant extracts.

8. *Chromatographic Column Variety*

   A wide range of chromatographic columns with different stationary phases is available for HPLC. This allows researchers to optimize separation conditions for specific anthocyanin compounds.

9. *Method Optimization*

   HPLC methods can be optimized to achieve the best separation and resolution of target anthocyanins. This includes adjusting mobile phase composition, gradient profiles, column temperature, and flow rates.

10. *Reliable Reproducibility*

    When properly validated and controlled, HPLC methods provide high reproducibility and robustness, ensuring consistent results across different laboratories and instruments.

11. *Comprehensive Data Analysis*

    Modern HPLC systems often come with advanced data analysis software that simplifies peak integration, quantification, and compound identification. This streamlines the analysis process and enhances result interpretation.

12. *Applications in Various Fields*

HPLC's accuracy and versatility make it suitable for various applications, such as food analysis, pharmaceutical research, environmental monitoring, and nutraceutical studies involving anthocyanin analysis.

27 Precision of HPLC in Detecting and Quantifying Anthocyanins

Precision refers to the reproducibility and consistency of results obtained from HPLC analyses. It encompasses two main aspects:

1. *Repeatability (Intra-day Precision)*

Repeatability assesses the variability in results when the same sample is analyzed multiple times within a short period under the same conditions. A high level of repeatability indicates that the HPLC method is producing consistent results for a given sample. In the context of anthocyanin analysis, consistent retention times, peak areas, and quantification results across repeated analyses are indicative of good repeatability.

2. *Intermediate Precision (Inter-day Precision)*

Intermediate precision evaluates the variability in results when the same sample is analyzed on different days, by different operators, or using different instruments. It assesses the robustness of the HPLC method over time and across variations in experimental conditions. Consistent quantification results between different instances of analysis suggest good intermediate precision. Achieving high precision in anthocyanin analysis is essential to ensure the reliability and credibility of results. It allows researchers to detect even small changes in anthocyanin concentrations and confidently distinguish differences between samples.

28 Sensitivity of HPLC in Detecting and Quantifying Anthocyanins

Sensitivity refers to the ability of an analytical method to detect and quantify compounds at low concentrations. Sensitivity is crucial in anthocyanin analysis, as these compounds can be present in trace amounts in certain samples. HPLC offers excellent sensitivity due to several factors:

1. *Detector Sensitivity*: HPLC detectors, such as UV-Vis and fluorescence detectors, are capable of detecting compounds with high sensitivity. Anthocyanins have characteristic absorption and emission properties that make them well-suited for detection using these methods.

2. *Sample Injection Volume*: HPLC allows precise control over the injection volume, enabling researchers to inject very small sample volumes for analysis. This is particularly useful when working with limited sample quantities.
3. *Concentration Techniques*: Pre-concentration techniques, such as solid-phase extraction (SPE), can be employed to concentrate anthocyanins from samples before injection, thereby enhancing sensitivity.
4. *Calibration Curve Construction*: By using a series of standard solutions with known concentrations of anthocyanins, calibration curves can be established. These curves allow accurate quantification of anthocyanins even at low concentrations.
5. *Column Efficiency and Resolution*: HPLC's high resolution and efficiency facilitate the separation of closely eluting compounds, reducing interference from matrix components and improving sensitivity.
6. *Detector Noise Reduction*: Modern HPLC systems incorporate noise-reduction features that minimize baseline noise and enhance the signal-to-noise ratio, increasing the detection sensitivity.

In anthocyanin analysis, sensitivity is essential when dealing with samples containing low concentrations of these compounds, such as in certain fruits, vegetables, or plant extracts. It ensures that even minor variations in anthocyanin content can be accurately detected and quantified.

29 Selectivity of HPLC in Differentiating Between Various Anthocyanin Compounds

The selectivity of High-Performance Liquid Chromatography (HPLC) is a critical aspect that enables the differentiation between various anthocyanin compounds in complex mixtures. Selectivity refers to the ability of the chromatographic system to separate compounds based on their chemical properties, leading to distinct and well-resolved peaks in the chromatogram [52]. Here's how HPLC achieves selectivity in differentiating between different anthocyanin compounds:

1. *Column Selection*

The choice of the HPLC column is pivotal for achieving selectivity. Different column types, such as reverse-phase C18 columns, normal-phase columns, and specialty columns, exhibit varying affinities for analytes based on their hydrophobic, polar, or specific interaction characteristics [53]. This helps separate anthocyanins with diverse structural features.

2. *Mobile Phase Composition*

The mobile phase, a mixture of solvents, plays a significant role in selectivity. By altering the composition of the mobile phase (e.g., changing the ratio of water to organic solvent), you can adjust the retention times and elution order of anthocyanin compounds. This allows the chromatographic system to separate compounds that may have similar chemical structures but different polarities.

3. *Gradient Elution*

Using gradient elution involves changing the composition of the mobile phase during analysis. This technique enhances selectivity by promoting the sequential elution of compounds based on their affinity for the changing solvent conditions. Gradient elution is particularly effective for resolving complex mixtures of anthocyanin compounds.

4. *Column Temperature*

Controlling the column temperature can affect selectivity by influencing the interactions between compounds and the stationary phase. Slight temperature adjustments can lead to changes in separation patterns, aiding in the differentiation of closely related anthocyanins.

5. *pH Adjustment*

For ionizable anthocyanins, pH adjustment of the mobile phase can impact their ionization states and interactions with the stationary phase. This can significantly alter the selectivity of the separation.

6. *Stationary Phase Chemistry*

Different stationary phases have unique chemical characteristics that can selectively interact with specific anthocyanin compounds. The choice of stationary phase can be tailored to enhance selectivity for certain classes of anthocyanins.

7. *Detector Specificity*

Coupling HPLC with specific detectors, such as UV-Vis, diode array, or mass spectrometry, adds an additional layer of selectivity. UV-Vis detectors, for example, can identify anthocyanins based on their characteristic absorption maxima, aiding in compound identification.

Through careful optimization of these parameters and the appropriate adjustment of method conditions, HPLC achieves a high degree of selectivity, enabling the differentiation between various anthocyanin compounds. This ability to separate and resolve compounds with similar structures is crucial in characterizing complex mixtures and understanding the composition of anthocyanin profiles in different samples [30].

30 How HPLC Enables Quantitative Analysis, Essential for Nutritional and Quality Assessments

High-Performance Liquid Chromatography (HPLC) is a powerful analytical technique that enables accurate and precise quantitative analysis, which is essential for nutritional and quality assessments of various products. HPLC's ability to quantify specific compounds within complex mixtures makes it a valuable tool for assessing the content and quality of nutrients, additives, contaminants, and bioactive compounds [30, 52]. Here's how HPLC facilitates quantitative analysis for nutritional and quality assessments:

1. *Calibration Curves and Reference Standards*

   HPLC quantification relies on the construction of calibration curves using reference standards of known concentrations. By analyzing standard solutions with known concentrations, a linear relationship between detector response (peak area or height) and analyte concentration is established. This calibration curve is then used to quantify unknown sample concentrations based on their detector responses.

2. *Accurate Compound Identification*

   HPLC's separation power and detector selectivity enable accurate compound identification, which is crucial for quantitative analysis. Knowing the exact compound being analyzed allows researchers to select appropriate reference standards and ensure accurate quantification.

3. *Precision and Reproducibility*

   HPLC provides high precision and reproducibility in quantification. Repeatability (intra-day precision) and intermediate precision (inter-day precision) ensure that results obtained from the same samples on different occasions are consistent and reliable.

4. *Sensitivity and Detection Limits*

   HPLC is sensitive and capable of detecting compounds at low concentrations. This is essential for nutritional and quality assessments, where trace amounts of essential nutrients, contaminants, or bioactive compounds can significantly impact the product's quality and safety.

5. *Selective Detection Methods*

   HPLC can be coupled with various detectors, including UV-Vis, fluorescence, and mass spectrometry. Each detection method has its own advantages for specific analytes. For instance, UV-Vis detectors are commonly used for quantifying compounds that absorb UV or visible light, such as vitamins and pigments.

6. *Complex Sample Analysis*

HPLC excels in analyzing complex samples like food, beverages, supplements, and plant extracts. Its ability to separate and quantify compounds in intricate matrices ensures accurate assessment of nutritional and quality attributes.

7. *Compliance and Validation*

HPLC methods used for nutritional and quality assessments can be validated and verified to ensure their accuracy, reliability, and compliance with regulatory standards. This is particularly important for industries such as food, pharmaceuticals, and environmental monitoring.

8. *Monitoring Changes over Time*

HPLC allows for longitudinal studies, tracking changes in the content of compounds over time. This is crucial for assessing product shelf life, stability, and the impact of processing or storage conditions on nutritional and quality attributes.

In nutritional and quality assessments, HPLC's quantitative capabilities provide valuable insights into the content and composition of compounds that influence the nutritional value and overall quality of products [54]. Whether it's determining vitamin content in food, identifying contaminants, analyzing bioactive compounds, or assessing the presence of additives, HPLC plays a fundamental role in ensuring accurate and reliable quantitative analysis.

31 Comprehensive Profiling of Multiple Anthocyanin Compounds Using HPLC

Comprehensive profiling of multiple anthocyanin compounds using High-Performance Liquid Chromatography (HPLC) involves the systematic separation, identification, and quantification of a wide range of anthocyanins present in a sample. This process allows researchers to gain a comprehensive understanding of the anthocyanin composition in complex mixtures, such as plant extracts, foods, and beverages. Here's how this profiling is achieved:

1. *Sample Preparation*

Collect and prepare the sample, ensuring it accurately represents the matrix of interest (e.g., plant tissue, fruit extract). Proper sample preparation involves extracting anthocyanins from the matrix while preserving their integrity.

2. *HPLC Instrumentation*

Set up an HPLC system with the appropriate columns, mobile phase, and detector for anthocyanin separation and analysis.

Columns with different stationary phases can be selected to optimize the separation of various anthocyanins based on their chemical properties.

3. *Method Optimization*

   Develop or optimize the HPLC method to achieve the best separation and resolution of anthocyanin compounds. This includes selecting the appropriate column, mobile phase composition, gradient program, column temperature, and flow rate.

4. *Chromatographic Separation*

   Inject the sample into the HPLC system. The chromatographic separation separates the anthocyanin compounds based on their interactions with the stationary phase and mobile phase components. Different anthocyanin compounds elute at different times, forming distinct peaks in the chromatogram.

5. *Compound Identification*

   Identify individual anthocyanin compounds by comparing their retention times and UV-Vis absorption spectra with those of reference standards. Mass spectrometry can also be used to confirm compound identities by providing additional structural information.

6. *Quantification*

   Quantify the concentration of each identified anthocyanin compound using calibration curves or reference standards with known concentrations. HPLC software assists in peak integration and calculation of concentrations based on peak areas or heights.

7. *Data Analysis*

   Analyze the obtained chromatograms, UV-Vis spectra, and quantification results. Software tools help process and interpret the data, allowing for accurate identification and quantification of multiple anthocyanin compounds.

8. *Comprehensive Profiling*

   Repeat the analysis with different sample matrices or conditions to comprehensively profile a wide range of anthocyanin compounds present. This approach provides a detailed overview of the anthocyanin composition and variability within samples.

9. *Report Generation*

   Generate reports summarizing the identified anthocyanin compounds, their concentrations, and any relevant trends or patterns observed in the data. This information contributes to a comprehensive understanding of the sample's anthocyanin profile.

Comprehensive profiling using HPLC is particularly valuable in studies where the anthocyanin composition varies across different plant species, growth conditions, or processing methods. It allows researchers to explore the diversity of anthocyanin compounds present in complex mixtures and gain insights into their potential health benefits, flavor contributions, and applications in various industries.

32 Implications and Applications

The comprehensive profiling of anthocyanin compounds using High-Performance Liquid Chromatography (HPLC) has far-reaching implications and applications across various fields due to the importance of anthocyanins in nutrition, health, and industry. Here are some of the key implications and applications:

1. *Food and Nutrition*

Anthocyanins contribute to the color, flavor, and nutritional value of foods [24]. Comprehensive anthocyanin profiling using HPLC allows researchers to study their presence, concentrations, and variations in different fruits, vegetables, and plant-based products [41]. This information helps in understanding the potential health benefits of these compounds, such as their antioxidant and anti-inflammatory properties.

2. Dietary Assessment

HPLC-based anthocyanin profiling contributes to dietary assessments by determining the anthocyanin content in various foods and beverages. This aids in evaluating the intake of these bioactive compounds and their potential contributions to human health.

3. *Plant Breeding and Agriculture*

In plant breeding, anthocyanin profiling can assist in selecting and developing crops with higher anthocyanin content. It can also help in understanding the genetic and environmental factors influencing anthocyanin accumulation, leading to the production of nutritionally enhanced crops [30].

4. *Quality Control in Food Industry*

Food manufacturers use HPLC-based anthocyanin profiling to ensure product quality and consistency. By quantifying anthocyanin content, they can verify that products meet label claims and maintain desired color and flavor profiles.

5. *Pharmaceuticals and Nutraceuticals*

Anthocyanins have been investigated for potential pharmaceutical and nutraceutical applications due to their health-promoting properties. Comprehensive profiling assists in identifying specific anthocyanins with potential therapeutic effects, leading to the development of functional foods and supplements [24].

6. *Environmental Monitoring*

Anthocyanins are also found in various environmental samples, such as water bodies and soil. Profiling these compounds using HPLC can provide insights into environmental health, pollution levels, and ecological changes.

7. *Authenticity and Adulteration Detection*

Comprehensive profiling can help identify the presence of specific anthocyanin markers in products, assisting in detecting adulteration and ensuring the authenticity of natural colorants and extracts.

8. *Flavor and Beverage Industry*

Anthocyanins contribute to the sensory characteristics of beverages, such as wines and fruit juices. Profiling their composition aids in understanding flavor attributes and guiding product development.

9. *Research and Academic Studies*

Comprehensive anthocyanin profiling using HPLC is essential for research studies investigating the bioactivity, stability, and structure-activity relationships of these compounds. It contributes to the advancement of scientific knowledge about anthocyanins and their role in various biological processes.

Thus, the comprehensive profiling of anthocyanin compounds using HPLC has broad implications and applications in fields ranging from food and nutrition to pharmaceuticals and environmental monitoring. It enables a deeper understanding of the presence, composition, and potential benefits of anthocyanins, influencing product development, quality control, research, and our understanding of their roles in various biological systems.

33 Implications of Onion Anthocyanin Analysis for Understanding Nutritional Content and Potential Health Benefits

Onion anthocyanin analysis has significant implications for understanding the nutritional content and potential health benefits of this commonly consumed vegetable. Onions (*Allium cepa*) are usually not associated with anthocyanin-rich foods, but certain red and purple varieties contain these pigments, which can have profound effects on their nutritional profile and potential health-promoting properties [55]. Analyzing onion anthocyanins using techniques like High-Performance Liquid Chromatography (HPLC) offers insights into their composition and potential health benefits:

1. *Nutrient Diversity*

Anthocyanin-rich onions provide an additional source of nutrients beyond the common nutritional components found in onions. By analyzing the anthocyanin content, researchers can uncover hidden nutritional diversity in onions, contributing to a more complete understanding of their value in the diet.

2. *Antioxidant Activity*

Anthocyanins are potent antioxidants with the ability to neutralize harmful free radicals in the body. Analyzing the anthocyanin content in onions helps establish their antioxidant capacity, which is important for reducing oxidative stress and its associated health risks [54].

3. *Potential Anti-Inflammatory Effects*

Anthocyanins have shown anti-inflammatory properties in various studies. By assessing the anthocyanin content of onions, researchers can explore the potential of these compounds to mitigate chronic inflammation, a contributing factor to several chronic diseases [12].

4. *Cardiovascular Health*

Some anthocyanins have been linked to cardiovascular health benefits, including the ability to improve lipid profiles, reduce blood pressure, and enhance endothelial function [35]. Analyzing anthocyanins in onions helps determine their potential role in supporting heart health.

5. *Gut Health*

Anthocyanins may influence gut health by promoting the growth of beneficial gut bacteria. Analyzing onion anthocyanins can provide insights into their prebiotic potential, supporting a healthy gut microbiome.

6. *Diabetes Management*

Certain anthocyanins exhibit potential anti-diabetic effects by improving insulin sensitivity and glucose metabolism. Analyzing anthocyanins in onions contributes to understanding their role in diabetes management.

7. *Anti-Cancer Properties*

Some anthocyanins have demonstrated anti-cancer effects by inhibiting the growth of cancer cells and promoting apoptosis (programmed cell death). Analyzing onion anthocyanins can shed light on their potential role in cancer prevention and treatment.

8. *Cognitive Health*

Emerging research suggests that anthocyanins may have positive effects on cognitive function and brain health. Analyzing onion anthocyanins can help uncover their neuroprotective potential [35].

9. *Dietary Recommendations*

Analyzing onion anthocyanin content adds to the body of knowledge informing dietary recommendations. Including anthocyanin-rich onions in dietary guidelines can promote a more diverse and healthful diet.

10. *Product Development*

Understanding the anthocyanin content of onions can influence product development in the food industry. Incorporating anthocyanin-rich onion varieties into food products can enhance their nutritional profiles and potential health benefits.

34 Role of HPLC in Ensuring Quality Control of Onion-Derived Products

High-Performance Liquid Chromatography (HPLC) plays a crucial role in ensuring the quality control of onion-derived products. Whether in the food, cosmetic, pharmaceutical, or nutraceutical industries, HPLC provides a powerful analytical tool for verifying the composition, authenticity, and consistency of products derived from onions [56]. Here's how HPLC contributes to quality control:

1. *Compound Identification and Authentication*

HPLC enables the identification and authentication of specific compounds in onion-derived products. By comparing the chromatographic profiles of products to reference standards or established profiles, HPLC confirms the presence of characteristic compounds, such as anthocyanins, flavonoids, and other bioactive constituents.

2. *Quantitative Analysis*

HPLC quantifies target compounds in onion-derived products. Accurate quantification ensures that products meet label claims and regulatory standards. For instance, HPLC can measure the concentration of key compounds like quercetin or anthocyanins, validating their presence and potency.

3. *Detection of Contaminants and Adulterants*

HPLC can detect contaminants, adulterants, or unwanted substances in onion-derived products. By comparing chromatographic profiles to authentic samples, deviations from expected patterns can indicate potential adulteration or contamination issues.

4. *Batch-to-Batch Consistency*

HPLC ensures consistency in product quality from batch to batch. By quantifying specific compounds, manufacturers can monitor the consistency of active ingredients and overall composition, maintaining uniform product quality over time.

5. *Verification of Extraction Processes*

HPLC verifies the efficiency of extraction processes used to obtain active compounds from onions. It ensures that the desired compounds are present in the product at optimal levels, reflecting the extraction efficiency.

6. *Stability Studies*

HPLC is used in stability studies to monitor changes in compound concentrations over time. This is crucial for assessing the shelf life of onion-derived products and ensuring that their quality is maintained throughout their intended use.

7. *Regulatory Compliance*

For industries subject to regulatory oversight, such as food and pharmaceuticals, HPLC helps ensure compliance with quality standards and labeling requirements. Accurate quantification of key compounds ensures that products meet regulatory expectations.

8. *Product Development*

HPLC aids in the development of new onion-derived products. By analyzing prototypes and formulations, researchers can fine-tune the composition and extract efficiency to achieve desired levels of specific compounds.

9. *Research and Innovation*

HPLC enables researchers to explore the potential of novel compounds in onion-derived products. By identifying and quantifying lesser-known compounds, new health benefits or applications can be discovered.

10. *Consumer Confidence*

Consistent quality control using HPLC enhances consumer confidence. Accurate labeling and reliable product composition foster trust in the brand and product efficacy.

35 Potential Applications of Onion Anthocyanins in the Food Industry

Onion anthocyanins offer several potential applications in the food industry due to their vibrant colors, potential health benefits, and versatility in various food products. While not as commonly associated with anthocyanins as berries or grapes, certain red and purple onion varieties contain these pigments. Here are some potential applications of onion anthocyanins in the food industry:

1. *Natural Food Coloring*

Onion anthocyanins can serve as natural food colorants, providing shades of red, purple, and blue to a variety of food products. They offer an alternative to synthetic colorants, catering to consumer demand for clean-label and natural ingredients.

2. *Enhancing Visual Appeal*

The vibrant colors of onion anthocyanins can enhance the visual appeal of foods, making them more attractive and appetizing to consumers. They are particularly valuable in products like salads, sauces, dressings, and beverages.

3. *Functional Foods*

Incorporating onion anthocyanins into functional foods, such as yogurt, smoothies, and energy bars, can provide an additional layer of health-promoting compounds. These compounds are recognized for their potential antioxidant and anti-inflammatory properties.

4. *Bakery and Confectionery*

Anthocyanin-rich onion extracts can be used in baked goods, pastries, and confectionery items to impart natural colors and contribute to the potential health benefits of these products.

5. *Beverages*

Onion anthocyanins can be incorporated into beverages, including fruit juices, teas, and sports drinks, to provide not only color but also potential health-enhancing properties.

6. *Dairy and Desserts*

Adding onion anthocyanins to dairy products like yogurt or ice cream can create visually appealing and potentially more nutritious options for consumers.

7. *Sauces and Condiments*

Anthocyanin-rich onion extracts can be used in sauces, ketchups, and condiments, enhancing both color and potential health value.

8. *Novel Food Products*

Food manufacturers can innovate by developing novel products that highlight onion anthocyanins. These could include anthocyanin-infused snacks, spreads, and even functional foods targeting specific health concerns.

9. *Ethnic and Gourmet Foods*

Onion anthocyanins can be integrated into ethnic and gourmet food products to add unique colors and flavors that cater to specific culinary preferences.

10. *Marketing and Differentiation*

Food products containing onion anthocyanins can be marketed as offering not only pleasing aesthetics but also potential health benefits. This can provide a competitive edge in a market focused on health and wellness.

11. *Research and Development*

Exploring the applications of onion anthocyanins may lead to the discovery of new food formulations and product concepts that capitalize on their unique attributes.

Therefore, it's important to note that the incorporation of onion anthocyanins in food products requires careful consideration of factors such as flavor compatibility, stability, and regulatory approval. Additionally, sourcing reliable anthocyanin-rich onion extracts is crucial to ensure consistent quality and color performance.

36 Challenges and Future Directions

While onion anthocyanins hold promise for various applications, there are also challenges and future directions that need to be addressed to fully harness their potential. Here are some key challenges and areas of focus for the future.

Challenges

1. *Variability in Anthocyanin Content*: The anthocyanin content in onions can vary based on factors such as variety, growing conditions, and storage. Ensuring consistent and standardized anthocyanin levels poses a challenge for product development and quality control.
2. *Stability Issues*: Anthocyanins are sensitive to factors like pH, temperature, light, and oxygen, which can lead to degradation and color changes. Developing stable formulations that retain anthocyanin content and color over time is a challenge.
3. *Extraction Efficiency*: Extracting anthocyanins from onions efficiently while preserving their integrity is essential. Optimizing extraction methods to yield high-quality extracts without compromising compound stability is a challenge.
4. *Taste and Flavor Considerations*: Anthocyanin-rich extracts can sometimes introduce new flavors or alter the taste of food products. Overcoming any potential negative taste impact while maintaining product acceptability is a challenge.
5. *Regulatory Approval*: Utilizing novel ingredients like onion anthocyanins in food products may require regulatory approval in different regions. Meeting regulatory requirements for safety, labeling, and claims is a challenge.
6. *Cost and Scalability*: Sourcing and producing consistent, high-quality anthocyanin extracts can be costly. Finding cost-effective methods for large-scale production without compromising quality is a challenge.

Future Directions

1. *Breeding and Genetic Enhancement*: Developing onion varieties with higher and more stable anthocyanin content through breeding and genetic engineering can ensure a more reliable supply of anthocyanin-rich raw materials.
2. *Stability Enhancement*: Research into stabilization techniques, such as encapsulation, microencapsulation, and novel delivery systems, can improve the stability of anthocyanins in various food matrices.
3. *Sustainable Sourcing*: Exploring sustainable methods for sourcing anthocyanin-rich onions, such as vertical farming or controlled environment agriculture, can ensure a consistent supply while minimizing environmental impact.
4. *Processing Innovations*: Developing innovative processing methods that minimize degradation and optimize anthocyanin retention during food production can lead to more stable and attractive products.

5. *Synergistic Formulations*: Investigating combinations of anthocyanins with other natural compounds that enhance stability, color, and potential health benefits can lead to synergistic formulations.
6. *Consumer Education*: Educating consumers about the potential health benefits of onion anthocyanins can increase demand for products containing these compounds, fostering greater acceptance and understanding.
7. *Functional Food Development*: Continued research into the bioactive properties of onion anthocyanins and their potential effects on human health can drive the development of functional foods with specific health claims.
8. *Regulatory Frameworks*: Collaborating with regulatory bodies to establish clear and consistent guidelines for the use of onion anthocyanins in food products can facilitate market access and consumer confidence.
9. *Partnerships and Collaboration*: Collaborations between academia, industry, and regulatory agencies can accelerate research, development, and adoption of onion anthocyanin applications.

In a nutshell, while onion anthocyanins offer exciting opportunities in the food industry, addressing challenges related to stability, quality, and regulatory considerations is essential. By focusing on innovations in sourcing, processing, and product formulation, the food industry can unlock the potential of onion anthocyanins and develop a new generation of visually appealing, nutritious, and functional food products.

37 Potential Challenges in Conducting Onion Anthocyanin Analysis Using HPLC

Conducting onion anthocyanin analysis using High-Performance Liquid Chromatography (HPLC) presents several challenges due to the unique characteristics of anthocyanins and the complex matrix of onion samples. These challenges can impact the accuracy, reliability, and reproducibility of the analysis. Here are some potential challenges:

1. *Sample Complexity*

Onion samples are complex matrices containing a variety of compounds, such as sugars, organic acids, and phenolic compounds, which can interfere with anthocyanin separation and quantification. Proper sample preparation is crucial to minimize interference.

2. *Compound Diversity*

Onion anthocyanins comprise various compounds with distinct structural features. Ensuring optimal separation of these compounds to obtain accurate identification and quantification can be challenging.

3. *Extraction Efficiency*

Extracting anthocyanins from onions while maintaining their stability and integrity can be difficult. Variability in extraction efficiency can impact the accuracy of quantification.

4. *Matrix Effects*

The matrix effect refers to interactions between compounds in the sample matrix and the chromatographic system, affecting retention times and peak shapes. Matrix effects can lead to inaccurate quantification and require thorough method validation.

5. *Compound Degradation*

Anthocyanins are sensitive to factors like light, temperature, and pH. Inadequate sample handling and storage conditions can lead to compound degradation, affecting the reliability of results.

6. *Calibration Curve Accuracy*

Creating accurate calibration curves for quantification requires a series of standard solutions with known concentrations. The accuracy of the calibration curve depends on the purity and stability of the reference standards.

7. *Isomer Separation*

Anthocyanins can have isomeric forms, which can co-elute or exhibit similar retention times. Proper chromatographic conditions are needed to separate and quantify these isomers accurately.

8. *Standardization of Methods*

Standardizing HPLC methods for onion anthocyanin analysis can be challenging due to variations in onion types, growing conditions, and processing methods. Developing a method that accounts for these variations while maintaining reproducibility is essential.

9. *Sensitivity and Detection Limits*

Some onion varieties contain lower anthocyanin levels, requiring sensitive detection methods to accurately quantify these compounds. Achieving low detection limits while maintaining precision can be challenging.

10. *Sample Throughput*

HPLC analysis can be time-consuming, especially when dealing with a large number of samples. Efficient sample preparation and optimization of chromatographic conditions are necessary to ensure reasonable throughput.

11. *Method Validation*

Thorough method validation is essential to ensure the accuracy, precision, specificity, and robustness of the HPLC method for onion anthocyanin analysis. Addressing all potential challenges in method validation enhances the reliability of results.

12. *Data Interpretation*

Interpreting complex chromatograms and identifying individual anthocyanin peaks can be challenging, especially when dealing with complex matrices containing multiple compounds.

To overcome these challenges, researchers must carefully design and optimize HPLC methods for onion anthocyanin analysis, conduct proper sample preparation, implement quality control procedures, and perform thorough method validation. Addressing these challenges ensures the accuracy, reproducibility, and reliability of the results, allowing for meaningful insights into onion anthocyanin composition and content.

38 Limitations of the Technique, Such as Variations in Sample Preparation and Matrix Effects

High-Performance Liquid Chromatography (HPLC) is a powerful analytical technique, but like any method, it has its limitations. Two significant limitations are variations in sample preparation and matrix effects:

1. *Variations in Sample Preparation*

Sample preparation is a critical step in HPLC analysis, and variations in this process can introduce errors and inconsistencies. Factors such as extraction methods, sample homogeneity, and handling can impact the accuracy and reproducibility of results. Different operators or laboratories might use slightly different sample preparation protocols, leading to variability in the obtained data.

2. *Matrix Effects*

Matrix effects occur when components of the sample matrix interact with the chromatographic system, affecting the separation and detection of target compounds. These interactions can lead to altered retention times, peak shape distortions, and changes in detector response. Matrix effects are particularly problematic in complex matrices like food, biological fluids, and environmental samples, where multiple compounds interact with the chromatographic system differently.

Matrix effects can arise due to:

- Ion Suppression or Enhancement: Components in the matrix can suppress or enhance the ionization of analytes, leading to inaccurate quantification.
- Column Fouling: Matrix components can accumulate on the HPLC column, affecting separation and causing drift in retention times.
- Masking Effects: Overlapping peaks or background noise from the matrix can mask the presence of analytes of interest, making their identification and quantification challenging.
- Detector Saturation: In complex matrices, the detector can become saturated due to the presence of high-concentration matrix components, leading to inaccurate measurements.

To address these limitations, researchers and analysts take several steps:

- Method Development and Validation: Careful method development and thorough validation help account for matrix effects and variations in sample preparation. Validated methods provide assurance of accurate and reproducible results.
- Internal Standards: Incorporating internal standards that are unaffected by matrix effects can help correct for variations during analysis, improving accuracy.
- Standard Addition Method: This method involves adding known amounts of the analyte to the sample to account for matrix effects. It helps determine the extent of interference from the matrix.
- Sample Cleanup Techniques: Pre-analytical sample cleanup techniques, such as solid-phase extraction (SPE) or liquid-liquid extraction (LLE), can reduce matrix effects by removing interfering components before analysis.
- Quality Control and Standardization: Implementing strict quality control procedures and standardized sample preparation protocols can minimize variations introduced during the analysis process.

While variations in sample preparation and matrix effects are challenges in HPLC, careful method development, validation, and the use of appropriate techniques can help mitigate these limitations and improve the accuracy and reliability of the results.

39 Emerging Trends and Advancements in HPLC Technology for More Accurate and Efficient Analysis

Emerging trends and advancements in High-Performance Liquid Chromatography (HPLC) technology are continually shaping the field, enabling more accurate, efficient, and comprehensive analysis. These advancements address challenges, improve analytical capabilities, and enhance the overall performance of HPLC systems. Here are some notable trends and advancements:

1. *Ultra-High Performance Liquid Chromatography (UHPLC)*

UHPLC represents a significant advancement in HPLC technology. It uses columns with smaller particle sizes and higher pressures to achieve faster separations and improved resolution. UHPLC systems offer higher sensitivity, reduced solvent consumption, and shorter analysis times compared to traditional HPLC.

2. *Column Technology*

Advancements in column technology include the development of novel stationary phases, surface modifications, and improved packing techniques. These innovations enhance selectivity, efficiency, and separation of complex mixtures, resulting in more accurate and reliable analyses.

3. *Hybrid Column Configurations*

Combining different column chemistries in a single analysis—known as hybrid column configurations—provides enhanced separation capabilities. This approach offers improved selectivity and efficiency for complex samples.

4. *Advanced Detectors*

New generations of detectors, such as high-resolution mass spectrometers and multidimensional detection systems, provide increased specificity, sensitivity, and information-rich data. These detectors enable identification of compounds with greater accuracy and facilitate structural elucidation.

5. *Miniaturization and Portability*

Miniaturized HPLC systems and portable instruments are becoming more prevalent. These systems offer advantages like reduced solvent consumption, faster analysis times, and the ability to perform on-site analysis, suitable for field applications.

6. *Automation and Robotics*

Integration of automation and robotics streamlines sample preparation, injection, and data processing. This leads to improved reproducibility, reduced human error, and increased sample throughput.

7. *Data Integration and Analysis Software*

Advanced software platforms facilitate data integration, processing, and interpretation. These tools offer features like automated peak detection, quantification, and compound identification, enhancing the efficiency and accuracy of analysis.

8. *Multidimensional Chromatography*

Multidimensional chromatography involves using multiple separation techniques in tandem to improve separation efficiency and resolve complex mixtures. This approach is particularly useful for challenging samples with overlapping peaks.

9. *Green HPLC*

Sustainable practices are gaining momentum in HPLC. Green HPLC techniques focus on reducing solvent consumption, energy usage, and waste generation. This trend aligns with environmental concerns and regulatory pressures.

10. *Real-Time Monitoring and Quality Control*

Advancements in sensor technology enable real-time monitoring of parameters such as pressure, temperature, and UV absorbance. This ensures consistency in analyses and allows for prompt adjustments if anomalies are detected.

11. *Internet of Things (IoT) Integration*

IoT integration in HPLC systems enables remote monitoring, data sharing, and even predictive maintenance. This connectivity enhances instrument efficiency and enables remote data access.

12. *3D Printing and Customization*

3D printing technology allows for custom fabrication of components and accessories, enabling researchers to design and optimize their own chromatography setups.

These emerging trends and advancements in HPLC technology collectively aim to enhance the accuracy, efficiency, and versatility of analytical methods. They enable researchers to tackle complex analytical challenges, reduce analysis times, and obtain higher-quality data, ultimately advancing various fields such as pharmaceuticals, food analysis, environmental monitoring, and more.

40 Conclusion

In conclusion, High-Performance Liquid Chromatography (HPLC) analytical modelling has proven to be an indispensable tool for measuring the onion anthocyanin profile. This sophisticated analytical technique allows for the separation, identification, and quantification of anthocyanin compounds in complex onion samples. Through careful selection of columns, mobile phases, detectors, and optimized chromatographic conditions, HPLC enables researchers to unravel the intricate composition of anthocyanins present in various onion varieties. The principles behind HPLC, involving the interactions between the mobile phase, stationary phase, and analytes, form the foundation for successful anthocyanin analysis. The mobile phase solvents are selected based on their ability to elute the compounds of interest, while the stationary phase provides the necessary separation based on chemical properties. UV-Vis detectors play a vital role in identifying and quantifying anthocyanins by measuring their absorption spectra.

However, conducting onion anthocyanin analysis using HPLC is not without its challenges. Variations in sample preparation and matrix effects can introduce errors and inconsistencies into the results. Overcoming these challenges requires meticulous method development, validation, and the use of appropriate corrective techniques, such as internal standards and standard addition methods. As HPLC technology continues to advance, emerging trends like ultra-high performance liquid chromatography (UHPLC), advanced detectors, and automation are improving the accuracy and efficiency of analysis. These advancements contribute to better separation, enhanced sensitivity, and faster analysis times, thereby facilitating more comprehensive and reliable onion anthocyanin profiling.

The implications of onion anthocyanin analysis extend beyond the laboratory, encompassing diverse fields such as food and nutrition, pharmaceuticals, cosmetics, and more. The insights gained from analyzing onion anthocyanin profiles offer opportunities for natural food coloring, functional food development, and even potential health benefits due to the compounds' antioxidant and anti-inflammatory properties. In a world where consumers seek transparency, quality, and natural ingredients, HPLC analytical modelling for measuring onion anthocyanin profiles stands as a valuable tool for the food industry, research, and innovation. By addressing challenges, embracing advancements, and exploring the multifaceted applications of onion anthocyanins, this analytical approach contributes to a deeper understanding of the nutritional, functional, and sensory aspects of this often-overlooked vegetable.

References

1. Bélanger JMR, Jocelyn Paré JR, Sigouin M (1997) High performance liquid chromatography (HPLC): principles and applications. In: Paré JRJ, Bélanger JMR (eds) Techniques and instrumentation in analytical chemistry. Elsevier, pp 37–59
2. Feruza A et al (2023) Chapter 1: Principles and applications of ultra-high-performance liquid chromatography. In: Oscar N et al (eds) High performance liquid chromatography. IntechOpen, Rijeka
3. Beekwilder J et al (2005) Antioxidants in raspberry: on-line analysis links antioxidant activity to a diversity of individual metabolites. J Agric Food Chem 53:3313–3320
4. Alvarez-Suarez J et al (2021) Novel approaches in anthocyanin research – plant fortification and bioavailability issues. Trends Food Sci Technol 117:92–105
5. Ahmed M et al (2023) Himalayan sources of anthocyanins and its multifunctional applications: a review. Foods 12(11):2203
6. He J, Giusti MM (2010) Anthocyanins: natural colorants with health-promoting properties. Annu Rev Food Sci Technol 1:163–187
7. Khoo HE et al (2017) Anthocyanidins and anthocyanins: colored pigments as food, pharmaceutical ingredients, and the potential health benefits. Food Nutr Res 61(1):1361779
8. Lila MA (2004) Anthocyanins and human health: an in vitro investigative approach. J Biomed Biotechnol 2004(5):306–313
9. Mannino G et al (2021) Anthocyanins: biosynthesis, distribution, ecological role, and use of biostimulants to increase their content in plant foods—a review. Agriculture 11(3):212
10. Chadorshabi S, Hallaj-Nezhadi S, Ghasempour Z (2022) Red onion skin active ingredients, extraction and biological properties for functional food applications. Food Chem 386: 132737
11. González-de-Peredo AV et al (2022) Extraction of antioxidant compounds from onion bulb (Allium cepa L.) using individual and simultaneous microwave-assisted extraction methods. Antioxidants (Basel) 11(5):846
12. Samota MK et al (2022) Onion anthocyanins: extraction, stability, bioavailability, dietary effect, and health implications. Front Nutr 9: 917617
13. Diaconeasa Z et al (2020) Anthocyanins, vibrant color pigments, and their role in skin. Cancer Prev 8(9):336
14. Ho WW, Smith SD (2016) Molecular evolution of anthocyanin pigmentation genes following losses of flower color. BMC Evol Biol 16(1):98
15. Santos-Buelga C, Mateus N, De Freitas V, Anthocyanins. (2014) Plant pigments and beyond. J Agric Food Chem 62(29): 6879–6884
16. Kovinich N et al (2015) Abiotic stresses induce different localizations of anthocyanins in Arabidopsis. Plant Signal Behav 10(7):e1027850
17. Dabravolski SA, Isayenkov SV (2023) The role of Anthocyanins in plant tolerance to drought and salt stresses. Plants (Basel) 12(13):2558
18. Mattioli R et al (2020) Anthocyanins: a comprehensive review of their chemical properties and health effects on cardiovascular and neurodegenerative diseases. Molecules 25(17):3809
19. Nassarawa SS, Belwal T, Javed M, Luo Z (2022) Influence of the red LEDs light irradiation on the quality and chemical attributes of postharvest table Grape (Vitis vinifera L.) during storage. Food Bioprocess Technol. https://doi.org/10.1007/s11947-022-02824-1
20. Sanusi SN, Bolade MK, Oluwalana, IB (2018) Nutrient retention capacity of white and red varieties of onion (Allium Cepa) bulbs as influenced by storage conditions. International Journal of Food Science and Biotechnology 3 (3):95. https://doi.org/10.11648/j.ijfsb.20180303.14
21. Nassarawa SS, Bao N, Zhang X, Ru Q, Luo Z (2024) Evaluation of light irradiation on anthocyanins and energy metabolism of grape (Vitis vinifera L.) during storage. Food Chem 431:137–141. https://doi.org/10.1016/j.foodchem.2023.137141
22. Nassarawa SS, Luo Z (2022) Effect of light irradiation on sugar, phenolics and GABA metabolism on postharvest grape (Vitis vinifera L.) during storage. Food Bioprocess Technol 15:1–20. https://doi.org/10.21203/rs.3.rs-1704313/v1
23. Gouda M, Shamsudeen S, Gupta SD, Isma N, Muhammad M (2023) Evaluation of carbon dioxide elevation on phenolic compounds and antioxidant activity of red onion (Allium cepa L.) during postharvest storage. Plant Physiol Biochem 200(23):107–752. https://doi.org/10.1016/j.plaphy.2023.107752
24. Tsuda T (2012) Anthocyanins as functional food factors—chemistry, nutrition and health promotion. Food Sci Technol Res 18(3): 315–324
25. Jolayemi OS, Shamsudeen Nassarawa S, Lawal OM, Sodipo MA, Oluwalana IB (2018) Monitoring the changes in chemical properties of

red and white onions (Allium cepa) during storage. J Stored Prod Postharvest Res 9 (7):78–86. https://doi.org/10.5897/JSPPR2018.0263
26. Nassarawa SS, Sulaiman SA (2019) Extending the shelf life of tomato and onion in Nigeria: A review. Int J Food Sci Nutr 4. www.foodsciencejournal.com
27. Sanusi SN, Gambo A (2019) Effect of storage methods on some selected mineral and ascorbic acid content of red and white onion (Allium cepa). J Food Nutr Agric 2(1):6–9. https://doi.org/10.21839/jfna.2019.v2i1.177
28. Sagar NA et al (2022) Onion (Allium cepa L.) bioactives: chemistry, pharmacotherapeutic functions, and industrial applications. Food Front 3(3):380–412
29. Lee J, Rennaker C, Wrolstad RE (2008) Correlation of two anthocyanin quantification methods: HPLC and spectrophotometric methods. Food Chem 110(3):782–786
30. Seyit Y et al (2023) Chapter 7: HPLC-MS(n) applications in the analysis of Anthocyanins in fruits. In: Oscar N et al (eds) High performance liquid chromatography. IntechOpen, Rijeka
31. Ayvaz H et al (2023) Anthocyanins: metabolic digestion, bioavailability, therapeutic effects, current pharmaceutical/industrial use, and innovation potential. Antioxidants (Basel) 12(1):48
32. Garcia-Oliveira P et al (2021) Identification, quantification, and method validation of Anthocyanins. Chem Proc 5(1):43
33. Konczak I, Zhang W (2004) Anthocyanins-more than nature's colours. J Biomed Biotechnol 2004(5):239–240
34. Naing AH, Kim CK (2021) Abiotic stress-induced anthocyanins in plants: their role in tolerance to abiotic stresses. Physiol Plant 172(3):1711–1723
35. Ockermann P et al (2021) A review of the properties of Anthocyanins and their influence on factors affecting cardiometabolic and cognitive health. Nutrients 13(8):2831
36. Burton-Freeman B, Sandhu A, Edirisinghe I (2016) Anthocyanins. In: Gupta RC (ed) Nutraceuticals. Academic, Boston, pp 489–500
37. Grace MH et al (2019) Simultaneous LC-MS quantification of anthocyanins and non-anthocyanin phenolics from blueberries with widely divergent profiles and biological activities. Food Chem 277:336–346
38. Valenta K et al (2020) Ultraviolet radiation changes plant color. BMC Plant Biol 20(1):253
39. Liu Y et al (2018) Anthocyanin biosynthesis and degradation mechanisms in Solanaceous vegetables: a review. Front Chem 6:52
40. Landi M, Tattini M, Gould KS (2015) Multiple functional roles of anthocyanins in plant-environment interactions. Environ Exp Bot 119:4–17
41. Winefield C, Davies K, Gould K (2009) Anthocyanins: biosynthesis, functions, and applications. Springer, New York
42. Mekapogu M et al (2020) Anthocyanins in floral colors: biosynthesis and regulation in chrysanthemum flowers. Int J Mol Sci 21(18):6537
43. Oteiza PI, Cremonini E, Fraga CG (2023) Anthocyanin actions at the gastrointestinal tract: relevance to their health benefits. Mol Asp Med 89:101156
44. Ma Z et al (2021) An insight into anti-inflammatory activities and inflammation related diseases of Anthocyanins: a review of both in vivo and in vitro investigations. Int J Mol Sci 22(20):11076
45. Kozłowska A, Dzierżanowski T (2021) Targeting inflammation by Anthocyanins as the novel therapeutic potential for chronic diseases: an update. Molecules 26(14):4380
46. Li S et al (2019) The anti-inflammatory effects of dietary anthocyanins against ulcerative colitis. Int J Mol Sci 20(10):2588
47. Ahles S, Joris PJ, Plat J (2021) Effects of berry anthocyanins on cognitive performance, vascular function and Cardiometabolic risk markers: a systematic review of randomized placebo-controlled intervention studies in humans. Int J Mol Sci 22(12):6482
48. Gardener SL et al (2021) Intake of products containing anthocyanins, flavanols, and flavanones, and cognitive function: a narrative review. Front Aging Neurosci 13:640381
49. Lin BW et al (2017) Effects of anthocyanins on the prevention and treatment of cancer. Br J Pharmacol 174(11):1226–1243
50. Hudlikar R et al (2020) Anthocyanins and cancer prevention. In: Pezzuto JM, Vang O (eds) Natural products for cancer chemoprevention: single compounds and combinations. Springer International Publishing, Cham, pp 351–373
51. Fujita R et al (2020) Evaluation of fruit anthocyanin composition by LC/MS in interspecific hybrids between Haskap (*Lonicera caerulea* subsp. *edulis* (Turcz. ex. Herder) Hultén) and Miyama-uguisukagura (*Lonicera gracilipes* Miq.). The Horticulture Journal 89(4):343–350
52. Vieira GS et al (2017) Determination of anthocyanins and non-anthocyanin polyphenols by

ultra performance liquid chromatography/electrospray ionization mass spectrometry (UPLC/ESI-MS) in jussara (Euterpe edulis) extracts. J Food Sci Technol 54(7):2135–2144

53. Zhang Z et al (2004) Comparison of HPLC methods for determination of anthocyanins and anthocyanidins in bilberry extracts. J Agric Food Chem 52:688–691
54. Zhang S-L et al (2016) Quantification and analysis of anthocyanin and flavonoids compositions, and antioxidant activities in onions with three different colors. J Integr Agric 15(9): 2175–2181
55. Zhao XX et al (2021) Recent advances in bioactive compounds, health functions, and safety concerns of onion (Allium cepa L.). Front Nutr 8:669805
56. González-de-Peredo AV et al (2021) Development of a rapid UHPLC-PDA method for the simultaneous quantification of Flavonol contents in onions (Allium cepa L.). Pharmaceuticals (Basel) 14(4):310

Chapter 8

Use of LC-Q-TOF-MS for Elucidating the Flavonoids Chemical Structures of Tea Leaves

Wenkai Zhang, Xiaoli Li, Xusheng Ji, and Mostafa Gouda

Abstract

Flavonoids are important secondary metabolites in tea plants, which contribute to the formation of high-quality tea leaves. Their structural variability and similitude increase the difficulty of compound identification. The powerful tool of LC-Q-TOF-MS, combining the separation capabilities of liquid chromatography (LC) with the structural analysis potential of quadrupole time-of-flight mass spectrometry (Q-TOF-MS), provides high selectivity and sensitivity, making it ideal for detecting flavonoids. In this chapter, we present a protocol for the comprehensive identification of flavonoids compounds in tea leaves based on LC-Q-TOF-MS/MS, which concludes the basic principles of LC-Q-TOF-MS, a data processing methodology, relative software operational essentials.

Key words Phytochemicals, Liquid chromatography, Flavonoids, Mass spectrometry, Catechins

1 Introduction

Tea leaves contain a considerable number of flavonoids, the main secondary metabolites in plants and the most important quality-related compounds. They account for 20–40% of the dry matter in tea leaves [1]. These flavonoids contribute to the formation of good color, taste, and aroma in tea leaves. Both green tea and black tea are rich dietary sources of flavonoids [2]. Existing evidence suggests that regular tea consumption can reduce the risk of cardiovascular diseases. The basic structure of flavonoids consists of 15 carbon atoms arranged in three rings, labeled A, B, and C. This structure is often referred to as the C6(A-ring)–C3(C-ring)–C6 (B-ring) skeleton, with two benzene rings (the A and B rings) connected by a three-carbon chain that forms a closed pyran ring (the C ring). The diversity of flavonoids comes from the many different ways this basic structure can be modified, such as through the addition of various functional groups (like hydroxyl groups), methylation, glycosylation, or other types of rearrangements and

Mostafa Gouda et al. (eds.), *Plant Chemical Compositions and Bioactivities*, Methods and Protocols in Food Science, https://doi.org/10.1007/978-1-0716-3938-2_8,

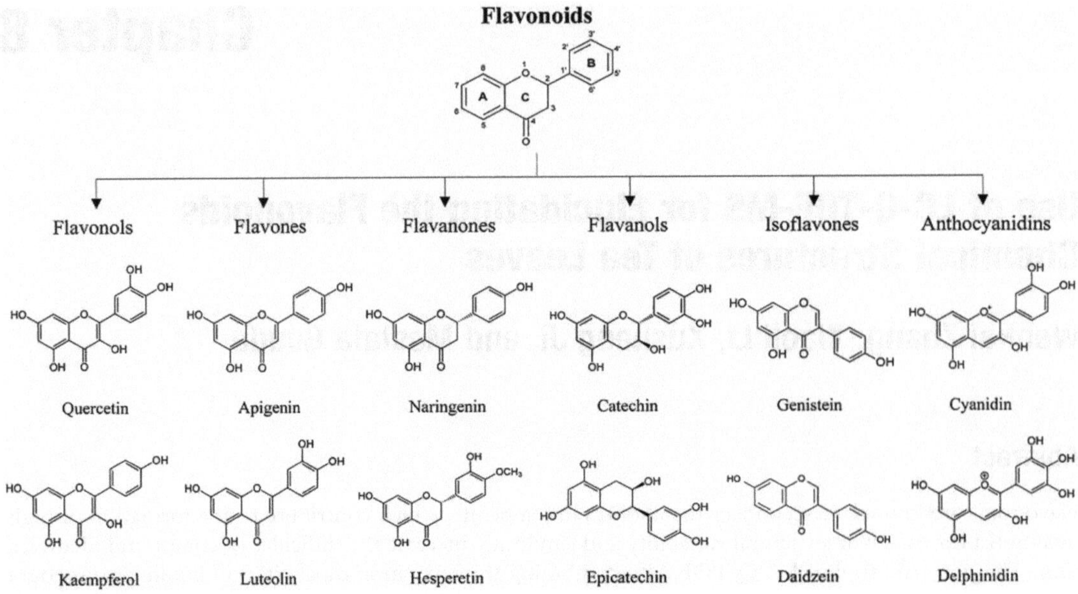

Fig. 1 Classification and examples of flavonoid in tea leaves [3]

conjugations. Thus, flavonoids are classified as flavonols, flavones, flavanones, flavanols (catechins and proanthocyanidins), isoflavones, anthocyanidins, according to heterocyclic ring's substitution pattern [3], as shown in Fig. 1. In tea leaves, the most abundant flavonoids are flavan-3-ols, also known as catechins. These include epicatechin (EC), epicatechin gallate (ECG), epigallocatechin (EGC), and epigallocatechin gallate (EGCG), with EGCG being the most abundant [4]. These compounds have a specific arrangement of hydroxyl groups on the B and C rings of the basic flavonoid structure, which is largely responsible for their bioactivity [5, 6].

Q-TOF-MS (quadrupole time-of-flight mass spectrometry), a type of high-resolution mass spectrometry (HRMS), is a powerful analytical technique that combines the capabilities of quadrupole precursor ion selection with high-resolution time-of-flight (TOF) mass analysis. It can separate ionized ions with different m/z values according to flight time [7]. Q-TOF-MS can be coupled with several preceding analysis and sample preparation techniques to detect substances with different properties, such as liquid chromatography (LC) [6, 8, 9], gas chromatography (GC) [10], capillary electrophoresis (CE) [8, 11], ion mobility spectrometry (IMS) [12], matrix-assisted laser desorption/ionization (MALDI) [13], and so on. When coupled to LC, it is an efficient flavonoid detection technique with high sensitivity [7]. With LC-Q-TOF-MS/MS, it is possible to detect flavonoids qualitatively and quantitatively. Establishing a correlation between the strength of mass signals and analyte quantity allows for high sensitivity. By comparing the analyte content in an unknown sample with that of a standard, this method can accurately determine analyte quantity [14]. By utilizing LC-Q-TOF-MS as either a target or nontarget metabolomic tool, it becomes possible to analyze the quality, origin, and cultivar of flavonoid-rich foods. Additionally, it allows for the evaluation of flavonoid quantities and chemical alterations that may occur during food processing [15].

This chapter will primarily introduce the basic principles of LC-Q-TOF-MS, and detail the specific steps and considerations for using this technology to detect flavonoid components in tea leaves.

2 Principle

To enhance the detection of flavonoids in tea leaves, it is crucial to grasp the fundamentals of LC-Q-TOF-MS/MS. Figure 2 illustrates the principle of LC-Q-TOF-MS, blending the power of liquid chromatography (LC) for component separation, and quadrupole time-of-flight mass spectrometry (Q-TOF-MS) for accurate mass analysis. LC separates sample components via a mobile and stationary phase, while the Q-TOF-MS filters ions of specific mass-to-charge ratios and measures their flight time in an electric field to determine their precise mass.

3 LC Separation

LC separation, including HPLC and UPLC, relies on analyte interactions with a stationary phase. The efficiency of column separation, as per Plate theory, is linked to the height equivalent to a theoretical plate (HETP, H). The van Deemter equation represents the relationship between HETP and mobile-phase linear velocity (μ) (Eq. 1) [7]:

$$H = A + \frac{B}{\mu} + C\mu = 2\lambda d_p + \frac{2\gamma DM}{\mu} + \frac{f(k) d_p^2 \mu}{D_M} \quad (1)$$

where A, B, and C are the eddy diffusion, axial diffusion, and overall solid-mobile phase mass transfer resistance, respectively. d_p and λ are the particle diameter and the structure factor of the column packing material, respectively. DM, γ, and k are the analyte diffusion coefficient, constant termed tortuosity, and retention factor for an analyte, respectively.

The separation in LC occurs due to varying affinities of compounds for the column material in a chosen mobile phase. The measure of LC's separation efficiency is resolution, computed as per Eq. 2 [7]:

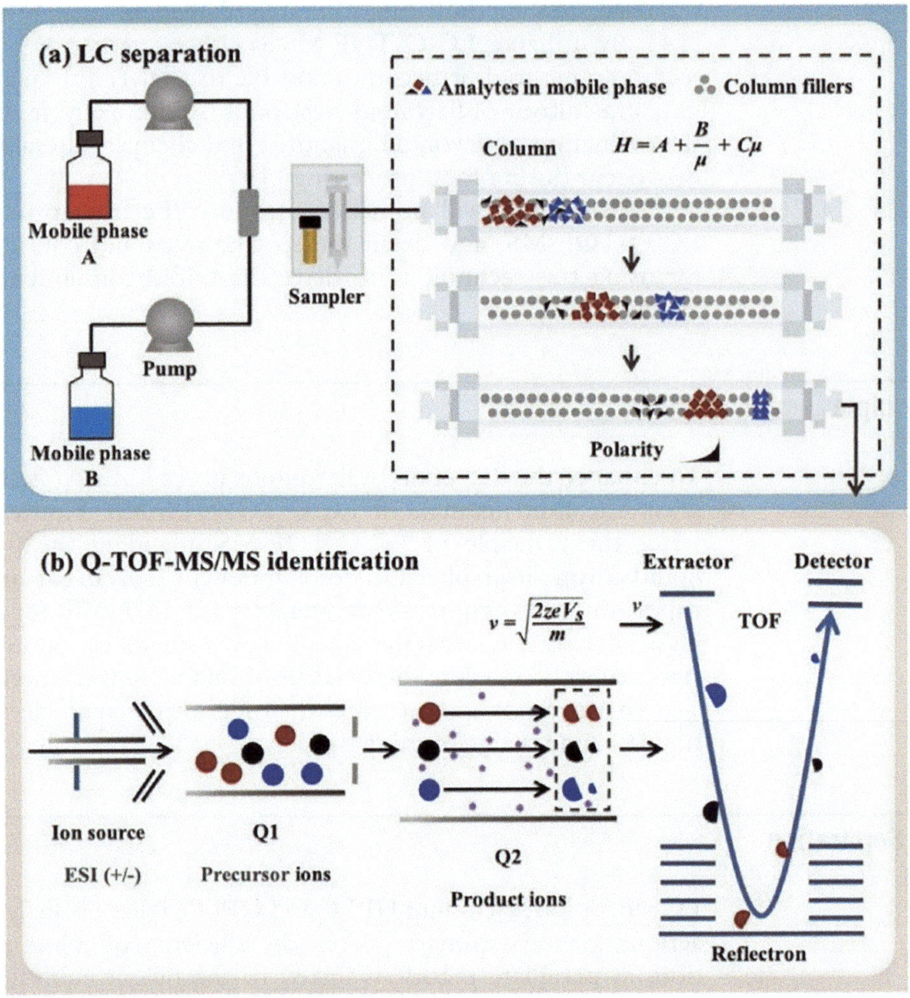

Fig. 2 Schematic illustration of the principle of LC-Q-TOF-MS/MS: (**a**) LC and (**b**) Q-TOF-MS/MS [7]

$$R_s = \frac{2(t_{r2} - t_{r1})}{W_{b2} + W_{b1}} = \left(\frac{\sqrt{N}}{4}\right)\left(\frac{\alpha - 1}{\alpha}\right)\left(\frac{k_2}{k_2 + 1}\right) \quad (2)$$

where R_s, t_r, W_b, k, N, and α are the resolution, retention time, peak width, capacity factor, theoretical plate number, and selectivity factor, respectively. k and α are mainly influenced by the mobile phase and pH value [16].

LC, widely used for food flavonoid separation, simplifies the process by avoiding complex pretreatments that could alter or degrade flavonoids [7]. UPLC, with 2 μm stationary phases, accelerates flavonoid analysis 3–9 times compared to HPLC, without compromising resolution or sensitivity. Its high selectivity, resolution, and repeatability make UPLC beneficial for food flavonoid separation [17].

4 Q-TOF-MS/MS Identification

Q-TOF-MS/MS analyzer comprises two stages. In the first stage, a quadrupole (Q) mass analyzer preselects ions, which are then dissociated in a collision cell filled with gas. The second stage involves a TOF analyzer, which analyzes the resultant product ions. These ions, given equal initial energy, have velocities (v) that are inversely related to their m/z values as per Eq. 3. The ions are subsequently accelerated toward the reflectron, altering direction by 90° to reach the final detector, which compensates for any discrepancies in their kinetic energy [18]. Ions possessing higher kinetic energy infiltrate the reflectron more deeply, resulting in the total flight time (t_{TOT}) being a summation of the flight time into and out of the reflectron (Eq. 4) [7]. The mass spectrum is then obtained by graphing the detector's output signal against time, which can subsequently be translated into an m/z value.

$$v = \sqrt{\frac{2zeV_S}{m}} \tag{3}$$

$$t_{TOT}^2 = \frac{m}{z}\frac{(L_1 + L_2 + 4d)^2}{2eV} \tag{4}$$

where m, z_e, and V_s are the mass of ions, total charge, and potential voltage, respectively. L_1, L_2, and d represent the distances covered before and after the reflectron, and the distance of ions penetrating the reflectron, respectively, and v represents velocity.

Q-TOF-MS/MS, due to its high spectral sensitivity, resolution, and mass accuracy, provides comprehensive structural information and enables higher order mass identification. When coupled with LC, it offers superior chromatographic resolution compared to other high-resolution mass spectrometry (HRMS) methods [19]. The TOF mass analyzer can detect all ions, including trace amounts of flavonoids, in each ionization pulse. These features make Q-TOF-MS/MS essential for distinguishing different flavonoids with similar mass. Additionally, its rapid data acquisition capacity allows simultaneous monitoring of multiple precursor ions from co-eluting peaks. Despite the challenge posed by the glycosidic forms of flavonoids, collision-induced dissociation (CID) in Q-TOF-MS/MS generates suitable fragment ions and facilitates structural analysis. Thus, Q-TOF-MS/MS is suitable for structural analysis of flavonoids in various complex matrices due to its high resolution, accurate mass determination, and CID detection.

5 LC Coupled with Q-TOF-MS/MS

LC-Q-TOF-MS/MS is a powerful tool combining the separation capabilities of liquid chromatography (LC) with the structural analysis potential of quadrupole time-of-flight mass spectrometry (Q-TOF-MS/MS) [7]. This combination provides high selectivity and sensitivity, making it ideal for detecting flavonoids in complex samples such as food and biological matrices. While LC simplifies flavonoid samples by reducing their complexity, some flavonoids, lacking chromophore moieties, present detection challenges using LC with ultraviolet or diode array detection [20]. Q-TOF-MS/MS complements this by providing detailed structural information through the calculation of molecular mass and elucidation of fragmentation pathways based on the m/z value. Direct detection through Q-TOF-MS/MS, however, necessitates complex sample pretreatment and may lead to loss of flavonoids. Furthermore, it does not allow for structural identification of compounds without known standards. The LC-Q-TOF-MS/MS method overcomes these limitations. In this method, LC separation aids in the ionization of flavonoids by MS, reducing the number of charged analytes entering the mass spectrometer simultaneously, and thus lessening competition for charge [21]. It enables structural identification by concurrently monitoring retention time and MS/MS spectral patterns, even without known standards. Consequently, the coupling of LC with Q-TOF-MS/MS enhances flavonoid detection.

In addition, qualitative information on the molecular weight and characteristic fragmentation pathways of flavonoids can be obtained by Q-TOF-MS/ MS as well as their retention behavior in LC. Meanwhile, quantitative information also can be obtained by using the matrix calibration curve-based approach. This review targets Tieguanyin (a type of Chinese Oolong tea) for the nontargeted analysis of its metabolites, which include flavonoids [6].

6 Materials

The used materials in this protocol are listed in Table 1. Adhere strictly to all pertinent regulations that guarantee safety when handling chemicals at work. Properly follow waste disposal guidelines to ensure careful and correct disposal of waste materials. Utilize ultrapure water, which is purified deionized water, for the preparation of all solutions and ensure the use of analytical grade reagents. Employ only the highest available purity of chemicals, for instance, HPLC or LC-MS grade. Unless directed otherwise, prepare and keep all reagents at room temperature.

Table 1
Reagent and standards

Methanol and acetonitrile	Chromatographic grade
Formic acid	Chromatographic grade
Deionized water	>18.2 MΩ
Leucine encephalin	ESI−: 554.2615, ESI+: 556.2771
Theaflavin (TF)	Purity > 98.0%
Vitexin	Purity > 98.0%
Kaempferol	Purity > 98.0%
Epigallocatechin gallate (EGCG)	Purity > 98.0%
Gallic acid (GA)	Purity > 98.0%
Epigallocatechin (EGC)	Purity > 98.0%
Epicatechin gallate (ECG)	Purity > 98.0%
Catechin (C)	Purity > 98.0%
Gallocatechin (GC)	Purity > 98.0%
Epicatechin (EC)	Purity > 98.0%
Gallocatechin gallate (GCG)	Purity > 98.0%
Catechin gallate (CG)	Purity > 98.0%
Scopoletin	Purity > 98.7%
Phloretin	Purity > 98.0%
Caffeic acid	Purity > 99.0%
Rutin, 3-caffeoylquinic acid	Purity > 98.0%
Apigenin	Purity > 98.0%
Quercetin	Purity > 98.0%
Theaflavin-3′-gallate (TF-3′-G)	Purity > 98.0%
Theaflavin-3-gallate (TF-3-G)	Purity > 98.0%
Theaflavin-3,3′-digallate (TF-3,3′-DIG)	Purity > 98.0%
Sucrose	Purity > 99.5%
Tartaric acid	Purity > 99.5%
Theanine	Purity > 98.0%
Malic acid	Purity > 99.5%
Citric acid	Purity > 99.5%
Ferulic acid	Purity > 98.0%
Myricetin	Purity > 98.0%
Naringoside	Purity > 98.0%

6.1 Reagent and Standards

6.2 Equipment

1. Ultraperformance liquid chromatography (UPLC) system (I-Class, Waters Corp., Manchester, UK).
2. Acquity UPLC HSS T3 C18 column (1.8 μm average particle size, 150 mm × 2.1 mm i.d., Waters Corp., Manchester, UK).

   The ACQUITY UPLC chromatographic column is packed with 1.8 um high-strength silica particles and can operate in a mechanically high-pressure environment. The column hardware can withstand a maximum pressure of 124,106 kPa (1241 bar, 18,000 psi). The column dimensions provide optimal flow rates suitable for MS, and the equipped outlet tubing minimizes the effect of extra column volume.
3. Quadrupole time-of-flight mass spectrometer (Q-TOF/MS) system (Xevo G2-S, Waters Corp., Manchester, UK).

7 Software

1. Masslynx (Waters Corp, Manchester, UK): Collecting raw data files acquired from UPLC-Q-TOF/MS analysis.
2. Markerlynx XS (Waters Corp, Manchester, UK): Background drift correction, peak extraction, and peak alignment.
3. Matlab (MathWorks, USA): Conducting *t*-test.
4. EZinfo 2.0 software (Umetrics AB, Umeå, Sweden).

8 Methods

8.1 Extract Solution Preparation

1. For fresh tea leaves, they should be fixed in an oven at 120 °C for 5 min after picking, and then continue to be dried at 75 °C for 120 min.
2. If the prepared sample is dry tea, it should be further ground into powder (Grindomix GM200, Retsch, Germany), ensuring it can pass through a 60-mesh sieve, and preserved at fridge with a temperature under −70 °C before extraction.
3. Prepare a mixed solution of water, methanol, and acetonitrile in a ratio of 3:2:1 (v/v). To 5 mL mixed solution is added 0.1 g of tea powder, then extracted with an ultrasonic instrument (DS-8510 DTH, Shanghai, China) for 30 min.
4. Before the untargeted metabolite analysis by UPLC-Q-TOF/MS, centrifuge the mixed solution at 15,000 × *g* for 3 min (Avanti J-E, Beckman Coulter, USA), and filter it through a 0.22 μm PTFE filter (F2513-4, Thermo Fisher Scientific, USA).

8.2 UPLC Method

1. Start the ACQUITY UPLC I-Class system workstation, then start the binary solvent manager and sample manager and open the MassLynx 4.0 software.
2. To calibrate the instrument before, use LC system (*see* **Note 1**).
3. Inspect the control panel parameters of the binary solvent manager, sample manager, detector, and chromatographic column manager from the chromatography data system.
4. The injection volume of sample was 5 μL.
5. Mobile phase: Samples were separated using the mobile phase consisting of water with 0.1% formic acid (phase A) and acetonitrile (phase B). The solvent gradient programming held at 5% phase B for 0.5 min in the beginning, then from 0.5 to 6 min; phase B was linearly increased from 5% to 40%, then linearly increased to 95% at 8 min; phase B was back to 5% at 10 min and maintained for 3 min. Thus, the total elapsed time required for a given chromatographic analysis was 13 min.
6. The flow rate was set 0.3 mL/min under the gradient elution procedure.
7. Use a column temperature of 25 °C.
8. Use a sequence consisting of multiple blanks at the beginning and end of the analysis. Ideally, inject at least three replicates per sample to ensure the reproducibility of the analyses.

8.3 Q-TOF-MS Method

1. Start the Xevo G2-XS Q-TOF system (*see* **Note 2**).
2. Make sure that the equipment is correctly calibrated (according to the manuals of each MS instrument) before testing.

The detailed conditions of the main Q-TOF-MS parameters are listed in Table 2.

9 Data Processing

1. Use Masslynx software (Waters Corp., Manchester, UK) to acquire raw data files from UPLC-Q-TOF/MS system.
2. Use Markerlynx XS (Waters Corp., Manchester, UK) to extract and detect peaks of raw data. Then a peak table, including retention time (RT), m/z, and MS intensity, was generated (schematic diagram as Fig. 3).
3. Process signals with a retention time (RT) of 0.8–11.0 min and m/z of 100–1000. The signal-to-noise threshold for peak detection is set to 3. The tolerances for peak pairing in retention time and mass number are set to 0.2 min and 0.01 Da, respectively.

Table 2
Q-TOF-MS parameters

Mode	Negative ionization mode
Capillary voltage	−2.5 kV
Sampling cone voltage	20 V
Temperature of the ion source	120 °C
Desolvation gas	400 °C
Flow rates of the cone gas	50 L/h
Desolvation gas	800 L/h
Mass scan range	100–1000 m/z
LockSpray reference ions	554.2615 m/z

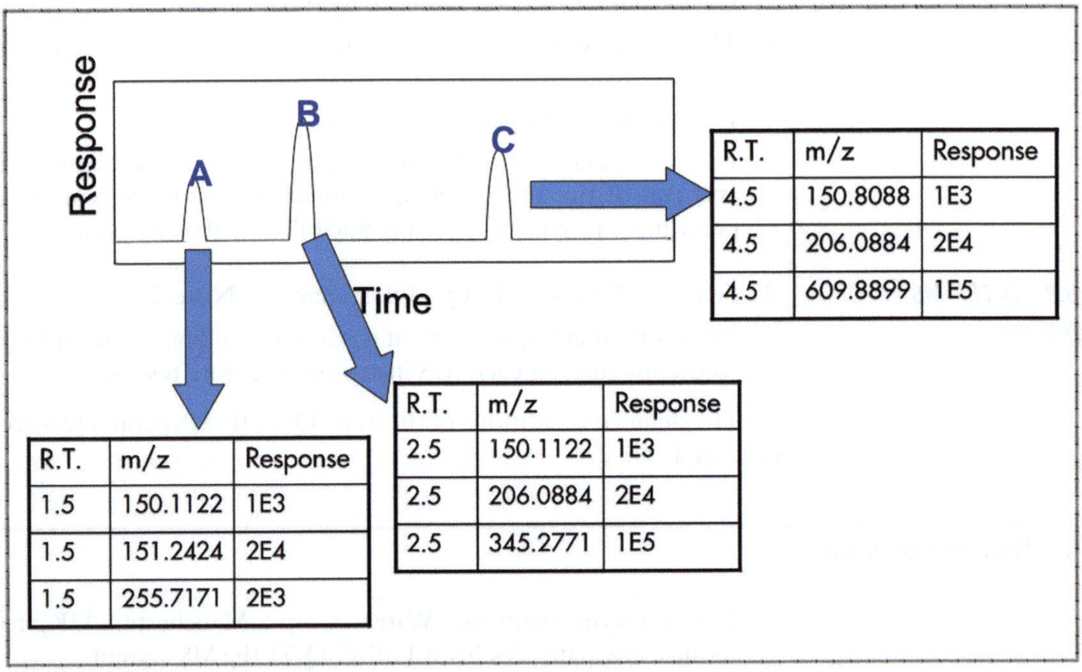

Fig. 3 Schematic diagram of peaks extraction

4. The identification of metabolite was carried out by using authentic standards, secondary mass spectra (https://metlin.scripps.edu), chemspider (http://www.chemspider.com), and Metlin (http://metlin.scripps.edu/).

10 Software Operation

10.1 Masslynx Operational Essentials

1. Open MassLynx 4.0 on the desktop to enter the mass spectrometry software (*see* **Note 3**).
2. Click on the mass spectrometer tuning icon (MS Tune) to enter the mass spectrometer tuning window (Fig. 4).
3. Select the "Ion" mode to be used in the mass spectrometer tuning window, navigate to the "Source" interface, and set the various parameters within the "Source" interface (Fig. 5).
4. Enter the "Analyser" interface and set the various parameters within the "Analyser" interface (Fig. 6).
5. Collect sample signals in the "Tune Page" interface. Click on the *gas icon* to turn on the nitrogen. Press the *operate button*. When the voltage is applied, the "Operate" icon will turn from red to green, confirming that the mass spectrometer is operational and that samples are entering the spectrometer. Clicking on the "*Acquire*" icon will bring up the following dialog box, and the signals would be collected after clicking the start button (Fig. 7).
6. Create the "Inlet Method." Click on the "Inlet Method" icon to enter the method editing interface and set the experimental parameters (Fig. 8).

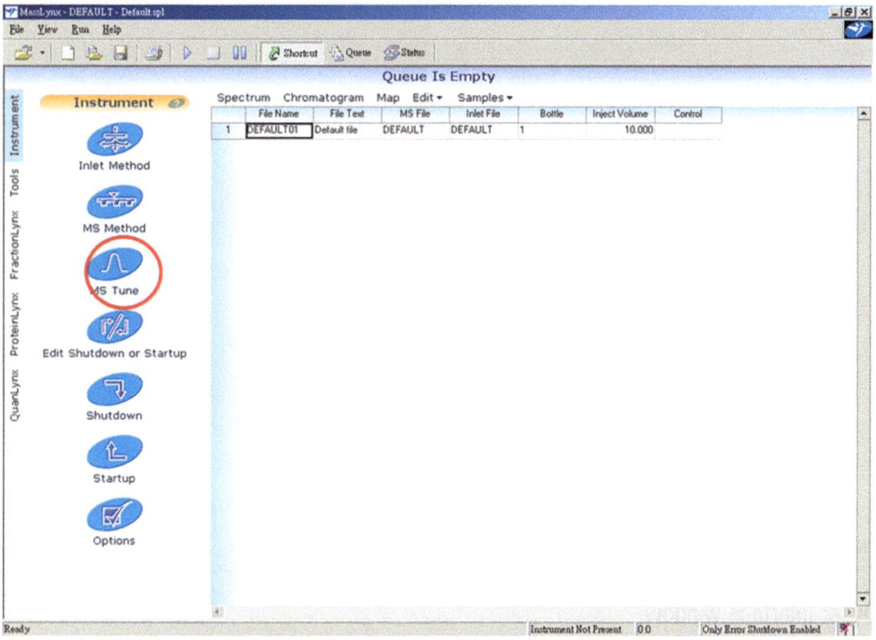

Fig. 4 Mass spectrometer tuning window

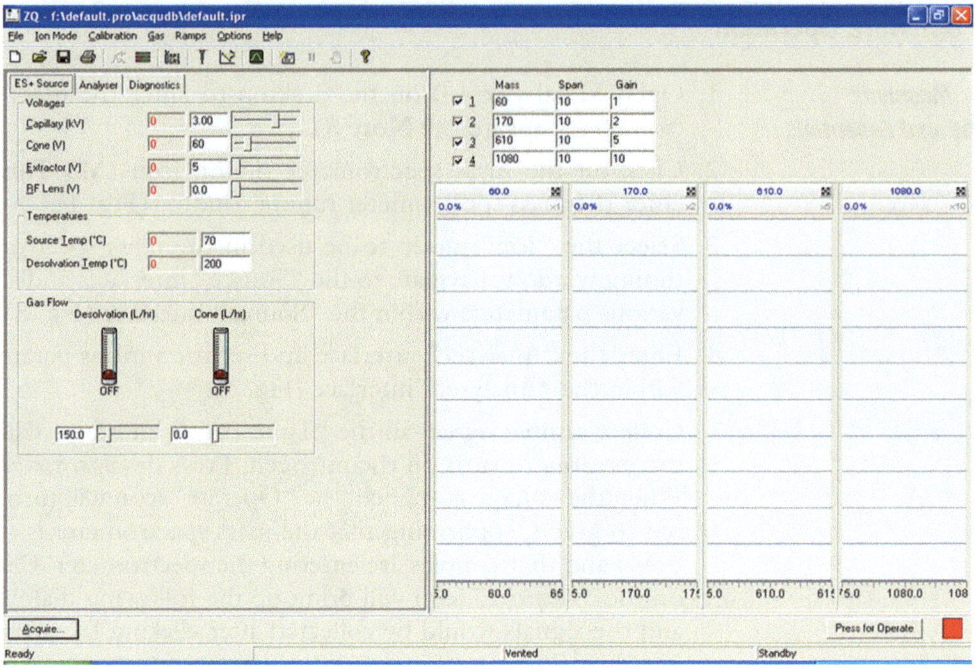

Fig. 5 Source interface window

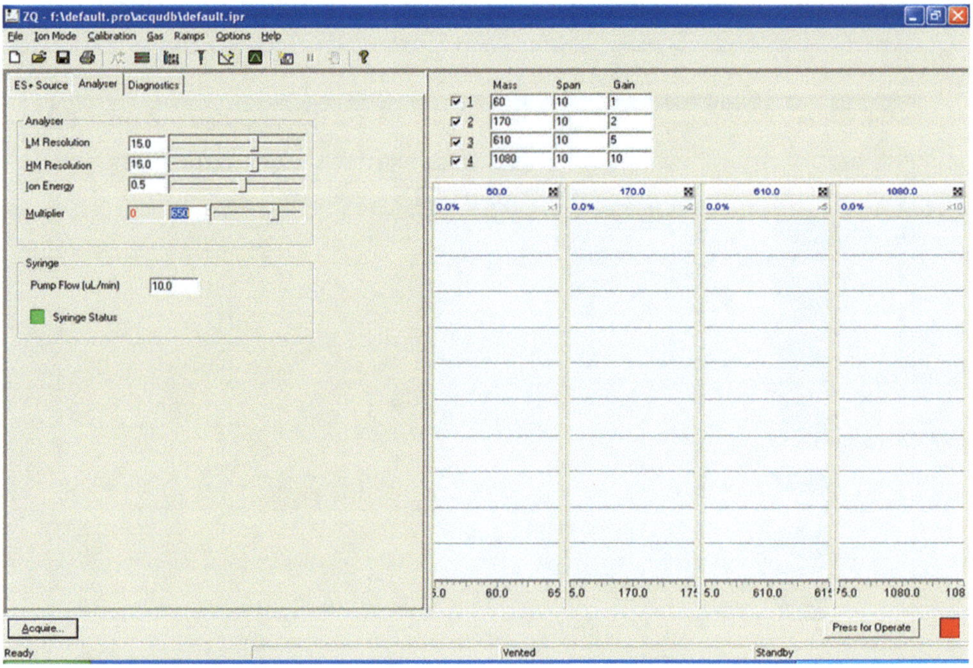

Fig. 6 Analyser interface window

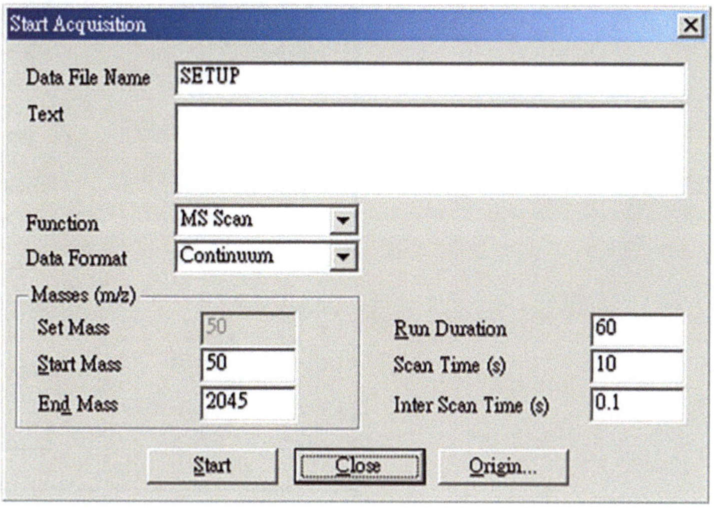

Fig. 7 Data acquisition dialog box

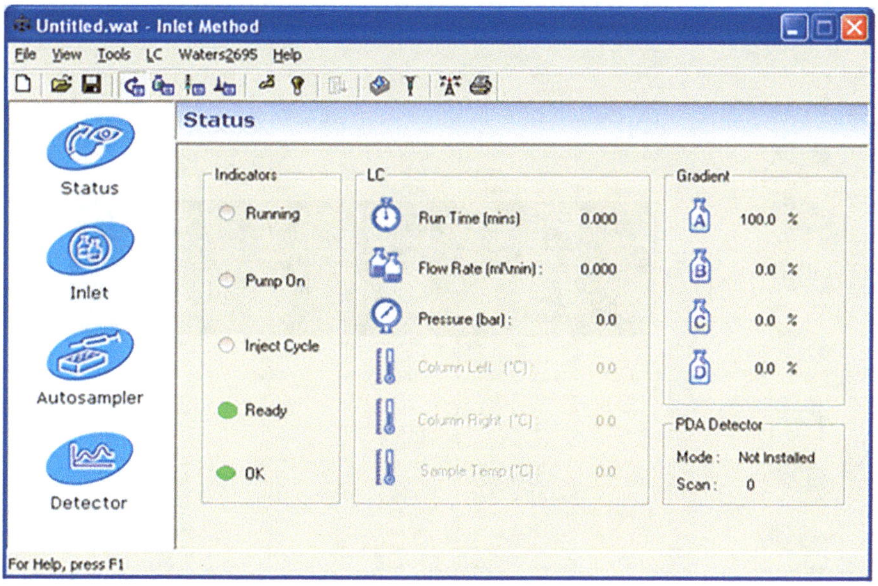

Fig. 8 Inlet method editing interface

7. Create the "MS Method." Click on the "MS Method" icon to enter the MS method editing interface and set the experimental parameters (Fig. 9).
8. View the chromatogram and mass spectrum. In the main MassLynx interface, select *File – Open* to open the relevant *Sample list* for the sample to be viewed (Fig. 10).
9. Close the LC-Q-TOF-MS system. Click *Press for standby* button to off mass spectrometry high voltage process. Then stop

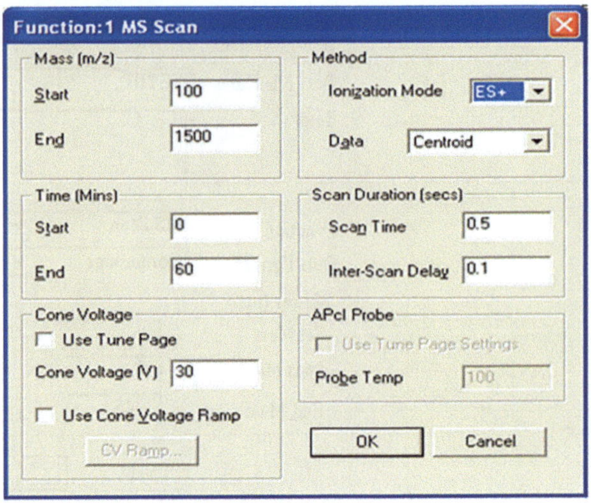

Fig. 9 MS method editing interface

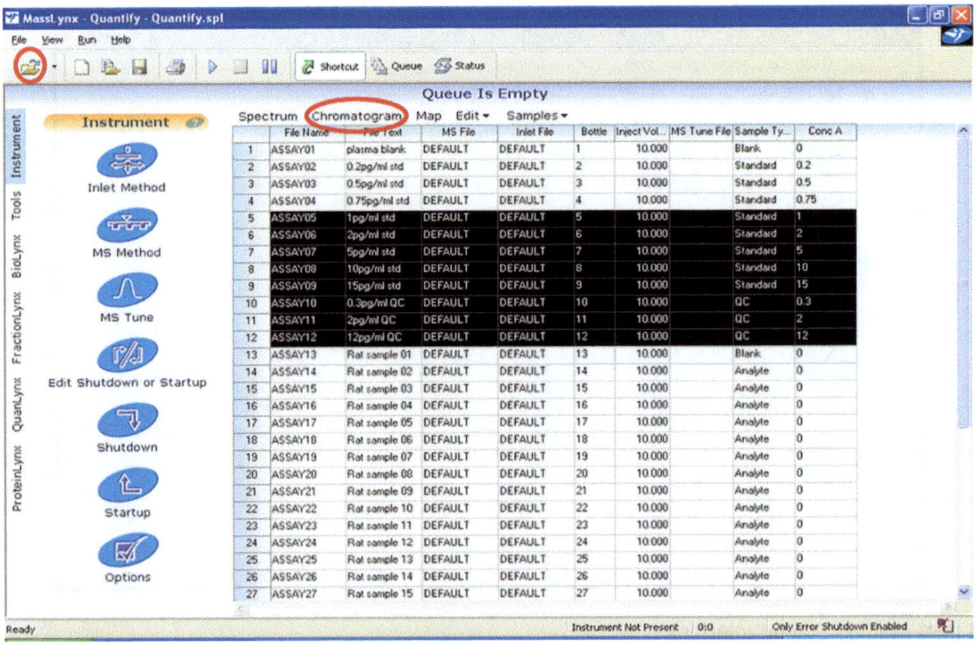

Fig. 10 Sample list

the liquid chromatography flow rate, remove the liquid chromatography tubing from the mass spectrometry to the waste bottle. Wait for the desolvation gas temperature (ESI) to drop to room temperature, then turn off the nitrogen gas. Select *Option/Vent*, the mass spectrometer will begin to vent, and the "Vacuum" status light on the ZQ front panel will start to blink. After a few minutes, the mechanical pump will stop running, and the mass spectrometer can be turned off.

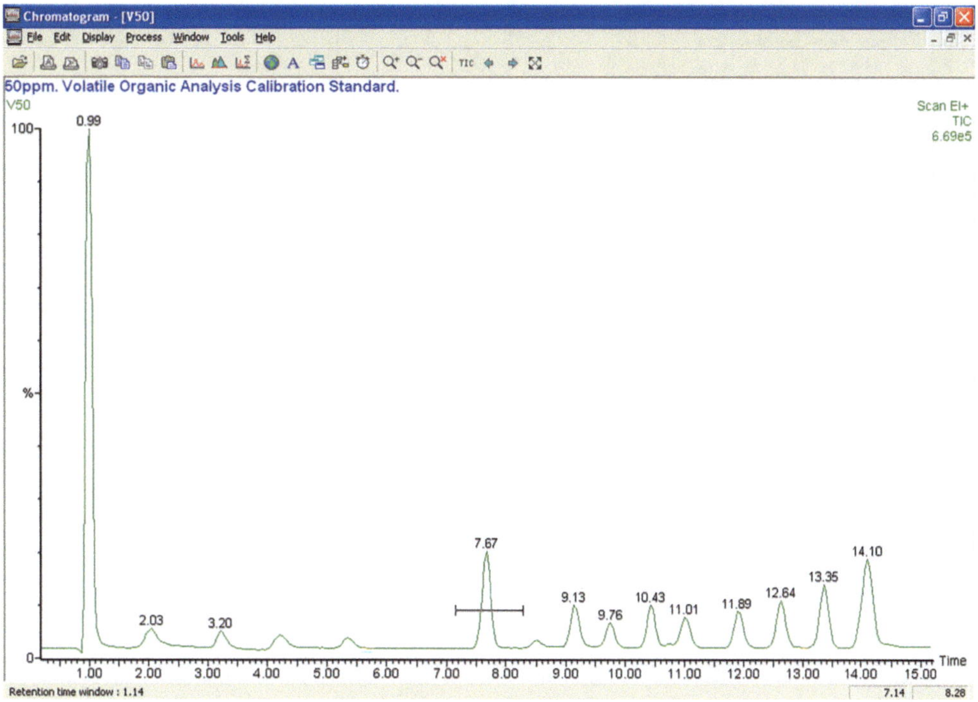

Fig. 11 Chromatogram window

10.2 Markerlynx XS Operational Essentials

The MarkerLynx Application Manager for MassLynx software accurately detects ions of interest from an MS sample set by extracting the relevant data in four automated processing steps:

1. Chromatographic peak detection (Fig. 11).
2. Ion intensity identification (Fig. 12).
3. Data matrix construction (Figs. 13 and 14).
4. PCA on the resulting dataset (Fig. 15).

11 Peak Detection and Compound Identification

Zhou et al. [6] compared the metabolic changes in Anxi Tieguanyin tea harvested in the autumn of 2012 and 2013. Figure 14 shows that a total of 107 metabolites, including the classification of catechins, alkaloids, flavone, organic acids, amino acids, and polyphenols, were tentatively extracted from datasets of UPLC-Q-TOF/MS. In order to evaluate the metabolic feature differences between Anxi Tieguanyin teas harvested in different seasons, an unsupervised PCA method based on UPLC-Q-TOF/MS analysis was used to distinguish Tieguanyin teas with seasonal variations. The 2D scatter plot of the scores for PC1 and PC2, as shown in Fig. 15,

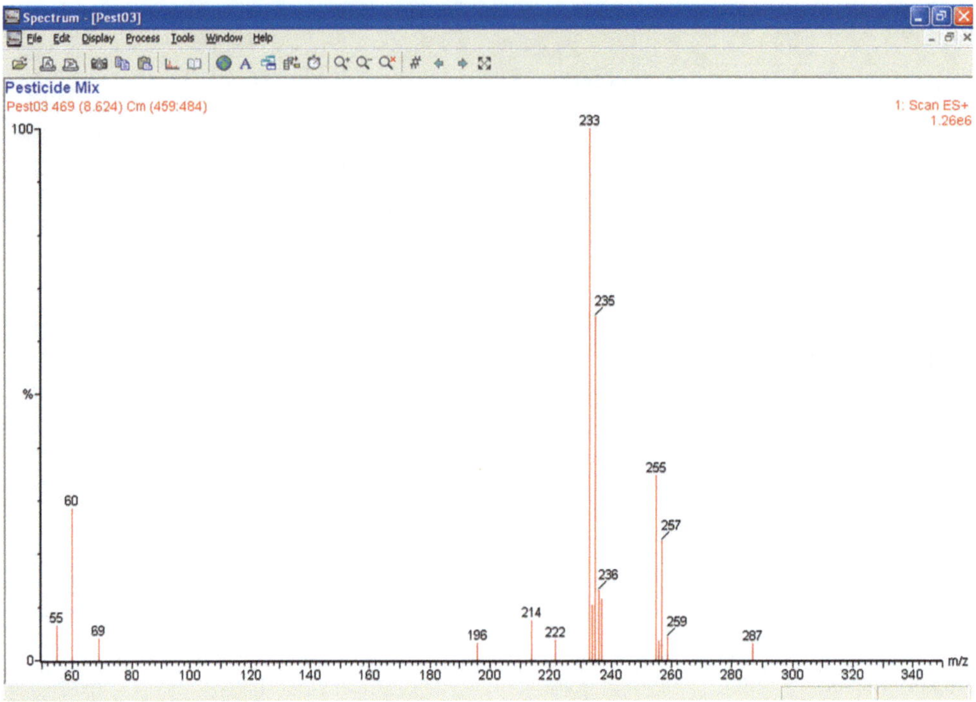

Fig. 12 Mass spectrum window

together explains 33.2% of the total variance (26.1% and 7.1%, respectively). It can be observed that the spring Tieguanyin tea and the autumn Tieguanyin tea can be completely classified and identified based on their metabolic spectra, which further demonstrates that the harvest season of Tieguanyin tea has a significant impact on the tea metabolome. More information concerning the identification of metabolites can be found in Table 3.

12 Notes

1. To calibrate the LC system, follow the acceptable calibration method that uses at least five standard samples to generate a standard curve. The concentration range of the standard samples should cover the entire range of QC samples, typical specimens, and atypical specimens.
2. Before starting the Mass analyzer, make sure the nitrogen source is connected to the instrument's nitrogen input connector. (The nitrogen must be dry and oil-free, with a purity of at least 95%. The gas supply pressure should be adjusted to a range of 650–700 kPa (6.5–7.0 bar, 94–102 psi).) In addition, make sure that the collision gas supply is connected to the instrument's collision cell gas inlet. (The collision gas must be

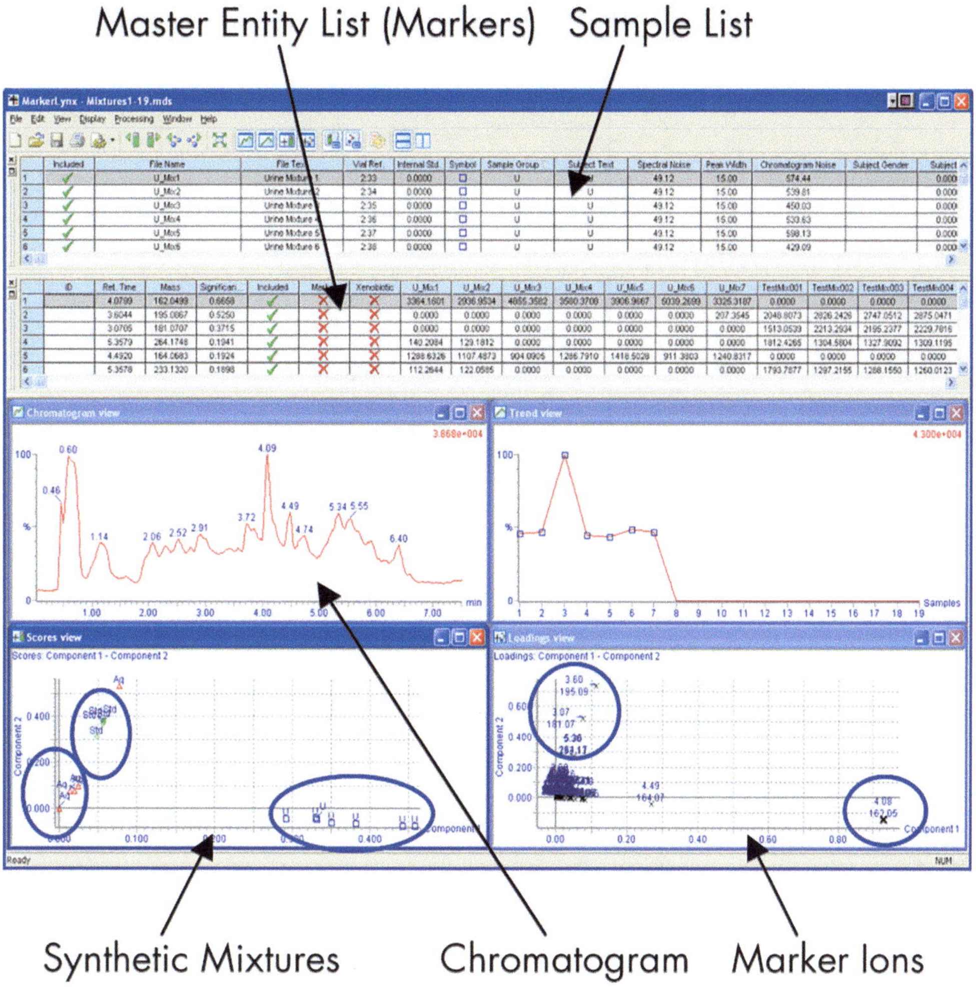

Fig. 13 MarkerLynx browser results of the synthetic mixture dataset [22]

dry, high-purity (99.997%) argon. Adjust the supply pressure to 50 kPa (0.5 bar, 7 psi).)

3. If the following message appears when entering the MassLynx software: "The embedded system is not responding, The system will run in standalone mode," it indicates that the communication connection between the mass spectrometer's built-in CPU (EPC) and the main computer has not been established. At this time, the mass spectrometer cannot be controlled. Please wait a while before reentering the software. However, if you are only opening the software to process data, this is not an issue. (The same message will appear if you enter the software when the mass spectrometer's main power is off.)

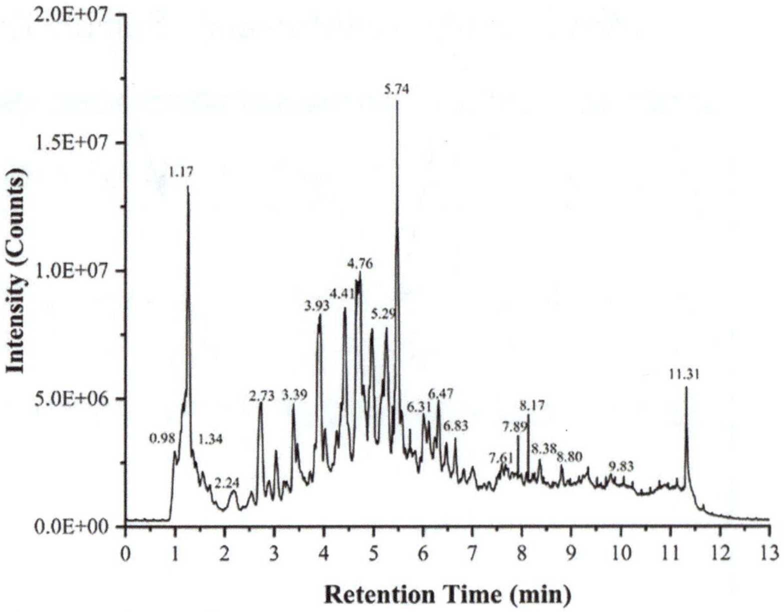

Fig. 14 Typical TIC chromatogram of Tieguanyin tea analyzed in the ESI-mode by UPLC-Q-TOF/MS

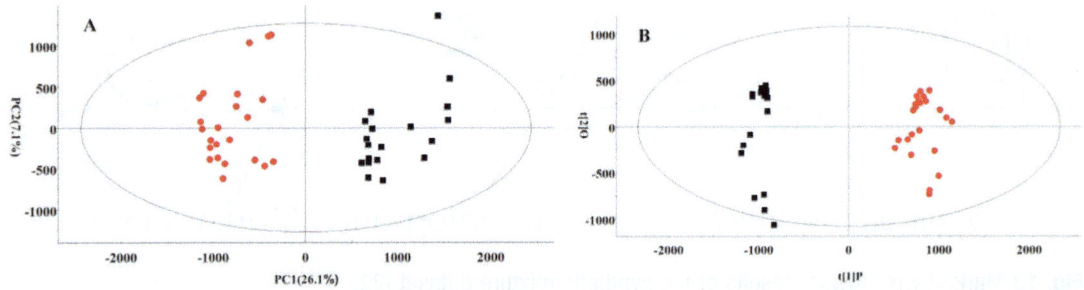

Fig. 15 Multivariate statistical analysis of spring (red circles) and autumn (black squares) Tieguanyin teas: (**a**) PCA score plot, $R^2X(cum) = 0.483$, $Q^2(cum) = 0.123$, (**b**) OPLS-DA score plot, R2Y(cum) = 0.974, Q2(cum) = 0.937 [6]

Table 3
Tentative identification of metabolites in Tieguanyin teas

Retention time (min)	Compounds	Molecular formula	ESI model	Theoretical mass (m/z)	Accurate mass (m/z)	Mass error (ppm)	Classification
1.13	Asparagine	$C_4H_8N_2O_3$	M-H	131.0457	131.0457	0.0	Amino acids
1.13	Aspartic acid	$C_4H_7NO_4$	M-H	132.0297	132.0296	−0.8	Amino acids
1.13	Glutamine	$C_5H_{10}N_2O_3$	M-H	145.0613	145.0614	0.7	Amino acids
1.13	Lactobionic acid	$C_{12}H_{22}O_{12}$	M-H	357.1033	357.1031	−0.6	Organic acids
1.17	Glutamic acid	$C_5H_9NO_4$	M-H	146.0453	146.0453	0.0	Amino acids
1.18	Gluconic acid	$C_6H_{12}O_7$	M-H	195.0505	195.0501	−2.1	Organic acids
1.20	Ribonic acid	$C_5H_{10}O_6$	M-H	165.0399	165.0398	−0.6	Organic acids
1.22	Sucrose[1]	$C_{12}H_{22}O_{11}$	M-H	341.1084	341.1080	−1.2	Saccharides
1.23	Inosine	$C_{10}H_{12}N_4O_5$	M-H	267.0729	267.0719	−3.7	Nucleosides
1.26	Quinic acid	$C_7H_{12}O_6$	M-H	191.0556	191.0562	3.1	Phenolic acids
1.26	Tartaric acid[1]	$C_4H_6O_6$	M-H	149.0086	149.0088	1.3	Organic acids
1.32	UMP	$C_9H_{13}N_2O_9P$	M-H	323.0280	323.0280	0.0	Nucleosides
1.35	Theanine[1]	$C_7H_{14}N_2O_3$	M-H	173.0926	173.0927	0.6	Amino acids
1.38	Malic acid[1]	$C_4H_6O_5$	M-H	133.0137	133.0139	1.5	Organic acids
1.56	Isocitric acid	$C_6H_8O_7$	M-H	191.0192	191.0196	2.1	Organic acids
1.69	Citric acid[1]	C_6H8O_7	M-H	191.0192	191.0193	0.5	Organic acids
2.19	Galloylglucose	$C_{13}H_{16}O_{10}$	M-H	331.0665	331.0667	0.6	Phenolic acids
2.39	Theasinensin C	$C_{30}H_{26}O_{14}$	M-H	609.1244	609.1233	−1.8	Catechins and relatives
2.72	Theogallin	$C_{14}H_{16}O_{10}$	M-H	343.0665	343.0664	−0.3	Phenolic acids

(continued)

Table 3 (continued)

Retention time (min)	Compounds	Molecular formula	ESI model	Theoretical mass (m/z)	Accurate mass (m/z)	Mass error (ppm)	Classification
2.75	Gallic acid[1]	$C_7H_6O_5$	M-H	169.0137	169.0140	1.8	Phenolic acids
2.75	Maltol	$C_6H_6O_3$	M-H	125.0239	125.0241	1.6	alcohols
3.37	Gallocatechin[1]	$C_{15}H_{14}O_7$	M-H	305.0661	305.0666	1.6	Catechins and relatives
3.38	2,6-Dihydroxybenzoic acid	$C_7H_6O_4$	M-H	153.0188	153.0192	2.6	Phenolic acids
3.47	Prodelphinidin B	$C_{30}H_{26}O_{14}$	M-H	609.1244	609.1234	−1.6	Catechins and relatives
3.58	Theasinensin B	$C_{37}H_{30}O_{18}$	M-H	761.1354	761.1341	−1.7	Catechins and relatives
3.63	5-Caffeoylquinic acid	$C_{16}H_{18}O_9$	M-H	353.0873	353.0872	−0.3	Phenolic acids
3.73	Theophyline	$C_7H_8N_4O_2$	M-H	179.0569	179.0576	3.9	alkaloids
3.79	Digalloylglucose	$C_{20}H_{20}O_{14}$	M-H	483.0775	483.0771	−0.8	Phenolic acids
3.86	Prodelphinidin A2 3′-gallate	$C_{37}H_{28}O_{18}$	M-H	759.1197	759.1178	−2.5	Catechins and relatives
3.90	Epigallocatechin[1]	$C_{15}H_{14}O_7$	M-H	305.0661	305.0668	2.3	Catechins and relatives
3.90	Salicylic acid	$C_7H_6O_3$	M-H	137.0239	137.0235	−2.9	Phenolic acids
4.02	Sanguiin H4	$C_{27}H_{22}O_{18}$	M-H	633.0728	633.0724	−0.6	Phenolic acids
4.02	Prodelphinidin B2 3′-gallate	$C_{37}H_{30}O_{18}$	M-H	761.1354	761.1341	−1.7	Catechins and relatives
4.07	Procyanidin dimer	$C_{30}H_{26}O_{12}$	M-H	577.1346	577.1344	−0.3	Catechins and relatives
4.13	3-*p*-Coumaroylquinic acid	$C_{16}H_{18}O_8$	M-H	337.0923	337.0924	0.3	Phenolic acids
4.16	3-Caffeoylquinic acid[1]	$C_{16}H_{18}O_9$	M-H	353.0873	353.0870	−0.8	Phenolic acids
4.22	4-Caffeoylquinic acid	$C_{16}H_{18}O_9$	M-H	353.0873	353.0870	−0.8	Phenolic acids
4.25	Catechina[1]	C_{15}H14O6	M-H	289.0712	289.0715	1.0	Catechins and relatives

4.26	Theasinensin A	$C_{44}H_{34}O_{22}$	M-H	913.1463	913.1432	−3.4	Catechins and relatives
4.39	Apigenin-6,8-C-diglucoside	$C_{27}H_{30}O_{15}$	M-H	593.1506	593.1505	−0.2	Flavonol and flavonol/ flavone glycosides
4.39	Chicoric acid	$C_{22}H_{18}O_{12}$	M-H	473.0720	473.0710	−2.1	Phenolic acids
4.43	Eriocitrin	$C_{27}H_{32}O_{15}$	M-H	595.1663	595.1660	−0.5	Flavonol and flavonol/ flavone glycosides
4.49	Gallic acid methyl ester	$C_8H_8O_5$	M-H	183.0293	183.0292	−0.5	Phenolic acids
4.60	Trigalloyl glucopyranose	$C_{27}H_{24}O_{18}$	M-H	635.0884	635.0883	−0.2	Phenolic acids
4.61	Myricetin 3-O-galactosylrutinoside	$C_{33}H_{40}O_{22}$	M-H	787.1933	787.1932	−0.1	Flavonol and flavonol/ flavone glycosides
4.62	Theacitrin C	$C_{44}H_{32}O_{22}$	M-H	911.1307	911.1297	−1.1	Catechins and relatives
4.64	Epicatechin[1]	$C_{15}H_{14}O_6$	M-H	289.0712	289.0712	0.0	Catechins and relatives
4.65	Epigallocatechin gallate[1]	$C_{22}H_{18}O_{11}$	M-H	457.0771	457.0775	0.9	Catechins and relatives
4.68	Assamicain A	$C_{44}H_{36}O_{22}$	M-H	915.1620	915.1598	−2.4	Catechins and relatives
4.73	Caffeic acid[1]	$C_9H_8O_4$	M-H	179.0344	179.0337	−3.9	Phenolic acids
4.75	5-*p*-Coumaroylquinic acid	$C_{16}H_{18}O_8$	M-H	337.0923	337.0925	0.6	Phenolic acids
4.75	Apigenin 6-C-glucosyl-8-C-arabinoside	$C_{26}H_{28}O_{14}$	M-H	563.1401	563.1398	−0.5	Flavonol and flavonol/ flavone glycosides
4.75	Shikimic acid	$C_7H_{10}O_5$	M-H	173.0450	173.0456	3.5	Phenolic acids
4.79	Myricetin 3-O-rhamnosylglucoside	$C_{27}H_{30}O_{17}$	M-H	625.1405	625.1394	−1.8	Flavonol and flavonol/ flavone glycosides
4.81	Gallocatechin gallate[1]	$C_{22}H_{18}O_{11}$	M-H	457.0771	457.0762	−2.0	Catechins and relatives
4.81	Procyanidin dimer B2 3′-gallate	$C_{37}H_{30}O_{16}$	M-H	729.1456	729.1430	−3.6	Catechins and relatives
4.86	Apigenin 6-C-arabinosyl-8-C-glucoside	$C_{26}H_{28}O_{14}$	M-H	563.1401	563.1404	0.5	Flavonol and flavonol/ flavone glycosides

(continued)

Table 3 (continued)

Retention time (min)	Compounds	Molecular formula	ESI model	Theoretical mass (m/z)	Accurate mass (m/z)	Mass error (ppm)	Classification
4.90	Catechin gallate[1]	$C_{22}H_{18}O_{10}$	M-H	441.0822	441.0806	−3.6	Catechins and relatives
4.91	Quercetin 3-O-galactosylrutinoside	$C_{33}H_{40}O_{21}$	M-H	771.1984	771.1981	−0.4	Flavonol and flavonol/flavone glycosides
4.92	Myricetin 3-O-galactoside	$C_{21}H_{20}O_{13}$	M-H	479.0826	479.0818	−1.7	Flavonol and flavonol/flavone glycosides
4.94	Myricetin 3-O-glucoside	$C_{21}H_{20}O_{13}$	M-H	479.0826	479.0813	−2.7	Flavonol and flavonol/flavone glycosides
4.96	Quercetin 3-O-glucosylrutinoside	$C_{33}H_{40}O_{21}$	M-H	771.1984	771.1992	1.0	Flavonol and flavonol/flavone glycosides
4.97	Kaempferol 3-O-rhamnosylgalactoside	$C_{27}H_{30}O_{15}$	M-H	593.1506	593.1500	−1.0	Flavonol and flavonol/flavone glycosides
5.07	Quercetin 3-O-dirhamnosylglucoside	$C_{33}H_{40}O_{20}$	M-H	755.2035	755.2023	−1.6	Flavonol and flavonol/flavone glycosides
5.12	4-*p*-Coumaroylquinic acid	$C_{16}H_{18}O_8$	M-H	337.0923	337.0918	−1.5	Phenolic acids
5.13	Vitexin-2-O-rhamnoside	$C_{27}H_{30}O_{14}$	M-H	577.1557	577.1566	1.6	Flavonol and flavonol/flavone glycosides
5.18	Rutin[1]	$C_{27}H_{30}O_{16}$	M-H	609.1456	609.1469	2.1	Flavonol and flavonol/flavone glycosides
5.18	Kaempferol 3-O-galactosylrutinoside	$C_{33}H_{40}O_{20}$	M-H	755.2035	755.2038	0.4	Flavonol and flavonol/flavone glycosides
5.23	Epiafzelechin	$C_{15}H_{14}O_5$	M-H	273.0763	273.0765	0.7	Catechins and relatives
5.23	Epigallocatechin 3-O-(4-methylgallate)	$C_{23}H_{20}O_{11}$	M-H	471.0927	471.0926	−0.2	Catechins and relatives

5.24	Methylepigallocatechin gallate	$C_{23}H_{20}O_{11}$	M-H	471.0927	471.0918	−1.9	Catechins and relatives
5.27	Kaempferol 3-O-glucosylrutinoside	$C_{33}H_{40}O_{20}$	M-H	755.2035	755.2039	0.5	Flavonol and flavonol/ flavone glycosides
5.32	Vitexin[1]	$C_{21}H_{20}O_{10}$	M-H	431.0978	431.0977	−0.2	Flavonol and flavonol/ flavone glycosides
5.33	Kaempfero 3-O-xylosylrutinoside	$C_{33}H_{40}O_{19}$	M-H	739.2086	739.2079	−0.9	Flavonol and flavonol/ flavone glycosides
5.38	Quercetin 3-O-galactoside	$C_{21}H_{20}O_{12}$	M-H	463.0877	463.0866	−2.4	Flavonol and flavonol/ flavone glycosides
5.40	Quercetin 3-glucoside	$C_{21}H_{20}O_{12}$	M-H	463.0877	463.0868	−1.9	Flavonol and flavonol/ flavone glycosides
5.41	Ellagic acid	$C_{14}H_6O_8$	M-H	300.9984	300.9986	0.7	Phenolic acids
5.48	Epicatechin gallate[1]	$C_{22}H_{18}O_{10}$	M-H	441.0822	441.0810	−2.7	Catechins
5.53	*p*-Coumaric acid	$C_9H_8O_3$	M-H	163.0395	163.0397	1.2	Phenolic acids
5.56	Kaempferol 3-O-rutinoside	$C_{27}H_{30}O_{15}$	M-H	593.1506	593.1498	−1.3	Flavonol and flavonol/ flavone glycosides
5.70	Kaempferol 3-O-galactoside	$C_{21}H_{20}O_{11}$	M-H	447.0927	447.0924	−0.7	Flavonol and flavonol/ flavone glycosides
5.80	Naringoside[1]	$C_{27}H_{32}O_{14}$	M-H	579.1714	579.1727	2.2	Flavonol and flavonol/ flavone glycosides
5.83	Kaempferol 3-O-glucoside	$C_{21}H_{20}O_{11}$	M-H	447.0927	447.0930	0.7	Flavonol and flavonol/ flavone glycosides
5.84	Kaempferol 3-O-acetyl-dirhamnosylhexoside	$C_{35}H_{42}O_{20}$	M-H	781.2191	781.2189	−0.3	Flavonol and flavonol/ flavone glycosides
5.90	Scopoletin[1]	$C_{10}H_8O_4$	M-H	191.0344	191.0341	−1.6	Flavonol and flavonol/ flavone glycosides
5.98	3,5-Dicaffeoylquinic acid	$C_{25}H_{24}O_{12}$	M-H	515.1190	515.1188	−0.4	Phenolic acids

(continued)

Table 3 (continued)

Retention time (min)	Compounds	Molecular formula	ESI model	Theoretical mass (m/z)	Accurate mass (m/z)	Mass error (ppm)	Classification
6.01	Methoxyepiafzelechine gallate	$C_{23}H_{20}O_{10}$	M-H	455.0978	455.0968	−2.2	Catechins and relatives
6.05	Epiafzelechin gallate	$C_{22}H_{18}O_9$	M-H	425.0873	425.0871	−0.5	Catechins and relatives
6.30	Ferulic acid[1]	$C_{10}H_{10}O_4$	M-H	193.0 501	193.0503	1.1	Phenolic acids
6.32	Kaempferol 3-O-*p*-coumaroylglucosyl rhamnosylglactoside	$C_{42}H_{46}O_{22}$	M-H	901.2402	901.2401	−0.1	Flavonol and flavonol/ flavone glycosides
6.39	Myricetin[1]	$C_{15}H_{10}O_8$	M-H	317.0297	317.0302	1.6	Flavonol and flavonol/ flavone glycosides
6.48	Quercetin[1]	$C_{15}H_{10}O_7$	M-H	301.0348	301.0345	−1.0	Flavonol and flavonol/ flavone glycosides
6.65	Theaflavin[1]	$C_{29}H_{24}O_{12}$	M-H	563.1190	563.1182	−1.4	Flavonol and flavonol/ flavone glycosides
6.65	Kaempferol 3-O-*p*-coumaroylrhamnosyldihexoside	$C_{42}H_{46}O_{22}$	M-H	901.2402	901.2391	−1.2	Flavonol and flavonol/ flavone glycosides
6.71	Theaflavate b	$C_{36}H_{28}O_{15}$	M-H	699.1350	699.1344	−0.9	Catechins and relatives
6.81	Theaflavin-3-gallate[1]	$C_{36}H_{28}O_{16}$	M-H	715.1299	715.1302	0.4	Catechins and relatives
6.83	Theaflavin-3,3′-digallate[1]	$C_{43}H_{32}O_{20}$	M-H	867.1409	867.1420	1.3	Catechins and relatives
6.89	Theaflavate a	$C_{43}H_{32}O_{19}$	M-H	851.1460	851.1453	−0.8	Catechins and relatives
6.93	Theaflavin-3′-gallate[1]	$C_{36}H_{28}O_{16}$	M-H	715.1299	715.1287	−1.7	Catechins and relatives
6.99	Apigenin 6,8-C-dipentoside	$C_{27}H_{30}O_{11}$	M-H	529.1710	529.1720	1.9	Flavonol and flavonol/ flavone glycosides

7.74	Phloretin[1]	$C_{15}H_{14}O_5$	M-H	273.0763	273.0771	2.9	Flavonol and flavonol/ flavone glycosides
7.81	Apigenin[1]	$C_{15}H_{10}O_5$	M-H	269.0450	269.0454	1.5	Flavonol and flavonol/ flavone glycosides
7.86	Kaempferol 3-O-2′,6′-di-*p*-coumaroylglucoside	$C_{39}H_{32}O_{15}$	M-H	739.1663	739.1663	0.0	Flavonol and flavonol/ flavone glycosides
7.86	Kaempferol[1]	$C_{15}H_{10}O_6$	M-H	285.0399	285.0404	1.8	Flavonol and flavonol/ flavone glycosides
9.40	Oleic acid	$C_{18}H_{34}O_2$	M-H	281.2481	281.2480	−0.4	Lipid
10.84	Linoleic acid	$C_{18}H_{32}O_2$	M-H	279.2324	279.2329	1.8	Lipid
10.30	Linolenic acid	$C_{18}H_{30}O_2$	M-H	277.2168	277.2166	−0.7	Lipid

References

1. Zhang Q, Shi Y, Ma L, Yi X, Ruan J (2014) Metabolomic analysis using ultra-performance liquid chromatography-quadrupole-time of flight mass spectrometry (UPLC-Q-TOF MS) uncovers the effects of light intensity and temperature under shading treatments on the metabolites in tea. PLoS One 9(11):e112572. https://doi.org/10.1371/journal.pone.0112572
2. Hodgson JM, Croft KD (2010) Tea flavonoids and cardiovascular health. Mol Aspects Med 31(6):495–502. https://doi.org/10.1016/j.mam.2010.09.004
3. Kamiloglu S, Tomas M, Ozdal T, Capanoglu E (2021) Effect of food matrix on the content and bioavailability of flavonoids. Trends Food Sci Technol 117:15–33. https://doi.org/10.1016/j.tifs.2020.10.030
4. Zhang Q, Liu M, Ruan J (2017) Metabolomics analysis reveals the metabolic and functional roles of flavonoids in light-sensitive tea leaves. BMC Plant Biol 17(1):64. https://doi.org/10.1186/s12870-017-1012-8
5. Yamasaki H, Sakihama Y (1997) Flavonoid-peroxidase reaction as a detoxification mechanism of plant cells against H2O2. Plant Physiol 115:1405–1412
6. Zhou P, Hu O, Fu H, Ouyang L, Gong X, Meng P et al (2019) UPLC–Q-TOF/MS-based untargeted metabolomics coupled with chemometrics approach for Tieguanyin tea with seasonal and year variations. Food Chem 283:73–82. https://doi.org/10.1016/j.foodchem.2019.01.050
7. Yang M, Li J, Zhao C, Xiao H, Fang X, Zheng J (2021) LC-Q-TOF-MS/MS detection of food flavonoids: principle, methodology, and applications. Crit Rev Food Sci Nutr 0(0):1–21. https://doi.org/10.1080/10408398.2021.1993128
8. Chen H, Shi P, Fan F, Tu M, Xu Z, Xu X, Du M (2018) Complementation of UPLC-Q-TOF-MS and CESI-Q-TOF-MS on identification and determination of peptides from bovine lactoferrin. J Chromatogr B 1084:150–157. https://doi.org/10.1016/j.jchromb.2018.03.022
9. Liu R, Chou J, Hou S, Liu X, Yu J, Zhao X et al (2018) Evaluation of two-step liquid-liquid extraction protocol for untargeted metabolic profiling of serum samples to achieve broader metabolome coverage by UPLC-Q-TOF-MS. Anal Chim Acta 1035:96–107. https://doi.org/10.1016/j.aca.2018.07.034
10. Zhang B, Zhang H, Du C, Ng QX, Hu C, He Y, Ong CN (2017) Metabolic responses of the growing Daphnia similis to chronic AgNPs exposure as revealed by GC-Q-TOF/MS and LC-Q-TOF/MS. Water Res 114:135–143. https://doi.org/10.1016/j.watres.2017.02.046
11. Wang Q-Y, Dong X, Yang J, Zhen X-T, Ye L-H, Chu C et al (2019) Solid acids assisted matrix solid-phase dispersion microextraction of alkaloids by capillary electrophoresis coupled with quadrupole time-of-flight mass spectrometry. J Sep Sci 42(23):3579–3588. https://doi.org/10.1002/jssc.201900647
12. Bauer A, Kuballa J, Rohn S, Jantzen E, Luetjohann J (2018) Evaluation and validation of an ion mobility quadrupole time-of-flight mass spectrometry pesticide screening approach. J Sep Sci 41(10):2178–2187. https://doi.org/10.1002/jssc.201701059
13. Xu C-F, Lu Y, Ma J, Mohammadi M, Neubert TA (2005) Identification of phosphopeptides by MALDI Q-TOF MS in positive and negative ion modes after methyl esterification * S. Mol Cell Proteomics 4(6):809–818. https://doi.org/10.1074/mcp.T400019-MCP200
14. Šulniūtė V, Pukalskas A, Venskutonis PR (2017) Phytochemical composition of fractions isolated from ten Salvia species by supercritical carbon dioxide and pressurized liquid extraction methods. Food Chem 224:37–47. https://doi.org/10.1016/j.foodchem.2016.12.047
15. Fraser K, Harrison SJ, Lane GA, Otter DE, Hemar Y, Quek S-Y, Rasmussen S (2014) Analysis of low molecular weight metabolites in tea using mass spectrometry-based analytical methods. Crit Rev Food Sci Nutr 54(7):924–937. https://doi.org/10.1080/10408398.2011.619670
16. Al-Sayah MA, Rizos P, Antonucci V, Wu N (2008) High throughput screening of active pharmaceutical ingredients by UPLC. J Sep Sci 31(12):2167–2172. https://doi.org/10.1002/jssc.200700594
17. Pyrzynska K, Sentkowska A (2015) Recent developments in the HPLC separation of phenolic food compounds. Crit Rev Anal Chem 45(1):41–51. https://doi.org/10.1080/10408347.2013.870027
18. El-Aneed A, Cohen A, Banoub J (2009) Mass spectrometry, review of the basics: electrospray, MALDI, and commonly used mass analyzers.

Appl Spectrosc Rev 44(3):210–230. https://doi.org/10.1080/05704920902717872

19. Ibáñez C, García-Cañas V, Valdés A, Simó C (2013) Novel MS-based approaches and applications in food metabolomics. TrAC Trends Anal Chem 52:100–111. https://doi.org/10.1016/j.trac.2013.06.015
20. Sut S, Poloniato G, Malagoli M, Dall'Acqua, S. (2018) Fragmentation of the main triterpene acids of apple by LC-APCI-MSn. J Mass Spectrom 53(9):882–892. https://doi.org/10.1002/jms.4264
21. Gathungu RM, Kautz R, Kristal BS, Bird SS, Vouros P (2020) The integration of LC-MS and NMR for the analysis of low molecular weight trace analytes in complex matrices. Mass Spectrom Rev 39(1–2):35–54. https://doi.org/10.1002/mas.21575
22. Stumpf CL, Goshawk J (n.d.) MarkerLynx application manager: informatics for mass spectrometric metabonomic discovery

Chapter 9

Gas Chromatography for Omega-3 (ω-3) Fatty Acid Profiles of Microalgae

Shuai Zhang, Zhihao Zhan, Xiaoli Li, and Mostafa Gouda

Abstract

There are various types of fatty acids, which are essential nutrients for humans and affect the balance of dietary structure, endocrine stability, and the occurrence of metabolic syndrome. The accurate determination of fatty acid content in food, serum, and metabolite samples is of great significance for food quality control, disease monitoring, and health assessment. Gas chromatography analysis, one of the important instrumental analysis methods, is the characteristics of high separation efficiency, fast analysis speed, high sensitivity, accurate qualitative and quantitative analysis results for complex multicomponent mixtures, easy automation, and high selectivity. It is widely used in the analysis of fatty acids. In this chapter, we detailed the method for detecting omega-3 (ω-3) fatty acids (eicosapentaenoic acid (EPA) and docosahexaenoic acid (DHA)) of microalgae, which also provide important high-throughput pathway for studying lipid metabolism in microalgae.

Key words Gas chromatography, Omega-3, Microalgae, Standard method

1 Introduction

Marine microalgae are rich in two types of omega-3 (ω-3) fatty acids, eicosapentaenoic acid (EPA) and docosahexaenoic acid (DHA), among which the former can help reduce cholesterol and triglyceride levels, promoting the metabolism of saturated fatty acids in the body, reduce blood viscosity, enhance blood circulation, increase tissue oxygen supply, and eliminate fatigue [1–5], and the latter is crucial for the growth and development of brain nerves. In addition, clinical studies have found that arachidonic acid (ARA) can synthesize substances that cause inflammation and allergy in the body [6]. Proper intake of DHA can avoid the formation of excessive ARA, reduce excessive inflammatory reaction and allergy, and enhance immunity. Therefore, it is very important to establish standard detection methods of the EPA and DHA.

Mostafa Gouda et al. (eds.), *Plant Chemical Compositions and Bioactivities*, Methods and Protocols in Food Science, https://doi.org/10.1007/978-1-0716-3938-2_9,

With the development of society, humans are paying more attention to dietary health and have developed a series of detection methods to detect important nutrients in food. Traditional Soxhlet extraction methods can only perform rough detection of fatty acids, which is time-consuming and laborious. At present, the main methods for analyzing fatty acids are spectroscopy and chromatography. Near-infrared spectroscopy and Raman spectroscopy techniques can quickly and nondestructive qualitatively detect fatty acids, which are widely used in edible oil adulteration analysis and identification of edible oil types. However, statistical models need to be established based on stoichiometric methods for analysis and identification, and it is difficult to accurately qualitatively and quantitatively determine each type of fatty acid [7–9]. Liquid chromatography is also commonly used in the study of fatty acids, but due to the lack of conjugated structures in fatty acids, UV absorption is weak. In addition, fatty acids also lack functional groups that generate fluorescence and other signals, and the fluorescence signal is also weak. Therefore, liquid chromatography cannot be directly detected using ultraviolet detectors, fluorescence detectors, and so on. It requires derivatization of fatty acids for detection [15]. The evaporative luminescence detector can be used for the analysis of long-chain fatty acids, but cannot analyze Short-chain fatty acid. Liquid chromatography–mass spectrometry is also commonly used for quantitative and qualitative analysis of fatty acids, and has the advantages of fast speed, high resolution, and high sensitivity. However, when analyzing saturated medium- and long-chain fatty acids, the sensitivity is relatively low [18–20]. Therefore, liquid chromatography–mass spectrometry has obvious limitations in fatty acid analysis [10]. However, the emergence of gas chromatography can simultaneously identify the rich nutrients in microalgae quantitative analysis of eicosapentaenoic acid (EPA) and docosahexaenoic acid (DHA) in ω-3 fatty acids. In addition, the fatty acid profile not only helps to evaluate the applicability of microalgae fatty acids in biological refining but also serves as a useful tool for classifying and characterizing microalgae or microbial resources [11].

In this chapter, we describe a method for determining two main types of ω-3 fatty acid-produced microalgae compounds eicosapentaenoic acid (EPA) and docosahexaenoic acid (DHA) by gas chromatographs. In detail, we introduced the required materials, instruments and equipment, measurement steps, result analysis, and precautions, and completed the quantitative analysis of the compound.

2 Materials

2.1. Ultrapure water: Milli-Q water system (Millipore, USA).

2.2. Methanol: (Sigma-Aldrich, Steinheim, Germany).

2.3. Hexane: HPLC grade (Sigma-Aldrich, Steinheim, Germany).

2.4. DHA: Purity ≥99.9% (Sigma-Aldrich, Steinheim, Germany).

2.5. EPA: Purity ≥99.9% (Sigma-Aldrich, Steinheim, Germany).

2.6. Heptadecanoic acid: (Sigma-Aldrich, Steinheim, Germany).

2.7. Potassium hydroxide methanol solution (0.8 mol/L): (Sigma-Aldrich, Steinheim, Germany).

Weigh 5.24 g of potassium hydroxide and dissolve it in 100 mL of anhydrous methanol, thoroughly and evenly.

2.8. Heptadecanoic acid internal standard stock solution (2.0 mg/mL):

Weigh 0.050 g heptadecanoic acid into a 25 mL volumetric flask, dilute to the scale, and shake well.

2.9. DHA and EPA standard solutions:

Weigh 0.010 g of DHA and EPA standard respectively into a 5 mL volumetric flask, add proper amount of *n*-hexane to dissolve, dilute to the scale, shake up, then take 0.1, 0.2, 0.4, 0.8, and 1.5 mL, respectively, transfer them into five 10 mL volumetric flasks, respectively, transfer 2.0 mL of heptadecanoic acid internal standard stock solution, add *n*-hexane to dissolve, dilute to the scale, and shake up.

2.1 Instruments and Equipment

1. iConnect™ Flame Ionization Detector (FID) for TRACE™ 1300 and 1600 Series GC (Thermo Fisher Scientific, USA).
2. Analytical balance, BSA124S-CW (Apresys, USA).
3. Grinding mill Tissuelyser-48 (Shanghai Jinxing Instrument Manufacturing Co., Ltd., China).
4. Vortex Mixer HS120210-1EA (Sigma-Aldrich, Steinheim, Germany).
5. Ultrasonic crusher (Shanghai Lanyi Industrial Co., Ltd., China).
6. Vacuum oven (Zhengzhou Yuhua Instrument Manufacturing Co., Ltd., China).
7. Thermostat water bath (Shanghai Jiecheng Experimental Instrument Co., Ltd., China).

3 Method [12]

3.1 Measurement Steps

3.1.1 Sample Preparation

Microalgae: Weigh 10.00 g of microalgae, grind, and weigh 1.00 g of a uniform sample for oil extraction.

3.1.2 Extraction

Take 1.00 g of microalgae dry powder sample, put it into a 50 mL polytetrafluoroethylene test tube, add 6.0 mL of distilled water, then add 22.5 mL of chloroform methanol (1:2, volume ratio) mixture, shake and mix it evenly, break it with an ultrasonic crusher for 2 min, add 7.5 mL of chloroform, shake and mix it evenly for 30 s, add 7.5 mL of distilled water, shake and mix it evenly for 30 s, put the separatory funnel still for layering, separate the oil layer, vacuum dry it, weigh it, and calculate the oil content.

3.1.3 Fatty Acid Methyl Esterification

Weigh 0.10 g of oil sample and place it in a 10 mL centrifuge tube with a lid. Add 2 mL of a mixture of ether *n*-hexane (2:1 by volume), shake well for 15 min, add 2 mL of 0.8 mol/L potassium hydroxide methanol solution, shake well, and react in a constant temperature water bath at 60 °C for 15 min. Add water along the wall to the 10 mL mark, let it stand, and layer. Transfer all the supernatant to a volumetric flask, dilute to 1 mL with *n*-hexane, dilute to a certain multiple, and analyze the sample.

3.1.4 Detection

The reference conditions for gas chromatography are as follows:

Reference Conditions for Gas Chromatography

1. Chromatographic column: Polydihydropropylsiloxane capillary column (0.32 mm × 0.25 μM × 30 m), or equivalent;
2. Carrier gas and flow rate: The carrier gas is nitrogen, with a flow rate of 1.5 mL/min;
3. Column temperature program: The initial temperature is 160 °C (maintained for 3 min), and it is increased to 230 °C at a rate of 6 °C/min for 15 min;
4. Injection port temperature: 250 °C;
5. Detector temperature: 280 °C;
6. Injection volume: 1 μ L;
7. Injection method: Split injection, with a split ratio of 50:1.

Drawing of Standard Curves

Take 1 mL of standard working solution and 1 mL of blank solution (*n*-hexane) of different concentrations and esterify them according to Subheading 3.1.3. Inject the samples in sequence, deduct the blank value, and determine the peak areas of DHA methyl ester, EPA methyl ester, and heptadecanoate methyl ester.

Draw a standard curve using the ratio of the mass concentration of standard working solutions DHA and EPA to the mass concentration of internal standard substance as the *x*-axis, and the peak area ratio of DHA methyl ester and EPA methyl ester to heptadecanoate methyl ester as the *y*-axis. Refer to Appendix Fig. 1 for the chromatogram.

3.2 Result Analysis

Calculate the content of DHA or EPA in the sample according to Eq. 1:

$$\omega = \frac{Vn(y-b)\rho f \times 100}{am} \quad (1)$$

where

ω: the content of DHA or EPA in the sample, in g/100 g;

V: final volume of the test solution, in milliliters (mL);

n: dilution ratio of the test solution;

y: the peak area ratio of DNA methyl ester or EPA methyl ester to heptadecanoate methyl ester in the sample;

b: intercept of standard curve;

ρ: the mass concentration of the final internal standard substance of the test solution, in milligrams per milliliter (mg/mL);

f: the conversion coefficient of f-DNA methyl ester to DHA (value 0.9590), or the conversion coefficient of EPA methyl ester to EPA (value 0.9557);

a: the slope of the standard curve;

m: mass of the sample, in milligrams (mg).

The calculation results are expressed as the arithmetic mean of the parallel measured values, with three significant figures reserved. The detection limit of this method is 10 mg/kg, and the quantitative limit is 33 mg/kg.

3.3 Repeatability

The absolute difference between two independent measurement results obtained under repeatability conditions shall not exceed 10% of the arithmetic mean (Fig. 1).

4 Notes

1.1. Working Principle.

Chromatography using gas as the mobile phase. It utilizes the difference in the distribution coefficients of substances between the mobile phase and the fixed phase. When the two phases move relative, the sample components are repeatedly distributed between the two phases. Even if

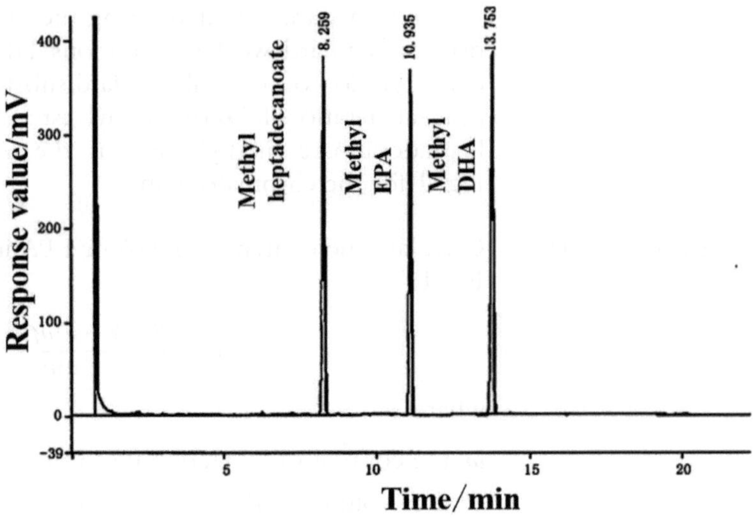

Fig. 1 Standard chromatograms of DHA, EPA, and heptadecanoic acid

there is only a slight difference in the distribution coefficients of each component, there can be distance as the mobile phase (gas) moves. Finally, the measured sample components are separated and determined.

1.2. Standard Samples.

The mass fraction of the chromatographic main body shall not be less than 99.9%.

1.3. Hydrogen and Nitrogen.

Its volume fraction shall not be less than 99.8%. Before use, a dehydration device, silica gel, molecular sieve, or activated carbon should be used for purification treatment.

1.4. Air.

It should be free of corrosive impurities and undergo deoiling and dehydration treatment before use.

1.5. Stability of the Entire Machine.

When hydrogen gas is used as the carrier (when using a thermal conductivity detector) or nitrogen gas as the carrier (when using a flame ionization detector), the white diatomaceous earth carrier (0.18–0.25 mm) is coated with dinionyl phthalate (20% mass fraction) as the stationary phase, with a column length of 2 m and a column temperature of 80 °C. The carrier gas flow rate should be selected appropriately, and the sensitivity of the instrument should be close to the sensitivity requirements of the entire machine. The baseline drift value of the instrument within 10 min shall not exceed 1% of full scale.

1.6. Overall Sensitivity.

Instruments using flame ionization detectors.

Using benzene as the sample, the experimental conditions are the same as 7.3. The sensitivity is calculated based on the detection limit D, and the value is expressed in g/s. The detection limit should not be greater than 5×10^{-10} g/s. Calculate according to Eq. 2:

$$D = \frac{2Nm}{A} \tag{2}$$

where

N: the value of baseline noise, in millivolts (mV);

m: the numerical value of m-benzene injection amount, in grams (g);

A: arithmetic mean value of benzene peak area, in millivolt seconds (mV s).

1.7. Quantitative Repeatability.

Using benzene as the sample, the test conditions are the same as 7.3. The injection volume is constant. After the instrument is stable, continuously inject 11 samples, and conduct the experiment after the relative standard deviation is stable.

2. Selection of Experimental Conditions.

Based on the characteristics and specifications of the product and the components to be tested, select the optimal conditions according to the following provisions:

(a) Detector: thermal conductivity detector or flame ionization detector;

(b) Carrier gas: carrier gas type and flow rate:

(c) Chromatographic column: chromatographic column (filled, capillary), column length, inner diameter;

(d) Fixed liquid: type of fixed liquid, carrier, mass fraction of fixed liquid, thickness of fixed liquid film;

(e) Temperature: column temperature, vaporization chamber temperature, detection chamber temperature;

(f) Separation degree: According to the requirements of method precision and accuracy, specify the separation degree of the tested component and it is difficult to separate substances (retain two significant digits);

(g) Asymmetric factor: According to the requirements of method precision and accuracy, the asymmetric factor of the main peak is specified;

(h) Effective plate height: On the basis of meeting the requirements of separation degree and asymmetry factor, the effective plate height of the chromatographic column is specified. The calculation method is shown in Appendix A (retain two significant digits);

(i) Relative retention value (rounded to two decimal places);

(j) Injection volume: It should be controlled within a linear response range, and each impurity peak and internal standard should be recorded clearly at this injection volume. When using the normalization method, the main peak height (or after attenuation) should be above 70% of the full scale on the recorder;

(k) Other instrument conditions: bridge flow, split ratio, and tail blowing;

(l) Quantitative methods.

Note: The separation and relative retention values of difficult to separate substances can be determined as needed. The carrier gas flow rate, column temperature, vaporization chamber temperature, split ratio, and tail blowing and injection volume conditions can be adjusted appropriately according to the specific instrument performance during operation.

3. Operation Methods.

3.1 Chromatographic Column.

3.1.1 Filled Columns.

3.1.1.1 Fixed Liquid Coating.

Dissolve the fixed solution in the solvent to form a homogeneous solution, immerse the carrier in the solution (heat and reflux if necessary), and gently stir or shake well. Do not crush the carrier. Place in a fume hood and let the solvent evaporate and dry under an infrared light.

The mass fraction of the fixed solution is calculated in w, and the value is expressed in %. Calculate according to Eq. 3:

$$w = \frac{m1}{m1 + m2} \times 100\% \tag{3}$$

where

$m1$: the numerical value of the mass of the fixed solution, in grams (g);

$m2$: the numerical value of the carrier mass, in grams (g).

3.1.1.2 Pretreatment of Empty Columns.

First, remove mechanical impurities from the column, wash with nitric acid solution (mass fraction 10%), and wash with water until neutral. Wash again with sodium hydroxide solution (100 g/L).

Finally, wash with water until neutral and dry.

3.1.1.3 Column Filling.

Plug one end of the pretreated empty column tightly with glass fiber and copper wire mesh, and connect it to a vacuum pump for pressure reduction and evacuation. Add a stationary phase at the other end and apply moderate vibration to ensure that the carrier is evenly and tightly packed into the chromatographic column.

3.1.1.4 Chromatographic Column Aging.

The aging chromatographic column should be slowly heated in an oxygen flow, and maintained for at least 4 h after reaching the maximum operating temperature of the low dry fixed solution (the temperature should not be too high to prevent loss). After aging, it should gradually cool down in the carrier gas.

3.1.2 Capillary Column.

At present, there are capillary columns available for different purposes on the market, and users can choose based on the properties of the samples.

3.2 Measurement of Carrier Gas Flow Rate.

3.2.1 Measurement of Carrier Gas Flow Rate with Thermal Conductivity Detector.

Connect the soap foam flowmeter to the carrier gas outlet and measure the carrier gas flow rate. The carrier gas flow rate is calculated in *F*, and the value is expressed in milliliters per minute (mL/min). Calculate according to Eq. 4:

$$F = \frac{V}{t} \tag{4}$$

where

V: the value of carrier gas volume, in milliliters (mL);

t: the numerical value of the measurement time, in minutes (min).

3.2.2 Determination of Carrier Gas Flow Rate Using Flame Ionization Detector.

The average linear velocity of the carrier gas flow rate of the flame ionization detector is in *u*, and the value is expressed in centimeters per second (cm/s). Calculate according to Eq. 5:

$$u = \frac{L}{t_m} \tag{5}$$

where.

L: the numerical value of the length of the chromatographic column, in centimeters (cm);

t_m: the numerical value of methane retention time, in seconds (s).

3.3 Injection Method.

Use a clean syringe (or other injection device) to extract the sample three times. After emptying the sample, take the specified amount of sample, quickly insert it into the vaporization chamber, push the sample once, and then extract the injector once.

3.4 Attenuation Ratio Calibration.

After the instrument stabilizes, adjust the zero occupation and recording position of the instrument to the zero position of the recording scale, calibrate the attenuation level before 16, and adjust the zero knob when using a conductivity detector. Adjust the zero knob, when using a flame ionization detector, use a basic current compensation knob, add a constant signal, and then attenuate step by step to record the specified position after attenuation. Calculate the attenuation ratio of each gear according to Eq. 6. When calibrating the attenuation range after gear 16, place the attenuation range at position 16. Repeat the above operation, calculate the attenuation ratio of each gear according to Eq. 6 and multiply it by the attenuation ratio of gear 16. The attenuation ratio is represented by A, and the value is calculated according to Eq. 6:

$$A = \frac{l1}{l2} \tag{6}$$

where

$l1$: the numerical value of the length of the signal on the recording scale, in millimeters (mm);

$l2$: the numerical value of the distance between the zero point and the pointer after attenuation, in millimeters (mm).

3.5 Peak Area Calculation.

The peak height and half peak width can be measured separately using measuring tools with an accuracy of 0.02 mm. The peak area is measured in A, and the value is expressed in square centimeters (cm^2). Calculate according to Eq. 7:

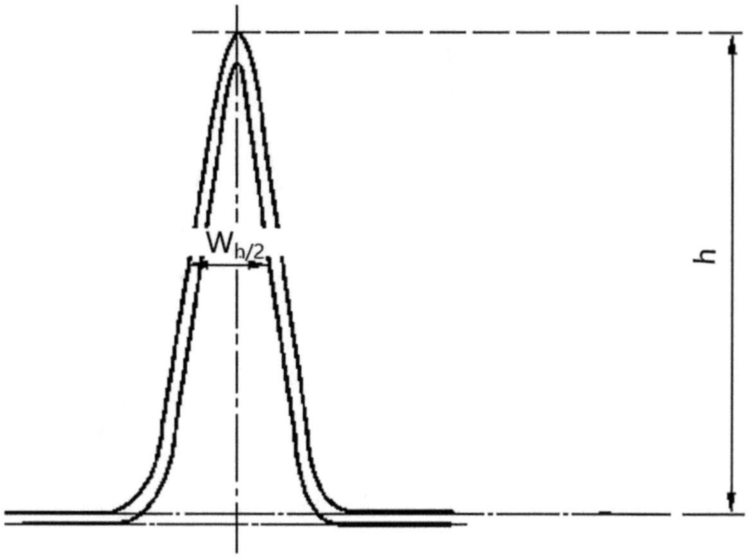

Fig. 2 Chromatographic peak distance calculation by using baseline to the vertex peak height

$$A = hW_{h/2} \quad (7)$$

where

H: the numerical value of peak height, in centimeters (cm);

$W_{h/2}$: the numerical value of the width at half peak, in centimeters (cm).

The value of peak area can also be directly provided by the data processing system.

3.5.1 Peak Height Measurement.

Draw a vertical line from the peak to the bottom of the chromatographic peak, and the distance from the intersection point along the baseline of the chromatographic peak to the vertex is the peak height h (Fig. 2).

3.6 Special Peak Pattern Processing.

3.6.1. When the separation of adjacent groups exceeds 1/2 of the small peak height, it can be handled according to Fig. 3.

3.6.2. When the separation between adjacent groups does not exceed 1/2 of the small peak height, it can be handled according to Fig. 4.

3.6.3. Trace magazine peaks on the leading or trailing lines of the main peak can be processed according to Fig. 5.

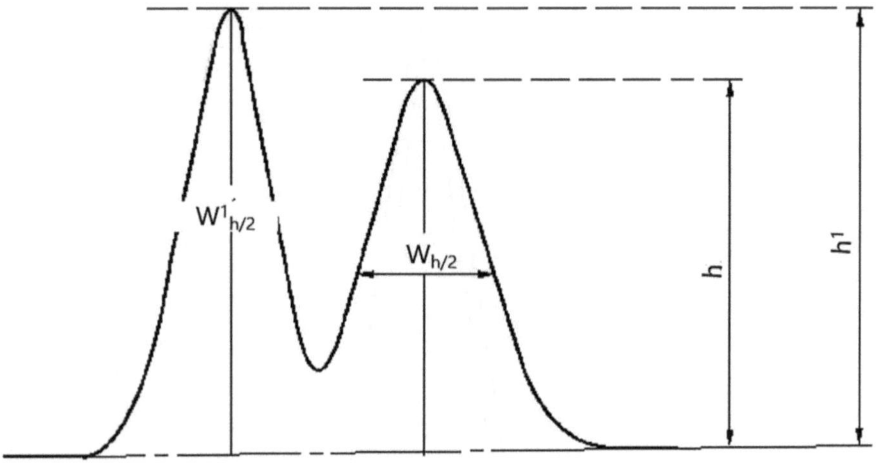

Fig. 3 Potential handling methods in case of separation of adjacent groups exceeds 1/2 of the small peak height

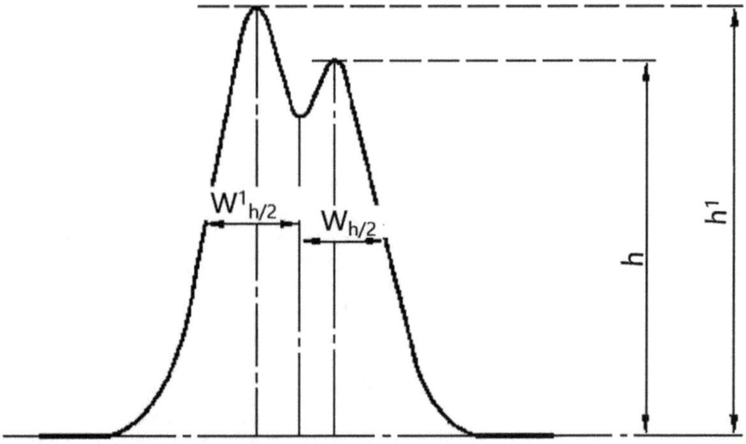

Fig. 4 Peak calculation when the separation between adjacent groups does not exceed 1/2 of the small peak height

4.1. Correction Factor.

This standard adopts the mass correction factor of component *i* relative to the main body.

4.1.1. Principle of Using Correction Factors.

For individual components included in technical indicators, correction factors are used regardless of their quality scores. Among the tested components, homologues with similar carbon numbers or substances with smaller differences in thermal conductivity can be corrected by adding a correction factor depending on the specific situation.

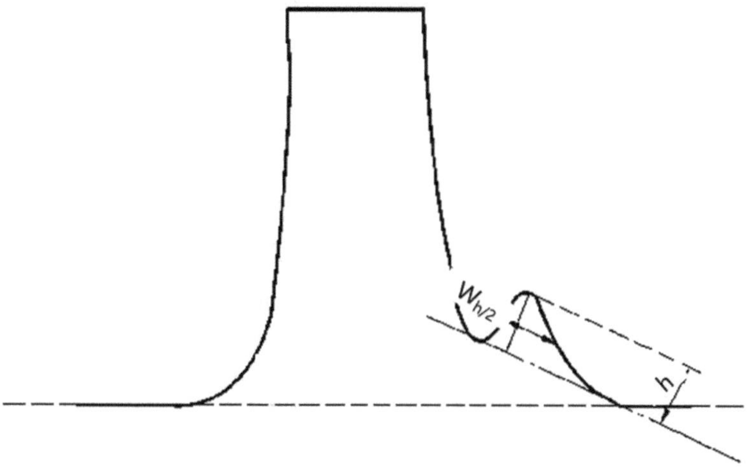

Fig. 5 Trace magazine peaks on the leading or trailing lines of the main peak data process

4.1.2. Determination of Correction Factors.

Prepare several standards with weighing method (accurate to 0.0001 g) that are similar to the corrected component indicators, and measure them according to the measurement conditions of the sample. The measurement results are rounded off with a confidence level of 95%, and the average value is calculated (retaining two significant digits).

The correction factor is represented by *f*, and the value is calculated according to Eq. 8:

$$f = \frac{As * mi}{Ai * ms} \tag{8}$$

where

As: the numerical value of the main peak area, in square centimeters (cm^2) or millivolts per minute (mV/min);

mi: the numerical value of the mass of component *i*, in grams (g);

Ai: the numerical value of the peak area of component *i*, in square centimeters (cm^2) or millivolts per minute (mV/min);

ms: the numerical value of the main body mass, in grams (g).

Note: When preparing standards, if there is no pure product, other methods can be used to measure the mass fraction for correction.

4.2. Internal Standard Method.

When using the internal standard method for quantification, the following requirements should be met:

(a) The internal standard should be completely miscible with the sample and not undergo chemical reactions. The internal standard should not contain impurities that interfere with the analysis;

(b) The retention value of the internal standard substance should be close to the retention value of the tested substance, without overlap or severe tailing, and there should be no other impurity peaks in the area;

(c) The amount of internal standard sample preparation should be close to or slightly greater than the amount of the measured component;

(d) The injection amount should be within the linear range of the detector, and all the components to be tested should flow out under test conditions.

The mass fraction of the components measured by the internal standard method is expressed in *w*, and the numerical value is expressed in %. Calculate according to Eq. 9:

$$w = \frac{msAifs \times i}{mAs} \times 100\% \quad (9)$$

where

ms: the numerical value of the mass of the added internal standard substance, in grams (g);

Ai: the numerical value of the peak area of component *i*, in square centimeters (cm^2) or millivolts per minute (mV/min);

fs × *i*: the value of the correction factor for component *i* compared to the internal standard;

m: the numerical value of the sample mass, in grams (g);

As: the numerical value of the peak area of the internal standard substance, in square centimeters (cm^2) or millivolts per minute (mV/min).

5. Environmental Requirements and Safety Precautions

5.1. Environmental Requirements

The instrument should be placed at a room temperature of 5–35 °C and a relative humidity of less than 85%. Prevent the invasion of corrosive gases, vibration, direct sunlight, and strong magnetic fields from damaging the instrument. The operating site should have good ventilation.

5.2. Safety Precautions

5.2.1. The instrument power supply of the device should be well grounded. When hydrogen is used as carrier gas or combustion gas, the pipeline should have no leakage. There should be no open flames in the instrument operating area.

5.2.2. The gas at the outlet of the chromatographic column should be discharged to a ventilated area.

5.2.3. Standardize the operation when using high-pressure steel cylinders.

References

1. Diprat AB, Menegol T, Boelter JF, Zmozinski A, Rodrigues Vale MG, Rodrigues E et al (2017) Chemical composition of microalgae Heterochlorella luteoviridis and Dunaliella tertiolecta with emphasis on carotenoids. J Sci Food Agric 97(10):3463–3468
2. Rezanka T, Vitova M, Lukavsky J, Nedbalova L, Kolouchova I (2019) Rapid screening of very long-chain fatty acids from microorganisms. J Chromatogr A 1605: 460365
3. Shao Y, Fang H, Zhou H, Wang Q, Zhu Y, He Y (2017) Detection and imaging of lipids of Scenedesmus obliquus based on confocal Raman microspectroscopy. Biotechnol Biofuels 10:300
4. Vendruscolo RG, Fagundes MB, Jacob-Lopes-E, Wagner R (2019) Analytical strategies for using gas chromatography to control and optimize microalgae bioprocessing. Curr Opin Food Sci 25:73–81
5. Vendruscolo RG, Xavier Facchi MM, Maroneze MM, Fagundes MB, Cichoski AJ, Zepka LQ et al (2018) Polar and non-polar intracellular compounds from microalgae: methods of simultaneous extraction, gas chromatography determination and comparative analysis. Food Res Int 109:204–212
6. Woo S, Yoo K, Lee J, Bang S, Lee M, On K et al (2012) Comparison of fatty acid analysis methods for assessing biorefinery applicability of waste water cultivated microalgae. Talanta 97:8
7. Basri KN, Hussain MN, Bakar J, Sharif Z, Khir M, Zoolfakar AS (2017) Classification and quantification of palm oil adulteration via portable NIR spectroscopy. Spectrochim Acta A Mol Biomol Spectrosc 173:335–342
8. Li BN, Wu YW, Wang Y, Zu WC, Chen SC (2014) Raman spectroscopy combined with pattern recognition methods for rapid identification of crude soybean oil adulteration. Spectroscopy and Spectral Analysis 34(10): 2696–2700
9. Xiang L, Zhu L, Huang Y, Cai ZW (2020) Application of derivatization in fatty acids and fatty acyls detection: mass spectrometry-based targeted lipidomics. Small Methods 4(8): 2000160
10. Martinez-Bisbal MC, Mestre NC, Martinez-Manez R, Bauza J, Fillol MA (2019) Microalgae degradation follow up by voltammetric electronic tongue, impedance spectroscopy and NMR spectroscopy. Sensors and Actuators B: Chemical 281:9
11. Reindl W, Northen TR (2010) Rapid screening of fatty acids using nanostructure-initiator mass spectrometry. Anal Chem 82(9): 3751–3755
12. GBT38095-2019.

Chapter 10

Metabolic Accumulation of Tea Theaflavins, Thearubicin, and Theabrownin Quality-Related Molecules by Surface-Enhanced Raman Spectroscopy (SERS)

Xuelun Luo, Xiaoli Li, and Mostafa Gouda

Abstract

The development of a rapid and simple method for the determination of theaflavins, thearubicin, and theabrownin quality-related molecules is very beneficial for the study of the fermentation process of black tea. This chapter takes the fermentation process of black tea as an example. Surface-enhanced Raman spectroscopy (SERS) and chemometrics method are used to detect theaflavins, thearubicin, and theabrownin quality-related molecules. The results show that this method can be used to determine the optimal fermentation stage of black tea. The important Raman peaks at 317.71, 619.59, 731.48, 956.08, and 1326.70 cm^{-1} were found for monitoring quality changes of black tea by density functional analysis and correlation analysis. The prediction r^2 for catechin (C) and epigallocatechin gallate (EGCG) reached 0.81 and 0.82, respectively, by integrated SERS with a one-dimensional convolutional neural network (1D-CNN). In conclusion, this chapter revealed the Raman fingerprint characteristics of key compounds associated with the fermentation quality of black tea, presenting an opportunity to quantify the quality changes of tea during fermentation using SERS data. With the monitoring method developed in this research, the optimal fermentation stage can be determined accurately, thus decreasing fermentation costs and improving tea quality.

Key words Black tea, Quality monitoring, Fermented tea detection, SERS

1 Introduction

Black tea is one of the most popular beverages in the world [1]. The ingredients of black tea have many health benefits. For instance, tea polyphenols (TP) and total catechins (CC) provide benefits, including cancer and diabetes prevention [2, 3]. The key step in forming these ingredients is the fermentation process of black tea. This process alters the content of the ingredients of the tea, thereby modifying its flavor, aroma, color, and nutritional value [4]. During fermentation, CC are oxidized by polyphenol oxidase (PPO) and peroxidase (POD) to produce theaflavin (TF) and thearubicin

Mostafa Gouda et al. (eds.), *Plant Chemical Compositions and Bioactivities*, Methods and Protocols in Food Science,
https://doi.org/10.1007/978-1-0716-3938-2_10,

(TR), which are subsequently oxidatively polymerized to form theabrownin (TB) [5]. Monitoring the changes in ingredients of black tea during fermentation will help optimize the fermentation process, resulting in a more nutritious and tasty black tea. However, it is still difficult to quickly analyze the quality of black tea and monitor the changes in the ingredients of black tea in real time during fermentation.

The traditional way to determine the quality of black tea is highly subjective and has low reproducibility, which cannot satisfy the requirements of industrialization of black tea [6]. Scholars have utilized high-performance liquid chromatography (HPLC) to assess fermentation samples of various tea leaves [7]; however, this method is difficult and time-consuming. Some scholars have proposed approaches for the evaluation of black tea fermentation stages based on the merging of near-infrared ray (NIR) and computer vision (CV) [6, 8]. However, due to the weakness of the molecular fingerprint of the near-infrared spectrum and the fact that images can only reflect color and texture information, the quantitative detection model based on NIR and CV is mostly empirical rather than mechanical, which makes it difficult to reveal the mechanism of black tea quality formation during fermentation with these techniques.

As a fast and accurate method, SERS technology has fast analysis speed, strong fingerprint, low cost, and generates less waste [9]. Huang et al. stated that SERS could reduce the detection limit for CC to a concentration level of 10^{-18} M [10]. Jing et al. used SERS to detect eight CC monomers and used DFT (density functional theory) to analyze the peaks [11]. However, these two investigations were focused on the CC standard rather than CC in tea. Zhang et al. used SERS to identify Pu-erh tea from diverse origins and fermentation years, but they did not conduct an in-depth examination of the SERS data using appropriate chemometrics methods [12]. Although SERS data have many advantages, their application in tea detection is still limited. This is because it is difficult to prepare an even SERS base, and an uneven SERS base will result in the instability of SERS data [13]. In this chapter, it was believed that the chemometrics method could effectively eliminate the instability of SERS data and deeply mine the characteristics of SERS data of black tea during fermentation.

In this chapter, SERS was adopted to quickly obtain quality information of black tea during fermentation. First, SERS data of black tea extract was obtained, and the true value of ten quality indicators of black tea was measured by the method of Yao et al. [15] and the national standard method. Then, the characteristic peak of SERS data was theoretically analyzed to confirm the attribution of the characteristic peak, and a quantitative model between SERS data and ten quality indicators of black tea was established.

2 Materials

2.1 Chemicals and Materials

All the chemicals below are purchased from Sigma-Aldrich Co. (St. Louis, MO, USA; www.sigmaaldrich.com). All chemicals were analytical grade. Ultrapure Milli-Q water was used throughout the experiments.

1. Silver nitrate ($AgNO_3$) (99%; PubChem CID: 24470).
2. Sodium citrate (99%; PubChem CID: 6224).
3. Sodium chloride (NaCl) (99%; PubChem CID: 5234).
4. Ethyl acetate (99.8%; PubChem CID: 8857).
5. *n*-Butanol (99%; PubChem CID: 263).
6. Ethanol (99.5%; PubChem CID: 702).
7. Theaflavin (99%; PubChem CID: 4263901).
8. Sodium bicarbonate (99%; PubChem CID: 516892).
9. Oxalic acid (99%; PubChem CID: 971).
10. Fengqing Daye (FQDY) tea leaves (Fengqing County, Yunnan Province (E: 99°31′; N: 24°13′), China).

2.2 Sample Preparation

1. The collected fresh leaves were fermented for 6 h according to the black tea fermentation process (*see* **Note 1**).
2. Samples were collected at ten time points (0 h, 1 h, 1.5 h, 2 h, 2.5 h, 3 h, 3.5 h, 4 h, 4.5 h, 5 h) during the fermentation process. The fermented samples were immediately frozen at −40 °C for 90 h for the downstream analyses.
3. At each fermentation time point, 1.2 g (±0.05 g) samples were weighed and ground into powder to obtain SERS (*see* **Note 2**).

2.3 Preparation of Silver Nanoparticles (AgNPs)

1. The preparation of AgNPs used for SERS followed the modified method of preparing AgNPs by Lee et al. [14] (*see* **Note 3**).

3 Methods

3.1 SERS Data Collection

1. To collect SERS data, the slides were first washed with 5% HNO_3 (v/v) solution.
2. 5 μL of AgNPs colloidal solution, 2.5 μL of 5% NaCl (w/v) solution, and 4 μL of tea extract were placed and well mixed on a glass slide.
3. The mixed droplet was dried and examined by a confocal Raman microscope (InVia Reflex, Renishaw plc, Gloucestershire, UK) (*see* **Note 4**).

3.2 Determination of Chemical Components in Tea

3.2.1 Determination of TF, TR, and TB Content

The contents of TF, TR, and TB were measured based on Yao et al. [15].

1. Four solutions (A, B, C, D) containing TF, TR, and TB were extracted by ethyl acetate, ethanol, *n*-butanol, and sodium bicarbonate.
2. With 95% ethanol as a reference, the absorbance values of A, B, C, and D were measured at 380 nm by the colorimetric method and recorded as EMA, EMB, EMC, and EMD.
3. TF (%) = (EMC*2.25)
4. TR (%) = 7.06*(2*EMA+2* EMD-EMC-2* EMB)
5. TB (%) = 2*EMB*7.06

3.2.2 Determination of TP and Catechin Monomer Content

The TP content was determined according to the Chinese standard method (GB/T8313-2018) [8]. The contents of epigallocatechin gallate (EGCG), epicatechin gallate (ECG), epigallocatechin gallate (EGC), epicatechin (EC), and catechin (C) were determined with the method proposed by Alireza et al. [16].

1. Before determine the TP content, the ground sample was sieved through a 40-mesh sieve.
2. The sieved tea powder of each sample (0.1 g) was weighed into a test tube, 25 mL of pure water was added, and the mixture was stirred.
3. The test tube was stored in a water bath at 85 °C for 20 min.
4. The supernatant was then filtered out with a 0.22 μm membrane filter and injected into the HPLC system.

3.3 Multivariate Data Analysis

3.3.1 Spectral Data Pretreatment

1. Wavelet transform (WT) and band ratio (BR) are two widely used spectral pretreatment methods for increasing the determination accuracy of spectroscopic technology in food quality detection [17–20].
2. Three preprocessing methods, including WT (*see* **Note 5**), BR (reference band = 860.71 cm^{-1}), and a combination of wavelet transform and band ratio (BRWT).
3. After testing the modeling effect, BRWT was confirmed as the best pretreatment method (Fig. 1).

3.3.2 SERS Analysis

1. The original Raman spectra of the sample (Fig. 2b) show considerable background fluorescence interference and no distinct peaks.
2. To reduce the background fluorescence of the spectra of tea extract, NaCl solution was added. The Raman peaks were more distinct after adding the AgNPs solution (Fig. 2c).

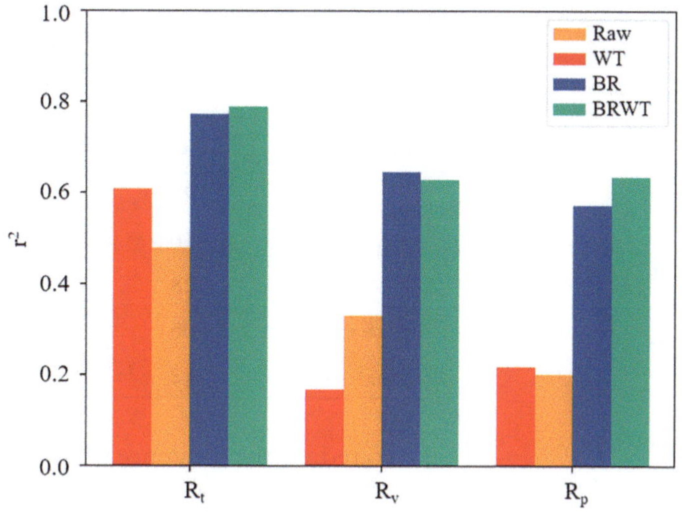

Fig. 1 PLS, LSSVM, and 1D-CNN were used to model ten kinds of tea quality indicators, and the modeling effects of the same pretreatment method were averaged, and the average effects (r^2) of four different pretreatment methods were obtained

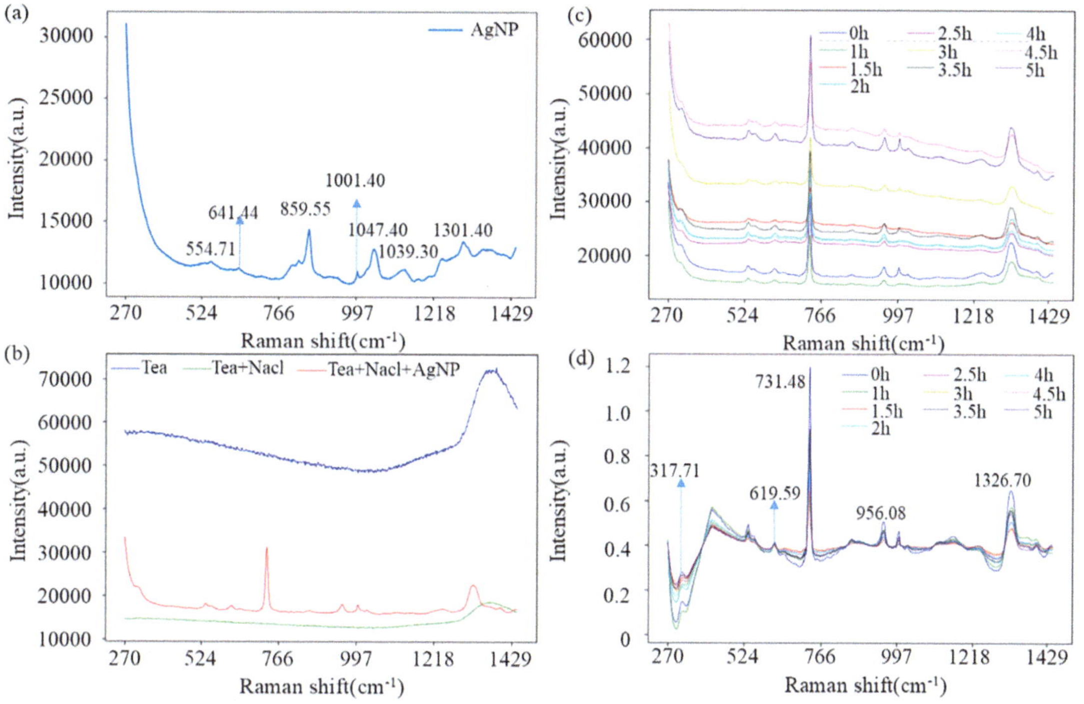

Fig. 2 (**a**) Raman spectra of AgNPs. (**b**) Raman spectra of tea extract; mixed droplets of tea extract and NaCl solution; mixed droplets of tea extract, NaCl solution, and AgNPs. (**c**) Original Raman spectra. (**d**) Raman spectrum after band ratio

Table 1
Position and intensity variances of important peaks to react the stability of Raman spectra

Peaks	Variances of Raman shift	Variances of intensity
317.71 cm^{-1}	1.998	7.633×10^7
619.59 cm^{-1}	1.446	1.097×10^8
731.48 cm^{-1}	1.528	1.203×10^8
956.08 cm^{-1}	1.406	1.330×10^8
1326.70 cm^{-1}	1.243	1.399×10^8

3. To reduce the influence of fluorescence and noise on the spectra, the original spectra were preprocessed with BRWT. Figure 2d shows the average spectra following pretreatment.
4. The Raman peaks that belong to the AgNPs solution (Fig. 2a) were removed from the peaks shown in Fig. 2d, and five primary Raman peaks were left: 317.71 cm^{-1}, 619.59 cm^{-1}, 731.48 cm^{-1}, 956.08 cm^{-1}, and 1326.70 cm^{-1}, all of which correspond to the FQDY extract. It is worth noting that the intensity of these primary peaks of the original spectra was unstable due to the fluorescence (Table 1). After pretreatment, the influence of fluorescence on the main peaks was significantly reduced.

3.3.3 Establishment of Determination Model

1. The PLS, LSSVM, and 1D-CNN were used to establish a quantitative model for the determination of ten chemical indicators of fermented tea based on SERS data (*see* **Note 6**).
2. Table 2 shows r^2 and RMSE of ten compounds under three modeling methods and four pretreatment methods. Table 3 and Fig. 4 show the best modeling method to establish the quantitative model results of ten tea quality indicators.
3. The 1D-CNN model was used, and network construction (Fig. 3) was based on the method of Zhang et al. [17] (*see* **Note 7**).

4 Notes

1. The collected fresh leaves were air-dried at room temperature (25 °C) to lose water at 38% of their fresh weight within 10 h. Then, the leaves were rolled by a 6CMK-55RN rolling machine (Changshaxiangfeng Ltd., China) for 50 min (light/heavy/light kneading for 15/20/15 min). After that, the fermentation process of leaves was carried out in a fermenter (6CMK-20/40FJC, Hangshaxiangfeng Ltd., China) at 28 °C and ≥90% humidity for 6 h.

Table 2
r^2 and RMSE of ten compounds under three modeling methods and four pretreatment methods

		Training set		Validation set		Prediction set	
Models	**Data type**	R_t^2	$RMSE_t$	R_v^2	$RMSE_v$	R_p^2	$RMSE_p$
C							
PLS	Raw	0.74	7.55E-04	0.57	2.19E-03	0.54	2.33E-03
	WT	0.47	3.22E-03	0.46	3.24E-03	0.27	6.06E-03
	BR	0.88	1.57E-04	0.67	1.12E-03	0.50	2.77E-03
	BRWT	0.74	7.91E-04	0.57	1.24E-03	0.64	1.48E-03
LSSVM	Raw	0.19	7.51E-03	0.20	7.52E-03	−0.01	1.15E-02
	WT	0.65	1.37E-03	0.20	7.29E-03	0.48	3.00E-03
	BR	0.69	1.07E-03	0.64	1.28E-03	0.68	1.82E-03
	BRWT	0.85	2.68E-04	0.68	6.73E-04	0.75	7.14E-04
CNN	Raw	0.95	7.59E-02	0.71	1.69E-01	0.75	1.64E-01
	WT	1.00	2.58E-03	0.34	2.28E-01	0.55	2.18E-01
	BR	0.98	5.12E-02	0.85	1.37E-01	0.81	1.43E-01
	BRWT	0.91	1.00E-01	0.79	1.58E-01	0.65	1.91E-01
TP							
PLS	Raw	0.66	9.00E-01	0.46	2.30E+00	0.44	2.44E+00
	WT	0.40	2.85E+00	0.39	2.91E+00	0.16	5.54E+00
	BR	0.74	5.34E-01	0.51	1.67E+00	0.19	5.06E+00
	BRWT	0.77	4.27E-01	0.46	1.94E+00	0.23	4.61E+00
LSSVM	Raw	−1.75	5.96E+01	−1.78	6.52E+01	−2.49	9.53E+01
	WT	0.21	4.90E+00	−2.08	8.56E+01	−1.15	3.62E+01
	BR	0.57	1.44E+00	0.49	1.73E+00	0.52	1.81E+00
	BRWT	0.77	4.11E-01	0.53	1.42E+00	0.50	1.97E+00
CNN	Raw	0.53	1.15E+00	0.21	1.40E+00	0.55	1.13E+00
	WT	0.33	1.37E+00	0.13	1.45E+00	0.18	1.52E+00
	BR	0.68	9.55E-01	0.64	1.07E+00	0.34	1.36E+00
	BRWT	0.72	8.92E-01	0.58	1.13E+00	0.43	1.26E+00
TB							
PLS	Raw	0.81	2.40E+00	0.64	9.23E+00	0.61	1.03E+01
	WT	0.53	1.53E+01	0.51	1.73E+01	0.36	2.76E+01
	BR	0.85	1.66E+00	0.70	5.61E+00	0.47	1.93E+01
	BRWT	0.84	1.88E+00	0.64	6.18E+00	0.56	1.35E+01
LSSVM	Raw	0.43	2.29E+01	0.46	2.05E+01	−0.41	2.35E+01
	WT	0.64	9.19E+00	0.26	4.04E+01	0.02	6.56E+01
	BR	0.69	6.94E+00	0.65	7.86E+00	0.68	6.84E+00
	BRWT	0.86	1.32E+00	0.75	3.09E+00	0.78	3.35E+00
CNN	Raw	0.80	1.28E+00	0.66	1.63E+00	0.71	1.55E+00
	WT	0.68	1.64E+00	0.38	1.91E+00	0.55	1.93E+00
	BR	0.88	1.00E+00	0.70	1.69E+00	0.80	1.29E+00
	BRWT	0.92	8.31E-01	0.83	1.25E+00	0.76	1.42E+00

(continued)

Table 2
(continued)

Models	Data type	Training set		Validation set		Prediction set	
		R_t^2	$RMSE_t$	R_v^2	$RMSE_v$	R_p^2	$RMSE_p$
TR							
PLS	Raw	0.79	4.04E-02	0.52	2.18E-01	0.48	2.47E-01
	WT	0.47	2.68E-01	0.42	3.13E-01	0.31	4.37E-01
	BR	0.75	5.99E-02	0.60	1.39E-01	0.45	2.79E-01
	BRWT	0.65	1.17E-01	0.51	1.43E-01	0.59	1.55E-01
LSSVM	Raw	0.38	3.59E-01	0.40	3.40E-01	−0.34	3.97E-01
	WT	0.63	1.31E-01	0.18	6.36E-01	−0.12	1.15E+00
	BR	0.60	1.50E-01	0.55	1.85E-01	0.60	1.47E-01
	BRWT	0.79	4.06E-02	0.63	7.81E-02	0.70	8.46E-02
CNN	Raw	0.66	5.75E-01	0.62	5.93E-01	0.55	6.59E-01
	WT	0.70	5.36E-01	0.34	7.06E-01	0.30	8.20E-01
	BR	0.83	4.02E-01	0.77	4.83E-01	0.67	5.61E-01
	BRWT	0.96	1.95E-01	0.68	5.88E-01	0.69	5.50E-01
TF							
PLS	Raw	0.73	8.90E-06	0.41	3.87E-05	0.36	4.84E-05
	WT	0.49	3.04E-05	0.35	5.42E-05	0.16	8.33E-05
	BR	0.67	1.26E-05	0.54	2.62E-05	0.51	2.77E-05
	BRWT	0.55	2.37E-05	0.42	2.71E-05	0.60	1.90E-05
LSSVM	Raw	0.16	8.33E-05	0.18	7.91E-05	−0.07	1.00E-04
	WT	0.50	2.93E-05	−0.32	2.07E-04	−0.66	3.23E-04
	BR	0.48	3.25E-05	0.48	3.23E-05	0.47	3.26E-05
	BRWT	0.72	9.49E-06	0.51	1.93E-05	0.66	1.33E-05
CNN	Raw	0.77	5.06E-02	0.42	7.69E-02	0.47	7.56E-02
	WT	0.91	3.18E-02	0.23	8.41E-02	0.38	8.18E-02
	BR	0.76	5.09E-02	0.66	6.23E-02	0.57	6.83E-02
	BRWT	0.84	4.22E-02	0.72	5.80E-02	0.67	5.93E-02
EC							
PLS	Raw	0.76	4.00E-07	0.59	1.20E-06	0.56	1.30E-06
	WT	0.48	1.80E-06	0.47	1.80E-06	0.28	3.40E-06
	BR	0.88	1.00E-07	0.67	6.00E-07	0.48	1.70E-06
	BRWT	0.73	5.00E-07	0.56	7.00E-07	0.63	9.00E-07
LSSVM	Raw	0.12	5.20E-06	0.13	5.20E-06	−0.13	8.30E-06
	WT	0.65	8.00E-07	0.16	5.00E-06	0.48	1.80E-06
	BR	0.68	7.00E-07	0.63	8.00E-07	0.66	7.00E-07
	BRWT	0.84	1.62E-07	0.67	3.92E-07	0.74	4.31E-07
CNN	Raw	0.90	1.64E-02	0.63	2.96E-02	0.67	2.88E-02
	WT	0.99	4.89E-03	0.39	3.40E-02	0.33	4.13E-02
	BR	1.00	1.50E-03	0.69	3.05E-02	0.73	2.62E-02
	BRWT	0.66	2.97E-02	0.72	2.81E-02	0.52	3.50E-02

(continued)

Table 2 (continued)

Models	Data type	Training set R_t^2	Training set $RMSE_t$	Validation set R_v^2	Validation set $RMSE_v$	Prediction set R_p^2	Prediction set $RMSE_p$
ECG							
PLS	Raw	0.77	8.76E-03	0.59	2.85E-02	0.56	3.13E-02
	WT	0.48	4.41E-02	0.46	4.64E-02	0.28	8.21E-02
	BR	0.88	2.44E-03	0.66	1.68E-02	0.48	4.26E-02
	BRWT	0.73	1.22E-02	0.57	1.87E-02	0.63	2.21E-02
LSSVM	Raw	0.13	1.24E-01	0.14	1.22E-01	−0.11	1.97E-01
	WT	0.65	2.01E-02	0.17	1.20E-01	0.48	4.31E-02
	BR	0.68	1.67E-02	0.62	2.02E-02	0.67	1.73E-02
	BRWT	0.84	4.15E-03	0.68	1.03E-02	0.74	1.06E-02
CNN	Raw	0.93	1.74E-01	0.66	3.53E-01	0.71	3.40E-01
	WT	0.92	1.76E-01	0.42	4.16E-01	0.54	4.30E-01
	BR	0.95	1.45E-01	0.83	2.81E-01	0.73	3.31E-01
	BRWT	0.96	1.30E-01	0.86	2.48E-01	0.79	2.86E-01
EGC							
PLS	Raw	0.06	2.19E-02	0.03	2.25E-02	0.05	2.28E-02
	WT	0.02	2.41E-02	−0.10	2.43E-02	−0.02	2.67E-02
	BR	0.05	2.25E-02	−0.03	2.43E-02	−0.04	2.72E-02
	BRWT	0.02	2.38E-02	−0.05	3.12E-02	0.06	2.26E-02
LSSVM	Raw	−3.20	4.39E-01	−3.08	3.98E-01	−3.68	5.57E-01
	WT	−0.49	5.51E-02	−5.18	7.56E-01	−3.43	5.00E-01
	BR	0.09	2.06E-02	0.00	2.42E-02	0.00	2.82E-02
	BRWT	0.40	8.85E-03	0.00	2.69E-02	0.14	1.88E-02
CNN	Raw	0.07	3.84E-01	0.00	3.97E-01	0.05	3.90E-01
	WT	0.23	3.49E-01	−0.18	4.67E-01	−0.07	4.13E-01
	BR	0.37	3.16E-01	−0.31	3.83E-01	−0.08	4.16E-01
	BRWT	0.22	3.50E-01	−0.03	4.15E-01	−0.08	4.14E-01
EGCG							
PLS	Raw	0.78	8.67E-05	0.55	3.55E-04	0.53	3.77E-04
	WT	0.46	4.99E-04	0.45	4.78E-04	0.27	8.99E-04
	BR	0.90	1.75E-05	0.70	1.25E-04	0.59	2.88E-04
	BRWT	0.76	1.03E-04	0.56	1.54E-04	0.68	1.71E-04
LSSVM	Raw	0.20	1.10E-03	0.20	1.14E-03	0.02	1.65E-03
	WT	0.66	2.05E-04	0.23	9.59E-04	0.48	4.54E-04
	BR	0.72	1.38E-04	0.66	1.72E-04	0.67	1.83E-04
	BRWT	0.87	3.13E-05	0.67	8.55E-05	0.78	8.22E-05
CNN	Raw	0.97	3.23E-02	0.69	1.10E-01	0.82	8.72E-02
	WT	1.00	1.86E-03	0.37	1.35E-01	0.58	1.31E-01
	BR	1.00	1.37E-02	0.90	7.18E-02	0.81	8.81E-02
	BRWT	0.98	2.59E-02	0.80	9.76E-02	0.82	8.60E-02

(continued)

Table 2
(continued)

Models	Data type	Training set R_t^2	$RMSE_t$	Validation set R_v^2	$RMSE_v$	Prediction set R_p^2	$RMSE_p$
CC							
PLS	Raw	0.72	2.40E-01	0.42	1.00E+00	0.43	9.77E-01
	WT	0.38	1.13E+00	0.36	1.09E+00	0.15	2.12E+00
	BR	0.82	9.91E-02	0.50	6.77E-01	0.28	1.55E+00
	BRWT	0.60	4.86E-01	0.37	7.55E-01	0.46	8.62E-01
LSSVM	Raw	−0.74	9.07E+00	−0.70	8.90E+00	−1.19	1.41E+01
	WT	0.41	1.02E+00	−0.84	1.05E+01	−0.06	3.30E+00
	BR	0.57	5.49E-01	0.50	6.82E-01	0.55	5.90E-01
	BRWT	0.75	1.83E-01	0.50	4.70E-01	0.57	5.51E-01
CNN	Raw	0.54	8.94E-01	0.36	1.01E+00	0.42	1.00E+00
	WT	0.73	6.83E-01	0.12	1.14E+00	0.29	1.11E+00
	BR	0.75	6.62E-01	0.65	7.96E-01	0.57	8.62E-01
	BRWT	0.71	7.06E-01	0.74	7.17E-01	0.58	8.54E-01

Table 3
Optimal prediction effect of 10 compounds

Tea compounds	Models	Data type	Training set R_t^2	$RMSE_t$	Validation set R_v^2	$RMSE_v$	Prediction set R_p^2	$RMSE_p$
C	CNN	BR	0.98	5.12E-02	0.85	1.37E-01	0.81	1.43E-01
TP	LS-SVM	BR + WT	0.77	4.11E-01	0.53	1.42E+00	0.50	1.97E+00
TB	CNN	BR	0.88	1.00E+00	0.70	1.69E+00	0.80	1.29E+00
TR	CNN	BR	0.83	4.02E-01	0.77	4.83E-01	0.67	5.61E-01
TF	CNN	BR + WT	0.84	4.22E-02	0.72	5.80E-02	0.67	5.93E-02
EC	CNN	BR	1.00	1.50E-03	0.69	3.05E-02	0.73	2.62E-02
ECG	CNN	BR + WT	0.96	1.30E-01	0.86	2.48E-01	0.79	2.86E-01
EGC	PLS	Raw	0.06	2.19E-02	0.03	2.25E-02	0.05	2.28E-02
EGCG	CNN	BR + WT	0.98	2.59E-02	0.80	9.76E-02	0.82	8.60E-02
CC	CNN	BR + WT	0.71	7.06E-01	0.74	7.17E-01	0.58	8.54E-01

2. 1 g (±0.005 g) of tea powder was added to a centrifuge tube and mixed with 10 mL alcohol (10%). After the mixture was shaken vigorously for 20 min, it was centrifuged at 223.6 × *g* and 4 °C for 5 min. The supernatant was collected as tea extract. In this

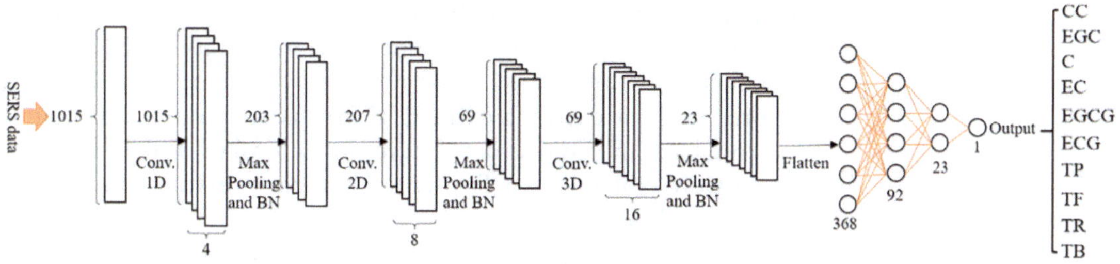

Fig. 3 Schematic diagram of the structure of the convolutional neural network

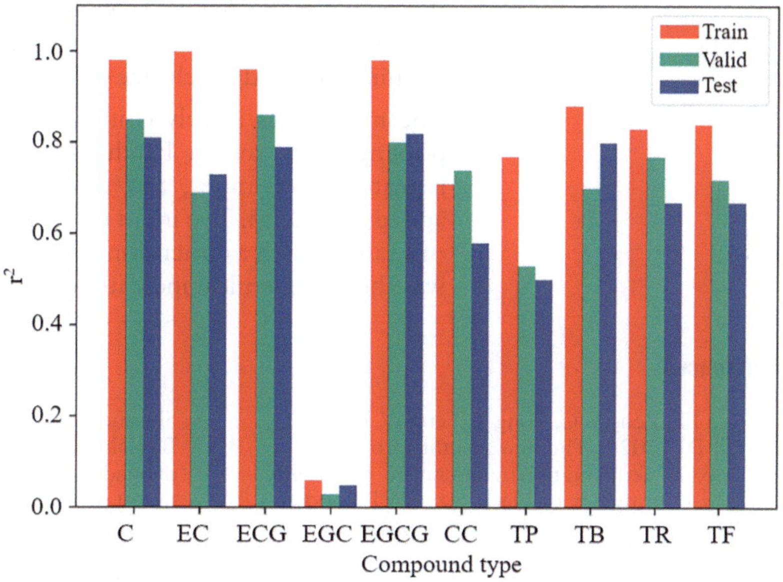

Fig. 4 Use the best modeling method to establish the quantitative model results of ten tea quality indicators

study, samples at each time point were used to prepare three tea extracts as described above, and a total of 30 tea extracts were prepared.

3. The preparation process of the substrate was as follows: 0.18 mg/mL silver nitrate ($AgNO_3$) in ultrapure water was heated to boiling at 100 °C. Then, 1% trisodium citrate was dropped on the boiling $AgNO_3$ solution every 5 s for up to 15 min. During the reaction, the solution changed from transparent to light brown and then to grayish green. After cooling at 25 °C, the above solution was centrifuged, and the supernatant was discarded. Then, ultrapure water was added, and the solution was shaken for 5 min. This process was repeated three times, and the obtained solution was stored in the dark.

4. The objective lens was 50×, the exposure time was 1 s, and the laser intensity was 1 mV. Since the instability of SERS may affect the robustness of the model, two mixed droplets were prepared for each tea extract. Then, six spectra were obtained from each mixed droplet. Therefore, a total of 360 SERS spectra were collected for fermented tea leaves extracts for further data analyses.
5. The chapter choose Biorthogonal wavelet as Wavelet basis.
6. Fivefold cross-validation was used to optimize the hyperparameters of the PLS and LSSVM models, including the number of principal components, the LSSVM kernel function type, and the punishment factor. Twenty percent of the data from the training set were defined as the validation set to determine the learning rate and batch size of the 1D-CNN.
7. The model was mainly composed of a convolutional layer, a pooling layer, and a batch norm layer. This structure was repeated three times, and three fully connected layers were added in the end. The input of the model was a single spectrum of a sample with a dimension of 1015. The output of the model was the ten chemical indicators of the sample.

References

1. Qu FF, Zeng WC, Tong X, Feng W, Chen YQ, Ni DJ (2020) The new insight into the influence of fermentation temperature on quality and bioactivities of black tea. LWT-Food Sci Technol 117:108646. https://doi.org/10.1016/j.lwt.2019.108646
2. Fu JY, Jiang CX, Wu MY, Mei RY, Yang AF, Tao HP, Chen XJ, Zhang J, Huang L, Zhao XF (2021) Theabrownin induces cell apoptosis and cell cycle arrest of oligodendroglioma and astrocytoma in different pathways. Front Pharmacol 12:664003. https://doi.org/10.3389/fphar.2021.664003
3. Li BR, Fu L, Kojima R, Yamamoto A, Ueno T, Matsui T (2021) Theaflavins prevent the onset of diabetes through ameliorating glucose tolerance mediated by promoted incretin secretion in spontaneous diabetic Torii rats. J Funct Foods 86:104702. https://doi.org/10.1016/j.jff.2021.104702
4. Tan JF, Dai WD, Lu ML, Lv HP, Guo L, Zhang Y, Zhu Y, Peng QH, Lin Z (2016) Study of the dynamic changes in the non-volatile chemical constituents of black tea during fermentation processing by a non-targeted metabolomics approach. Food Res Int 79:106–113. https://doi.org/10.1016/j.foodres.2015.11.018
5. Sharma P, Ghosh A, Tudu B, Sabhapondit S, Baruah BD, Tamuly P, Bhattacharyya N, Bandyopadhyay R (2015) Monitoring the fermentation process of black tea using QCM sensor based electronic nose. Sens Actuat B Chem 219:146–157. https://doi.org/10.1016/j.snb.2015.05.013
6. Jin G, Wang YJ, Li MH, Li TH, Huang WJ, Li LQ, Deng WW, Ning JM (2021) Rapid and real-time detection of black tea fermentation quality by using an inexpensive data fusion system. Food Chem 358:129815. https://doi.org/10.1016/j.foodchem.2021.129815
7. Wu QJ, Dong QH, Sun WJ, Huang Y, Wang QQ, Zhou WL (2014) Discrimination of Chinese teas with different fermentation degrees by stepwise linear discriminant analysis (S-LDA) of the chemical compounds. J Agric Food Chem 62:9336–9344. https://doi.org/10.1021/jf5025483
8. Jin G, Wang YJ, Li LQ, Shen SS, Deng WW, Zhang ZZ, Ning JM (2020) Intelligent evaluation of black tea fermentation degree by FT-NIR and computer vision based on data fusion strategy. LWT-Food Sci Technol 125:109216. https://doi.org/10.1016/j.lwt.2020.109216

9. Oliveira PK, Tosato MG, Alves RDS, Martin AA, Fávero PP, Raniero L (2012) Skin biochemical composition analysis by Raman spectroscopy Análise da composição bioquímica da pele por espectroscopia Raman. Rev Bras Eng Bioméd 28:278–287. https://doi.org/10.4322/rbeb.2012.032
10. Huang CC, Chen WL (2018) A SERS method with attomolar sensitivity: a case study with the flavonoid catechin. Microchim Acta 185:120. https://doi.org/10.1007/s00604-017-2662-9
11. Xia J, Wang D, Liang P, Zhang D, Du XQ, Ni DJ, Yu Z (2020) Vibrational (FT-IR, Raman) analysis of tea catechins based on both theoretical calculations and experiments. Biophys Chem 256:106282. https://doi.org/10.1016/j.bpc.2019.106282
12. Zhang L, Zhao YP, Feng YD (2013) Identification of pu'er ripe teas with different origins and fermentation years by surface-enhanced Raman spectroscopy. Spectrosc Spectr Anal 33:1575–1580. https://doi.org/10.3964/j.issn.1000-0593(2013)06-1575-06
13. Qi X, Wang X, Dong Y, Xie J, Gui X, Bai J, Duan JL, Liu J, Yao HJ (2022) Fast synthesis of gold nanostar SERS substrates based on ion-track etched membrane by one-step redox reaction. Spectrochim Acta A 272:120955. https://doi.org/10.1016/j.saa.2022.120955
14. Lee PC, Meisel D (1982) Adsorption and surface-enhanced Raman of dyes on silver and gold sols. J Phys Chem 86:3391–3395. https://doi.org/10.1021/j100214a025
15. Yao LH, Jiang YM, Caffin N, D'Arcy B, Datta N, Liu X, Singanusong R, Xu Y (2006) Phenolic compounds in tea from Australian supermarkets. Food Chem 96:614–620. https://doi.org/10.1016/j.foodchem.2005.03.009
16. Sanaeifar A, Huang XY, Chen MY, Zhao ZF, Ji YF, Li XL, He Y, Zhu Y, Chen X, Yu XX (2020) Nondestructive monitoring of polyphenols and caffeine during green tea processing using Vis-NIR spectroscopy. Nutr Food Sci 8:5860–5874. https://doi.org/10.1002/fsn3.1861
17. Zhang C, Wu WY, Zhou L, Cheng H, Ye XQ, He Y (2020) Developing deep learning based regression approaches for determination of chemical compositions in dry black goji berries (Lycium ruthenicum Murr.) using near-infrared hyperspectral imaging. Food Chem 319:126536. https://doi.org/10.1016/j.foodchem.2020.126536
18. Lopes GVD, Moraes GR, Issicaba D, Dotta D (2022) WAMS-based two-level robust detection methodology of power system events. Sustain Energy Grids 31:100689. https://doi.org/10.1016/j.segan.2022.100689
19. Doxaran D, Froidefond JM, Castaing P (2002) A reflectance band ratio used to estimate suspended matter concentrations in sediment-dominated coastal waters. Int J Remote Sens 23:5079–5085. https://doi.org/10.1080/01431160210000009912
20. Wang YC, Jin YT, Wang XN, Liao QH, Gu XH, Zhao ZH, Yang XF (2018) Quantitative inversion of organic matter content based on interconnection traditional spectral transform and continuous wavelet transform. Spectrosc Spectr Anal 38:2571–2577. https://doi.org/10.3964/j.issn.1000-0593(2018)08-2571-07

Chapter 11

Terahertz for Tracking the Cadmium Heavy Metal Accumulation During Pak Choi (*Brassica chinensis*) Food Production Processes

Yinglei Zhao, Mostafa Gouda, Lei Lin, and Yong He

Abstract

Detecting heavy metal–phytochelatin complexes in plants would simplify identification of cultivars with both high tolerance and enrichment abilities for heavy metals. In this study, we introduced a novel approach for the characterization and detection of Cd–PC_2 complexes using terahertz spectroscopy, density functional theory, circular dichroism, and chemometrics, where the phytochelatin2 (PC_2) was innovatively used as a natural probe to amplify terahertz signals of the trace Cd. Results showed that PC_2 chelates Cd^{2+} to form $Cd(PC_2)_2$ complex; the Cd connected to thiols of the PC_2 and changed β-turn and random coil of the PC_2 peptide chain to β-sheet, which presented as terahertz vibrations of the PC_2 around 1.03 and 1.71 THz being suppressed. This study demonstrated that the terahertz spectroscopy can be utilized to visualize and quantify chelated Cd in pak choi (*Brassica chinensis*) leaves, with a detection limit down to 1.151 ppm, and suggests that terahertz information of the complexes can be used as a simpler alternative to classical indices for phytoextraction performance evaluation and cultivar identification.

Key words Terahertz spectroscopy, Circular dichroism, Phytochelatin, Cadmium complexes

1 Introduction

Cadmium, a well-known hazardous heavy metal, is rarely 100% bioavailable in soil [1]. Thus, phytoextraction that involves planting various accumulation plants in Cd-contaminated soil, has been developed as a promising strategy to extract the contaminants from soil [2]. When Cd^{2+} enter plants, the ions are primarily chelated by thiols of the phytochelatins (PCs) to form stable and nonionic complexes (–S–Cd– bonding) that deposit in plants [3]. The Cd–PCs complexes are commonly detected using reverse-phase high-performance liquid chromatography (RP-HPLC) and inductively coupled plasma mass spectrometry (ICP-MS) [4]. However, these procedures are time-consuming and involve acidification and derivatization, which lead to dissociations of the complexes.

Mostafa Gouda et al. (eds.), *Plant Chemical Compositions and Bioactivities*, Methods and Protocols in Food Science,
https://doi.org/10.1007/978-1-0716-3938-2_11,

Furthermore, both the proportion of the PCs chelating Cd and the molecular weight of the complexes are poorly characterized [5].

Terahertz (THz) radiation employs electromagnetic waves from 0.1 to 10.0 THz to capture intermolecular interactions, lattice vibrations, skeletal vibrations, and rotational transitions of various vibrational frequencies. The terahertz spectral bands of analytes provide unique fingerprint information that can be harnessed to quantify the analytes and identify conformation of the analytes. As low energy photon of the terahertz radiation (1–10 me/V) does not cause any ionization damage to biomolecules, the terahertz-time domain spectroscopy (THz-TDS) provides a simple sample preparation, label-free, and nondestructive spectroscopic method for the characterization and visualization of biomolecules [6].

Despite the potential of terahertz spectroscopy, current materials used for terahertz radiation generator and receptor produce signals with high noise rates. The noise leads to deviation and deterioration of the fingerprint spectra, and negatively influences the chemometric analysis. Interpretation of the terahertz spectra remains a challenging task. Density functional theory (DFT) has been reported as an effective computational method for simulating the low-frequency vibration of biomolecules under radiational stimulation, aiding in identification and calibration of fingerprint peaks that enables more practical measurements.

Biomacromolecules are characterized by their complex structures and various chemical bonds, which contain vast amounts of terahertz information. Heavy metal atoms chelated by the biomacromolecules exhibit more distinctive spectral features. In this study, PC_2 was utilized as a natural probe to stabilize the chemical status of trace cadmium atoms, and amplify terahertz signals of the atoms for more precise detection. Specifically, structures and spectral features of the Cd–PCs complexes were analyzed using terahertz spectral information, DFT simulation, and circular dichroism. This study provided a general method for using terahertz spectrometry to analyze structures of the heavy metal–phytochelatin complexes, and precisely detect the complexes in plants for phytoextraction performance evaluation.

2 Materials

2.1 Chemicals

1. Cadmium chloride (CAS: 10108-64-2, Aladdin Corp., Shanghai, China).
2. PC_2 (CAS: 95014-75-8, Yeyuan Corp., Shanghai, China).
3. Polyethylene powders (CAS: 9002-88-4, Sigma-Aldrich Corp., USA).

2.2 Pak Choi Cultivars

1. Hangyoudong (HY), Shanghaiqing (SH), and Suzhouqing (SU) are the native pak choi cultivars in China, which are selected for the test.

2.3 Equipment

1. Tablet machine.
2. CCT-1800 terahertz time-domain spectrometer system (China Communication Technology Co., Ltd., Shenzhen, China) was used. The terahertz radiation enables the spectrometer to output reliable photoelectric absorption signals with a spectral resolution of 20 GHz in 0.1–5 THz.
3. The J-1500 circular dichroism spectrometer (Jasco, Japan) was used for acquiring secondary structure information.
4. The Agilent Infinity 1260 II SEC-HPLC system (Agilent Technologies, Santa Clara, CA, USA) was used to separate the Cd–PC_2 complex.
5. The TSK-GEL G3000 (5 μm particles; TOSOH, City, Japan) and TSK-GEL Guard SWXL (7 μm particles; TOSOH) were used as the chromatographic and guard columns, respectively.
6. The Agilent Technologies 7800 ICP-MS system (Agilent Technologies, Santa Clara, CA, USA) was used to detect target atoms.

3 Methods

3.1 Preparation of Standard Samples

1. 36.6 mg cadmium chloride and 5 mg PC_2 standards were dissolved in 1 mL and 43.1 μL deionized water, respectively, to 0.2 mol/L, which were defined as the 10^6 stock standard. The water was used to dilute the standard to 10^n stock standards, which were defined as nx standard samples ($n = 0, 1, 2, 3, 4$, and 5).
2. The 4x $CdCl_2$ standard was mixed with the 4× PC_2 standard in the specific volumes listed in Table 1, to obtain Cd–PC_2 mixture samples.
3. 10 μL of the samples were mixed with 0.1 g polyethylene (PE) powders, dried in 4 °C, and were pressed into pellet with 0.5 mm thick by the tablet machine with 30 MPa.

3.2 Preparation of Plant Samples

1. $CdCl_2$ was added to peat moss to 10 mg/kg.
2. The pak chois were raised in the peat moss in greenhouse by routine management. Pak chois without $CdCl_2$ stress were raised for the control set.
3. The 20-, 30-, and 40-day-old leaves were harvested and dried at 30 °C for 8 h. The leaves were ground into powders (pure leaf powders, PLP). 0.1 g of the powders was mixed with 10 μL

Table 1
Experimental design for the Cd–PC_2 mixture samples

Ratio ($PC_2/CdCl_2$ = *n*/*n*)	Water for dilution (μL)	Standards added to mixture (μL)	
		PC_2	$CdCl_2$
4:1	5.00	4.00	1.00
3:1	4.67	4.00	1.33
2:1	4.00	4.00	2.00
1:1	2.00	4.00	4.00
1:1.5	0.00	4.00	6.00

of the 0×, 1×, 2×, 3×, and 4× Cd–PC_2 complex standard (ST), respectively, and were dried and pressed into pellets by the tablet machine with 30 MPa to simulate leaf samples.

3.3 Density Functional Theory Computation

1. Molecule structures of $CdCl_2$ and PC_2 (supported by m.chemicalbook.com) were simulated in Gaussview 6.0.
2. PC_2 was simulated by the DFT with the base set of B3LYP/6-311+g(d,p), and $CdCl_2$ was simulated with the base set of pbepbe/lanl2dz in Gaussian16 (Semichem Inc., Shawnee Mission, KS, USA) [7].

3.4 Terahertz Spectrum and Image Acquisition

1. For transmission scanning the tablet samples, the sample thickness was set to 0.5 mm and accumulation, to 20 times, in control software.
2. For transmission imaging pak choi samples, the leaves were fixed on polyethylene film and dried in airflow oven at 30 °C for 8 h, the sample thickness was set to 0.05 mm and scan step, to 0.2 mm, accumulation, to 20 times [8].

3.5 Spectral Data Extraction and Processing

1. Raw absorbance data were developed by loading scanned data into Tdsanalyzer software and were loaded into Unscrambler X software (CAMO AS, Norway) for noise reduction by using the gaussian filter smoothing (GFS) function.
2. Parameter segment size (number of points) was set to 9 for the GFS. Relative actions were operated in the Unscrambler X software.

3.6 Chemical Distribution Reimaging and Content Prediction

1. Average spectrum of leaf and content of Cd–PC_2 complexes in the leaf were calculated by Eqs. 1 and 2, respectively, where the A, m, and S represent the average spectrum of selected leaf area, number of pixels of the area, and spectrum for each pixel in the area, respectively. The C, w, and a represents the Cd–PC_2 content of the leaf area, dry weight of the area, and molecular

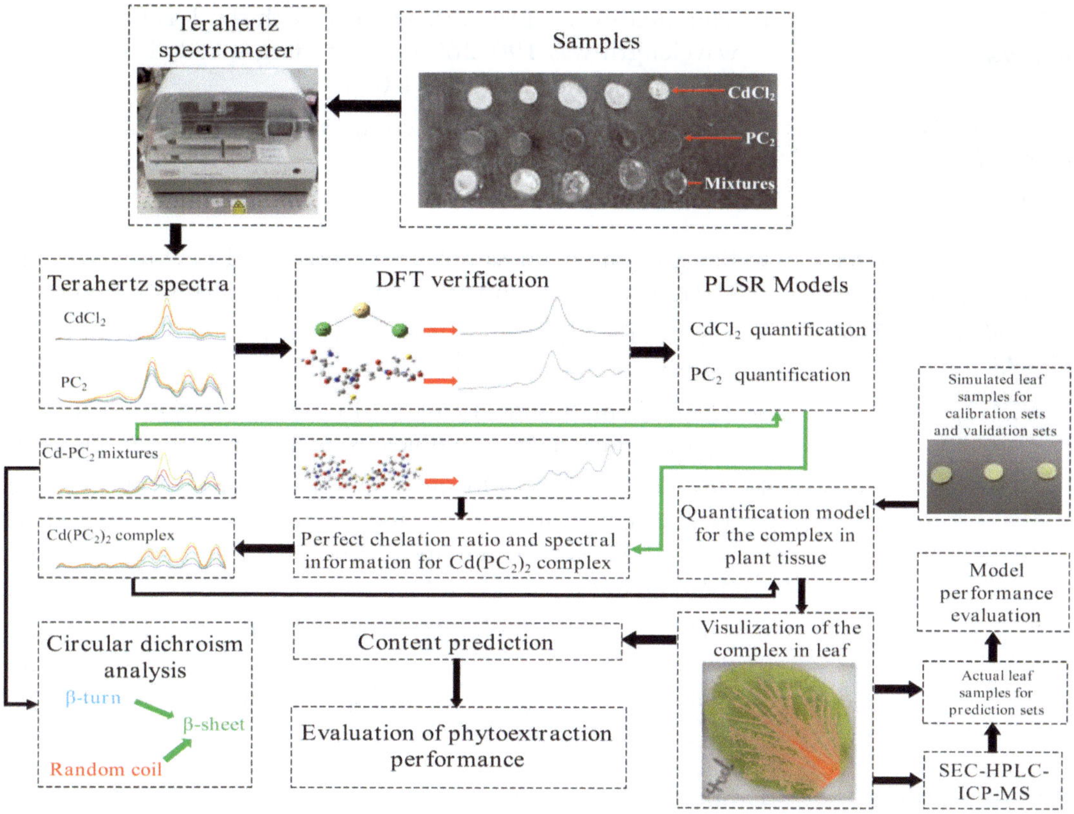

Fig. 1 The flowchart of analysis protocols

weight of the Cd–PC_2 complex, respectively. The n values for each of the pixels were calculated by the optimal Cd–PC_2 model, and the x is the mean value of the n. The b is the weight of simulated leaf samples and is 0.1 g here; the c is the concentration of the 10^6 stock Cd–PC_2 standards and is 0.2 mol/L here; and the d is the volume of the Cd–PC_2 mixture samples and is 10 μL here. The distribution of Cd–PC_2 in pak choi leaf was visualized by heat mapping the n values at pixel level, and relative actions were operated in Origin 2018 software. Fig. 1 shows the protocol overall flowchart.

$$A = \frac{\sum S_m}{m} \quad (1)$$

$$C = \frac{w \times a \times c \times d \times 10^{x-6}}{b} \quad (2)$$

3.7 Circular Dichroism Spectrometry for Assessing Structures of the Cd–PC2 Mixtures

1. The mixture samples were scanned with the J-1500. Scanning wavelength was 190–260 nm; scanning speed, 100 nm/min; DIT, 1 s; and band width, 1.0 nm.
2. The secondary structure composition was calculated using Yang's algorithm [9].

3.8 Detection for the Cd–PC2 Complex and Total Cd by SEC-HPLC-ICP-MS

1. 0.5 g of the plant samples was suspended in 5 mL Tris–HCl (10 mM, pH 7.5) and was centrifuged for 10 min at 10,000× *g*.
2. 1 mL of the supernatant was filtered by using a 0.20 μm syringe membrane filter.
3. The Agilent Infinity 1260 II SEC-HPLC system was used to separate the Cd–PC_2 complex using Tris–HCl as mobile phase.
4. The TSK-GEL G3000 and TSK-GEL Guard SWXL were used as the chromatographic and guard columns, respectively [10]. Peaks with a retention time of 8.8–9.1 min belong to the target complex [11].
5. The Agilent Technologies 7800 ICP-MS system was used to detect ^{111}Cd in the separated complexes. In addition, the standard curve was established for determining Cd of the Cd–PC_2 complex [12].

3.9 Data Analysis

3.9.1 Multivariate Modeling

1. Spectra data of the standard solution samples were used for calibrating partial least square regression (PLSR) models for analyzing terahertz characteristics of the Cd–PC_2 mixture samples.
2. Spectra data of the simulated leaf samples were used for calibrating PLSR models for detecting and visualizing the Cd–PC_2 complex in the pak choi leaves.
3. Successive projections algorithm (SPA) was implemented to identify signature variables from full spectra [13]. Relative actions were operated in the Unscrambler X (CAMO AS, Oslo, Norway) software.

3.9.2 Evaluation of Model Performance and Accuracy

1. Performance of the PLSR models were evaluated by examining R_V^2 and $RMSE_V$ coefficients of the validation sets and R_P^2 and $RMSE_P$ coefficients of the prediction sets [14].
2. Metaboanalyst software (MetaboAnalystR, CA) was used for analyzing variable importance in projection (VIP) and prediction (Q2) by a PLS-DA model [15].

4 Notes

1. Curve of absorbance was defined as the terahertz spectrum of sample.
2. Base signal of the PE powders and film were loaded as background.
3. Transmission scanning and imaging for the samples was conducted at 20 °C in a > 90% nitrogen environment.
4. The experimental spectra were divided into calibration set and validation set by a ratio of 4:1.
5. Models with high R_V^2 and R_P^2 values while low $RMSE_V$ and $RMSE_P$ values are considered as well performing.
6. The PLS-DA model shows a good prediction if Q2 > 0.5.

References

1. Hamid Y, Tang L, Sohail MI, Cao X, Hussain B, Aziz MZ, Usman M, He Z-L, Yang X (2019) An explanation of soil amendments to reduce cadmium phytoavailability and transfer to food chain. Sci Total Environ 660: 80–96
2. Mahar A, Wang P, Ali A, Awasthi MK, Lahori AH, Wang Q, Li R, Zhang Z (2016) Challenges and opportunities in the phytoremediation of heavy metals contaminated soils: a review. Ecotoxicol Environ Saf 126:111–121
3. Su H, Zou T, Lin R, Zheng J, Jian S, Zhang M (2020) Characterization of a phytochelatin synthase gene from Ipomoea pes-caprae involved in cadmium tolerance and accumulation in yeast and plants. Plant Physiol Biochem 155:743–755
4. Delafiori J, Ring G, Furey A (2016) Clinical applications of HPLC–ICP-MS element speciation: a review. Talanta 153:306–331. https://doi.org/10.1016/j.talanta.2016.02.035
5. Dennis KK, Uppal K, Liu KH, Ma C, Liang B, Go Y-M, Jones DP (2019) Phytochelatin database: a resource for phytochelatin complexes of nutritional and environmental metals. Database 2019:baz083
6. Zhao YL, Zhang JN, Gouda M, Zhang CH, Lin L, Nie PC, Ye HB, Huang W, Ye YX, Zhou CQ, He Y (2022) Structure analysis and non-invasive detection of cadmium-phytochelatin2 complexes in plant by deep learning Raman spectrum. J Hazard Mater 427
7. Khajehzadeh M, Moghadam M (2017) Molecular structure, FT IR, NMR, UV, NBO and HOMO-LUMO of 1-(3-(dimethylamino)-propyl)-1-(4-fluorophenyl)-1,3-dihydroisobenzofuran-5-carbonitrile by DFT/B3LYP and PBEPBE methods with LanL2DZ and 6-311++G(d,2p) basis sets. Spectrochim Acta A Mol Biomol Spectrosc 180:51–66
8. Nie P, Cai C, Qu F, Lin L, He Y (2019) Study of 2,4-D spectral characteristics and its detection in Zizania Latifolia using terahertz time-domain spectroscopy. Appl Sci 9:2248
9. Zhao Y, Hu M, Gao Z, Chen X, Huang D (2018) Biological mechanisms of a novel hydro-electro hybrid priming recovers potential vigor of onion seeds. Environ Exp Bot 150: 260–271
10. Munoz-Almagro N, Rico-Rodriguez F, Villamiel M, Montilla A (2018) Pectin characterisation using size exclusion chromatography: a comparison of ELS and RI detection. Food Chem 252:271
11. Yang H-X, Liu W, Li B, Wei W, Zhang H-J, Chen D-Y (2009) Speciation analysis of cadmium in Indian mustard (brassica Juncea) by size exclusion chromatography-high performance liquid chromatography-inductively coupled plasma mass spectrometry. Chin J Anal Chem 37:1511–1514. https://doi.org/10.1016/S1872-2040(08)60137-1
12. Pinheiro FC, Babos DV, Barros AI, Pereira-Filho ER, Nóbrega J (2019) Microwave-assisted digestion using dilute nitric acid solution and investigation of calibration strategies for determination of As, Cd, Hg and Pb in dietary supplements using ICP-MS. J Pharm Biomed Anal 174:471–478

13. Feng X, Yu C, Shu Z, Liu X, Yan W, Zheng Q, Sheng K, He Y (2018) Rapid and non-destructive measurement of biofuel pellet quality indices based on two-dimensional near infrared spectroscopic imaging. Fuel 228:197–205. https://doi.org/10.1016/j.fuel.2018.04.149
14. Jiang Y, Ge H, Zhang Y (2020) Quantitative analysis of wheat maltose by combined terahertz spectroscopy and imaging based on Boosting ensemble learning. Food Chem 307:125533
15. Yang S, Wang X, Duan C, Zhang J (2020) A novel approach combining metabolomics and molecular pharmacology to study the effect of Gei Herba on mouse hematopoietic function. Biomed Pharmacother 129:110437

Chapter 12

Use of Inductively Coupled Plasma Mass Spectrometry (ICP-MS) to Assess the Levels of Phosphorus and Cadmium in Lettuce

Kamel Mohamed Eltohamy and Mohamed Farag Taha

Abstract

Inductively coupled plasma mass spectrometry (ICP-MS) is a highly sensitive and reliable technique widely employed for the analysis of trace elements in complex matrices. While traditional detection methods such as atomic absorption and atomic emission are still used in some scientific circles, there has been a gradual transition toward the utilization of ICP-MS in plant analysis over the past decade. This chapter presents a comprehensive protocol outlining the application of ICP-MS for the assessment of phosphorus (P) and cadmium (Cd) levels in lettuce samples. The protocol encompasses crucial steps such as sample preparation, digestion, calibration, instrument setup, sample analysis, quality control, and data analysis. By employing this method, the levels of P and Cd in lettuce and other food samples can be monitored effectively, ensuring both safety and nutritional considerations. This chapter aims to provide researchers and analysts with a valuable resource for the implementation of ICP-MS in plant analysis and contribute to the advancement of food safety and quality assessment.

Key words ICP-MS, Cadmium, Phosphorus, Lettuce, Trace elements, Food analysis

1 Introduction

Food safety and quality are major concerns for consumers and public health agencies worldwide [1, 2]. The presence of trace elements in food samples can have both positive and negative effects on human health [3]. While some elements are essential nutrients that are required for proper bodily functions, others are toxic and can lead to adverse health effects when consumed in excess [3, 4].

Lettuce (*Lactuca sativa* L.) is a nutritious vegetable widely included in a healthy diet due to its abundance of vitamins and minerals, including vitamin C, iron, potassium, and calcium [5–7]. However, the quality and freshness of lettuce can rapidly deteriorate, making it essential for lettuce industries to focus on

Mostafa Gouda et al. (eds.), *Plant Chemical Compositions and Bioactivities*, Methods and Protocols in Food Science,
https://doi.org/10.1007/978-1-0716-3938-2_12,

improving its shelf life. Compared to cereals, grasses, and legumes, fast-growing leafy plants like lettuce have a higher tendency to accumulate metallic elements [8]. As a result, lettuce is highly sensitive to the presence of trace elements in the soil, such as phosphorus (P) and cadmium (Cd), which can accumulate in the edible parts [9–11]. The contamination of lettuce with these trace elements poses potential health risks that must be addressed. Therefore, it becomes imperative to quantitatively measure the total content of each component present in lettuce to ensure food safety and quality.

Phosphorus is an essential nutrient crucial for numerous physiological processes, including energy metabolism, bone mineralization, and DNA synthesis [12–14]. While P is necessary for proper bodily functions, excessive intake can have negative consequences, such as impaired kidney function, cardiovascular disease, and bone loss [15, 16]. On the other hand, Cd poses a significant health risk [17]. Its presence in food originates from the transfer of Cd from the soil to plants [18]. Additionally, various human and industrial activities contribute to both P and Cd exposure, including the use of nickel–cadmium batteries, application of manure and phosphate fertilizers, mining activities, recycling of electronic waste, and sewage sludge disposal [17–20]. Research studies have shown that excessive P and Cd intake can lead to adverse health effects, including impaired kidney function, cardiovascular disease, apoptosis, and bone loss [19, 21–23]. For instance, a study by Da et al. [23] specifically investigated the correlation between dietary P intake and kidney function in non-dialysis-dependent patients with chronic kidney disease, and the results demonstrated that each 1 mg/dL increase in serum P level independently raised the risk of kidney failure and mortality. Moreover, a study estimating the mean dietary exposure to Cd in Chinese individuals reported an average intake of 12.8 ± 4.2 μg/day, with vegetables, seafood, and rice identified as the primary sources responsible for 86.3% of the total exposure, and vegetables alone accounting for 40.2% [21]. These findings underscore the importance of monitoring and managing the intake of P and Cd to mitigate potential health risks. Following the Food and Agriculture Organization/World Health Organization (FAO/WHO) guidelines for both P and Cd intake can help ensure the safety and well-being of individuals [18].

Inductively coupled plasma mass spectrometry (ICP-MS) is a highly sensitive and robust analytical technique that has been widely used for the determination of multielemental composition (qualitatively), concentration (quantitatively), large analytical range (*see* Fig. 1), and isotopic abundances of plants and food samples [24]. Generally, an ICP-MS instrument consists of (1) a sample introduction system, (2) an ion source (inductively coupled plasma, ICP), (3) the electrostatic lenses, (4) an interface, (5) a mass spectrometer, and (6) a detector. Figure 2 depicts a typical

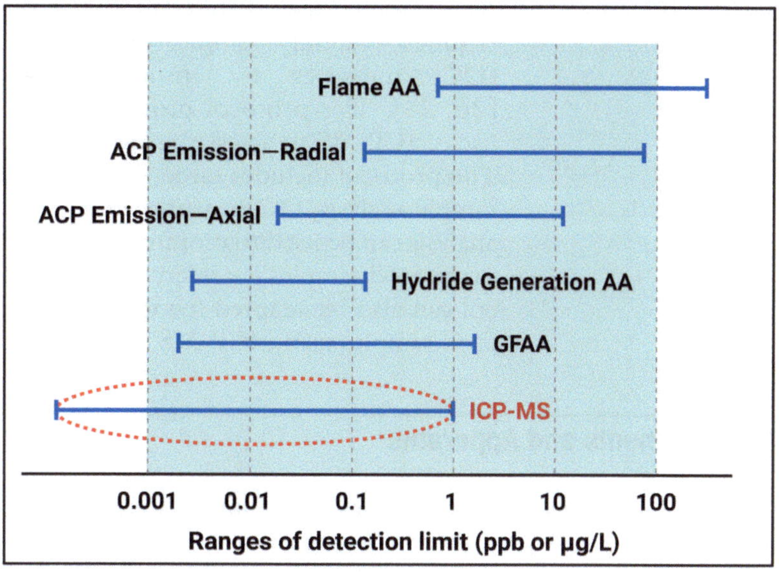

Fig. 1 Detection limit ranges of the various atomic spectroscopy techniques

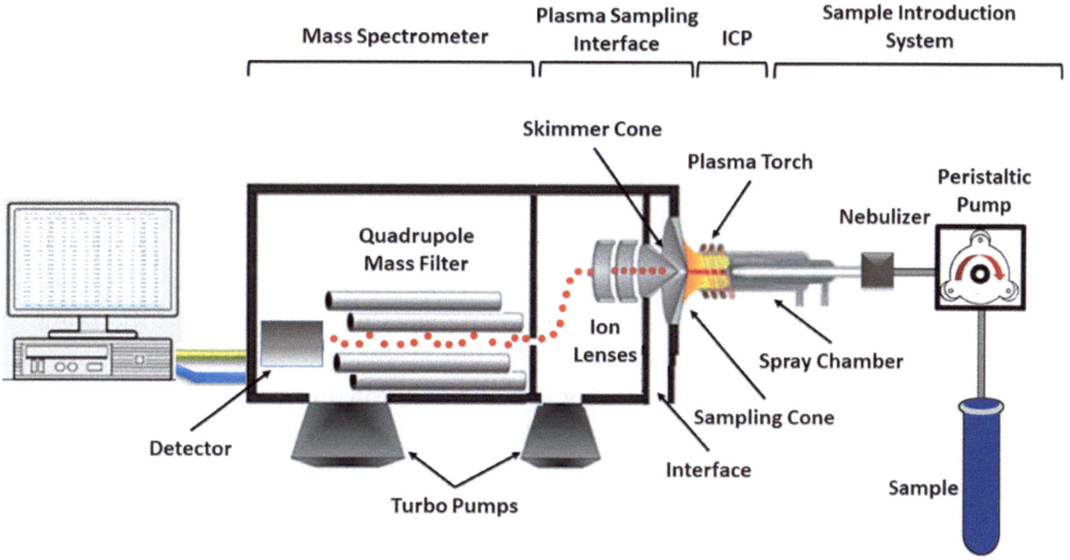

Fig. 2 Schematic diagram of the main components of an ICP-MS [27]

schematic representation of the basic components of ICP-MS. This technique involves the use of a plasma source to ionize the sample, which is then separated and detected by a mass spectrometer [24, 25]. ICP-MS has been used to determine the levels of Cd and P in various food samples, including lettuce [26].

However, the use of ICP-MS for the analysis of food samples requires careful sample preparation and quality control (QC) measures to ensure accurate and precise results [26, 28]. This protocol provides a standardized method for the use of ICP-MS to assess the levels of P and Cd in lettuce samples. The protocol includes sample preparation, instrument calibration, sample analysis, QC measures, and data analysis. The results of this analysis can be used to monitor the levels of P and Cd in lettuce and other food samples for safety and nutritional purposes. This protocol can also be adapted for the analysis of other trace elements in food samples using ICP-MS.

2 Instruments and Apparatus

Note: Equivalent equipment with comparable specifications may be used as a substitute.

1. *Analytical balance*: An analytical balance with a precision of 0.1 mg is used to accurately measure the mass of samples and reagents.
2. *Glassware*: Various glassware items such as flasks and beakers are employed for sample preparation and acid digestion.
3. *Microwave digestion system*: A microwave digestion system is utilized to perform the mineralization process on the lettuce samples. This system employs microwave energy to rapidly heat the samples, facilitating organic matter breakdown and sample matrix digestion. The microwave digestion system should have programmable settings for precise control of digestion parameters.
4. *PTFE digestion vessel*: A digestion vessel made of polytetrafluoroethylene (PTFE) material is necessary to hold the homogenized lettuce sample during the digestion process.
5. *Polymeric volumetric flasks*: Polypropylene or polymethylpentane volumetric flasks (50, 100, and 500 mL) are required for transferring the digested sample solutions after the digestion process is complete, as well as for the preparation of calibration standards and fortification standards.
6. *Timer*: A timer is used to accurately monitor the duration of the digestion process. The recommended digestion time typically ranges from 30 to 60 min, depending on the microwave system type and digestion program used.
7. *Filter paper*: Whatman No. 42 or an equivalent filter paper is utilized for the filtration of the digested samples.
8. *Vortex mixer*: A vortex mixer is used to ensure homogeneous mixing of the samples and reagents.

9. *Pipettes and micropipettes*: Pipettes and micropipettes are utilized for accurate measurement of volumes during sample and standard solution preparation.
10. *Polypropylene bottles*: Polypropylene bottles are used for storage and handling of reagents and solutions.
11. *Polypropylene centrifuge tubes*: Polypropylene centrifuge tubes are employed for sample processing and separation.
12. *Stirring rods (optional)*: Teflon or polypropylene stirring rods may be used for mixing purposes, if needed.
13. *ICP-MS instrument*: An ICP-MS instrument with an appropriate sample introduction system is utilized for the analysis of P and Cd in the lettuce samples. It incorporates ionization, mass discrimination, and detection processes to determine the elemental composition. The optimal operating conditions for the ICP-MS are presented in Table 1 [26, 29, 30].
14. *Computer software*: Computer software is used to set up and control the instrument parameters, such as ion lens voltages, mass range, and dwell time. It also facilitates data analysis and interpretation of the instrument response.

3 Reagents and Acids

Note: All reagents and acids should be of high purity, preferably ultrapure grade, to minimize the introduction of impurities during the digestion process. Proper handling and storage of these chemicals should be followed to ensure safety and accuracy in the analysis.

- Nitric acid (HNO_3), 69–70%, instra quality, J.T. Baker, conc (specific gravity 1.41), analytical grade.
- Hydrochloric acid (HCl), conc. (specific gravity 1.19), analytical grade.
- Perchloric acid ($HClO_4$), 70%, analytical grade.
- Hydrogen peroxide (H_2O_2), 30%, analytical grade.
- P standard solution, 1000 mg/L, certified reference material (CRM).
- Cd standard solution, 1000 mg/L, CRM.
- Internal standard solution, 10 μg/L, Yttrium (Y) in 2% HNO_3.
- Ultrapure water, Milli-Q (resistivity 18.2 MΩ/cm) or equivalent will be used for all dilutions.

Table 1
Operating conditions for ICP-MS analysis [26, 29, 30]

Perkin Elmer Elan DRC II	
Spray chamber	Cyclonic
Nebulizer	Meinhard®
Radiofrequency (RF) power (W)	1100
RF frequency (MHz)	27.12
RF matching (V)	1.72
Nebulizer pump (rps)	0.1
Sample intake (mL/min)	0.5
Spray chamber temperature (°C)	2
Extract 1 (V)	3.3
Extract 2 (V)	−110
Cell entrance (V)	−30
QP focus (V)	−10
Cell exit (V)	−40
Background	<5 cps (9 amu)
Isotope ratio precision (CeO/Ce) (%)	<3%
Short-term stability	<3% RSD
Long-term stability	<4% RSD/2 h
Sensitivity (cps/ppm)	80.10^{689}Y
Ar nebulizer gas flow (L/min)	0.6–0.9 (optimized daily)
Cones sampler and interface—Skimmer (machine)	Pt
Signal	Area peak
Plasma gas flow rate (L/min)	15
Auxiliary gas flow rate (L/min)	0.9
Measures	
Scan mode	Peak hopping
Resolution (amu)	0.7 (10% peak height)
Replicate time (s)	1
Reading/replicates	55
Dwell time (s)	50
Sweeps/reading	20
Integration time (s)	1000
Replicates	3
Isotopes	^{111}Cd, P-31

4 Methods

4.1 Preparation of Standard Solution

Note: It is recommended to store the standard solutions in inert containers, such as polypropylene, to minimize potential contamination.

4.1.1 Internal Standard (ISTD) Solution

To ensure accuracy and compensate for any variations during analysis, an internal standard solution (ISTD) with a concentration of 5000 μg/L is prepared. The ISTD is typically an element that is not present in the samples but behaves similarly to the analytes of interest (P and Cd). Follow these steps to prepare the ISTD solution:

1. Mix 50 mL of the stock ISTD solution with 10.0 mL of HNO_3 in a 500 mL volumetric flask.
2. Bring the volume to the mark with Milli-Q water, ensuring complete dissolution and homogeneity.
3. Use a suitable polypropylene or glass container to store the prepared ISTD solution.
4. Ensure proper labeling and identification.

4.1.2 Calibration Standards

Calibration standards are used to establish a calibration curve, relating the instrument response to known concentrations of P and Cd. Prepare a series of calibration standards with different concentrations to cover the expected range of analyte concentrations in the lettuce samples. The calibration standards should include concentrations similar to those found in the samples. Follow these steps to prepare the calibration standards:

(i) MS intermediate standard 1 (1000 μg/L):

1. Mix 1.0 mL of the MS mixed standard stock solution with 2.0 mL of HNO_3 in a 100 mL volumetric flask.
2. Bring the volume to the mark with Milli-Q water, ensuring thorough mixing.

(ii) MS intermediate standard 2 (100 μg/L):

1. Mix 5.0 mL of the 1000 μg/L MS intermediate standard 1 with 1.0 mL of HNO_3 in a 50 mL volumetric flask.
2. Bring the volume to the mark with Milli-Q water, ensuring thorough mixing.

Note: Prepare calibration standards by diluting the appropriate intermediate standards with 2% HNO_3. Add sufficient 5000 μg/L ISTD to achieve a final ISTD concentration of 5 μg/L. Use polymeric volumetric flasks for the preparation of these standards.

4.1.3 Fortification Standard

The preparation of a fortification standard is essential for assessing the accuracy and reliability of the analytical method by spiking the control tissue with a known concentration of the analyte, typically ranging from 1 to 2 times the maximum level of acceptance (MLA) based on a specified sample weight. To prepare the fortification standard, follow these steps:

1. Mix 2.0 mL of the MS fortification standard stock solution with 2.0 mL of HNO_3 in a 100 mL volumetric flask.
2. Fill the flask with Milli-Q water up to the mark to achieve the desired volume, ensuring thorough mixing.
3. The fortification standard should be added to the control tissue at a concentration ranging from 1 to 2 times the MLA, based on a 0.5 g sample weight.
4. The final concentrations of the fortification standard should be calculated considering the dilution of the stock standard.

By following these steps, you will have prepared the necessary standard solutions for the analysis of P and Cd in lettuce samples using ICP-MS. Ensure proper labeling and storage of the standard solutions to maintain their integrity and use appropriate containers to minimize contamination.

4.2 Sample Preparation

1. Collect fresh lettuce samples from the field or the store.
2. Wash the lettuce samples thoroughly with Milli-Q water to remove any surface contamination.
3. Dry the samples using an oven at 60 °C for 24 h to remove any moisture.
4. Once the samples are dry, cut the lettuce leaves into small pieces using a clean knife or scissors; it is better to grind it.
5. Homogenize the lettuce pieces using a blender or mortar and pestle until a uniform consistency is achieved.

4.3 Microwave Digestion

1. Weigh 0.5 g of the homogenized lettuce sample into a PTFE digestion vessel.
2. Add 10 mL of concentrated ultrapure HNO_3 to the sample and allow it to stand for 30 min.
3. Add 1.0 mL of H_2O_2 to each sample.
4. Add 2 mL of HCl and 2 mL of $HClO_4$ to the sample.
5. Expose the samples in the digestion vessel to mineralization by microwave digestion (30 min) and close the lid securely.
6. Program the microwave digestion system according to the manufacturer's instructions and start the digestion process. The digestion process typically takes 30–60 min, depending on the type of microwave system and the digestion program

Table 2
Recommended parameters for the digestion microwave [31]

No.	Parameter	Value
1	Power	1600 W[a]
2	Ramp time	12 min
3	Final temperature	180 °C
4	Temperature hold time	10 min
5	Cool down time	10 min

[a]The wattage can be reduced if the number of digestion vessels is less than eight

used. Table 2 lists the recommended parameters for the digestion microwave. These parameters vary according to the type of samples and differences between individual instruments.

7. After the digestion process is complete, allow the samples to cool to room temperature.
8. Filter the digested sample solution to remove any solid residues using a suitable filter paper or membrane.
9. Add 2 mL of Milli-Q water to each digestion vessel to dissolve any remaining salts and transfer the solution to a 50 mL volumetric flask.
10. Dilute the solution with Milli-Q water to the mark on the volumetric flask.
11. The prepared samples can now be used for analysis by ICP-MS.

Note: Temperature control requires a temperature sensor in one vessel during complete decomposition. It is preferable to provide the following conditions during the digestion process: power (W/time in min): 1000/28, ventilation 20 min. The internal temperature should be 240 °C during the last step and ventilation.

4.4 Instrument Calibration

1. Prepare the standard solutions of P and Cd by diluting the stock solution (1000 mg/L) with Milli-Q water to obtain a series of standard solutions with known concentrations (e.g., 0.1, 0.5, 1, 5, and 10 μg/L).
2. Turn on the ICP-MS instrument and allow it to warm up and stabilize for at least 30 min.
3. Assemble the spray chamber, nebulizer, and pump tubing according to the manufacturer's instructions.
4. Load the standard solutions into the sample introduction system of the ICP-MS instrument.

5. Using the computer software, set up the instrument parameters, such as the ion lens voltages, mass range, and dwell time.
6. Run the instrument calibration by analyzing the standard solutions one by one, starting with the lowest concentration and working up to the highest concentration.
7. Record the instrument response for each standard solution, typically in counts per second (cps).
8. Use the calibration curve, which relates the known concentration of the standard solutions to the instrument response, to determine the concentration of P and Cd in the lettuce samples.
9. Validate the instrument calibration by analyzing a QC sample, which is a standard solution with a known concentration that is analyzed together with the samples (*see* Subheading 5).
10. If the instrument response for the QC sample falls within an acceptable range, the instrument calibration is considered valid, and the analysis of the lettuce samples can proceed.

4.5 Sample Analysis

Within the ICP-MS analysis, the sample undergoes a series of sequential steps to determine its elemental composition. These steps include the introduction of the sample into the instrument, nebulization using a nebulizer to create an aerosol, dissociation of compounds into characteristic ionic species, ionization of elements within a high-temperature argon plasma, mass discrimination in the quaternary analyzer, and final detection based on the mass-to-charge ratio (m/z) followed by data analysis. The overall principle of ICP-MS is illustrated in Fig. 3, providing a visual representation of these key steps and processes involved in the analysis, as follows:

1. Load the grounded, dried, and digested lettuce samples into the sample introduction system of the ICP-MS instrument.
2. Set up the instrument parameters, such as the ion lens voltages, mass range, and dwell time, using the computer software.
3. Run the instrument calibration using the standard solutions of P and Cd, as described above.
4. Analyze the lettuce samples by running the instrument with the same instrument parameters as the calibration, and record the instrument response for each sample, typically in counts per second (cps).
5. Calculate the concentrations of P and Cd in the lettuce samples using the calibration curve and the instrument response for each sample.
6. Validate the accuracy of the analysis by analyzing the QC standard solution together with the samples and comparing

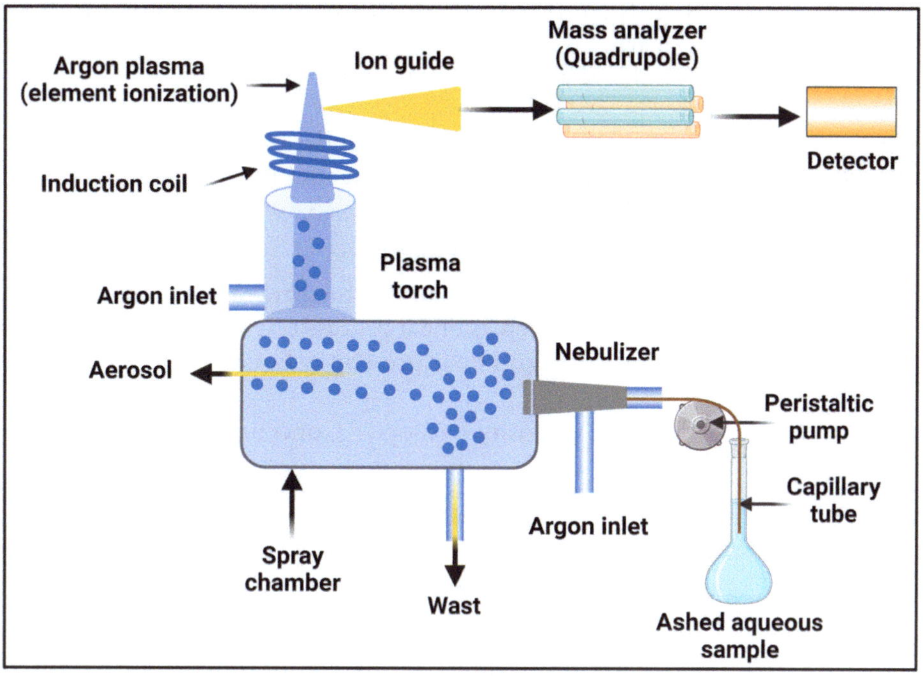

Fig. 3 Inductively coupled plasma mass spectrometry (ICP-MS) principle [32]

the instrument response to the expected value (*see* Subheading 5).

7. If the instrument response for the QC standard solution falls within an acceptable range, the results for the lettuce samples are considered valid.
8. Report the concentrations of P and Cd in the lettuce samples in appropriate units (e.g., μg/g or mg/kg).

5 Quality Assurance

Note: Quality assurance is of paramount importance in guaranteeing the integrity and credibility of analytical measurements. In the context of determining P and Cd concentrations in lettuce samples using ICP-MS, implementing rigorous quality assurance measures is crucial to ensure the validity and significance of the obtained results. This section provides a detailed overview of the quality assurance procedures employed in the analysis of P and Cd.

5.1 Calibration

5.1.1 Preparation of Calibration Standards

1. Prepare a series of calibration standards containing known concentrations of P and Cd.
2. Use CRMs or commercially available standards to ensure accuracy.

3. Prepare the standards by diluting stock solutions of P and Cd with a suitable matrix solution, such as HNO_3 solution, to achieve a range of concentrations.

5.1.2 Calibration Curve Using the ICP-MS Instrument

1. Analyze the calibration standards in triplicate and measure the corresponding signal intensities for P and Cd.
2. Plot the signal intensities against the known concentrations of the calibration standards.
3. Fit a regression line to the data using a suitable regression method, such as linear regression. The equation of the calibration curve can be expressed as Eq. 1:

$$\text{Signal intensity} = (\text{Slope} \times \text{Concentration}) + \text{Intercept}. \quad (1)$$

5.2 Validation

5.2.1 Accuracy and Precision

1. Analyze the CRMs with known concentrations of P and Cd in triplicate.
2. Calculate the average signal intensities and compare them to the certified values.
3. Assess the accuracy of the method by calculating the recovery rate using the following equation (Eq. 2):

$$\text{Recovery } (\%) = \frac{\text{Average measured concentration}}{\text{Certified concentration}} \times 100 \quad (2)$$

4. Calculate the percent relative standard deviation (%RSD) as a measure of precision using Eq. 3:

$$\%\text{RSD} = \frac{\text{Standard deviation}}{\text{Average measured concentration}} \times 100. \quad (3)$$

5.2.2 Limit of Detection (LOD) and Limit of Quantification (LOQ)

Determine the LOD and LOQ for P and Cd to establish the method's sensitivity. The LOD can be calculated as the signal corresponding to three times the standard deviation of the blank divided by the slope of the calibration curve (*see* Eq. 4). The LOQ is typically calculated as ten times the LOD (*see* Eq. 5).

$$\text{LOD} = 3 \times \frac{\text{Standard deviation of blank}}{\text{Slope of calibration curve}} \quad (4)$$

$$\text{LOQ} = 10 \times \text{LOD}. \quad (5)$$

5.3 Quality Control

Detail the QC measures implemented throughout the analysis. This includes the use of blank samples, spike recovery experiments, and regular analysis of reference materials to ensure the accuracy and precision of the analytical results.

5.3.1 Blank Samples

To correct for background interferences:

1. Analyze blank samples, consisting of reagents and matrix components without the analytes of interest, in triplicate.
2. Calculate the average signal intensity and subtract it from the sample measurements.

5.3.2 Spike Recovery

1. Perform spike recovery experiments to evaluate the method's accuracy and matrix effect.
2. Spike known amounts of P and Cd to the plant samples and analyze them in triplicate.
3. Calculate the spike recovery using Eq. 6:

$$\text{Spike recovery (\%)} = \frac{\text{Measured concentration after spiking}}{\text{Expected concentration}} \times 100. \tag{6}$$

5.4 Instrument Stability and Drift

1. Monitor the stability and drift of the ICP-MS instrument throughout the analysis.
2. Regularly analyze blank samples and reference materials to identify any instrumental drift or background interferences.

5.5 Method Validation

1. Perform method validation to assess the accuracy, precision, and reliability of the analytical method. This involves analyzing CRMs and comparing the obtained results with the certified values.
2. Validate the method for parameters such as linearity, precision, accuracy, specificity, and robustness.

5.6 Quality Assurance Reporting

Report the measured concentrations of P and Cd in the lettuce samples along with their respective uncertainties or confidence intervals. Include the calibration curve equation, LOD, LOQ, recovery rates, %RSD, and any additional relevant information in the quality assurance report. By incorporating all the necessary information into the quality assurance report, a comprehensive assessment of the analytical method for determining P and Cd in the samples using ICP-MS can be achieved. This documentation ensures transparency, reproducibility, and facilitates comparisons with other studies, ultimately enhancing the scientific rigor and confidence in the research outcomes.

6 Data Analysis

After all the samples have been analyzed, the data should be compiled and analyzed. The concentration of P and Cd in each sample should be reported in units of milligrams per kilogram (mg/kg). The data should also be compared to regulatory limits or guidelines, if available, to assess the safety and quality of the lettuce samples.

1. Export the raw data from the ICP-MS instrument to the computer software for data analysis.
2. Use the software to perform background correction, which subtracts the signal from blank samples or reagents from the sample and standard signals.
3. Create a calibration curve by analyzing standard solutions of known concentrations of P and Cd with the ICP-MS instrument.
4. Use the calibration curve to convert the instrument response (cps) for each sample into the corresponding concentrations of P and Cd.
5. Calculate the average concentrations and standard deviation of P and Cd for each sample.
6. Compare the concentrations of P and Cd in the lettuce samples to the acceptable limits set by regulatory agencies.
7. Evaluate the potential health risks associated with the observed levels of P and Cd in the lettuce samples.

Interpret the data and draw conclusions about the levels of Cd and P in the lettuce samples.

7 Data Preprocessing

Prior to analysis, it is important to preprocess the raw data to ensure accurate and reliable results. This involves several steps, including background correction, internal standardization, and calibration curve fitting.

1. Background correction is used to correct any background noise in the data, which can be caused by the instrument or the sample matrix. Several methods can be used for background correction, including subtracting a blank measurement or using mathematical algorithms to estimate and correct the background signal.
2. Internal standardization involves adding a known amount of an element that is not present in the sample to the sample matrix

prior to analysis. This helps correct any variations in the sample matrix and improves the accuracy and precision of the results.

3. Calibration curve fitting is used to establish a linear relationship between the measured signal and the concentration of the element of interest. This involves analyzing a series of standards with known concentrations of the element and plotting the resulting signal against the concentration. The resulting curve is then used to interpolate the concentration of the element in the samples.

Overall, data preprocessing is a crucial step in ensuring the accuracy and precision of the results obtained from ICP-MS analysis.

8 Notes

1. Prior to conducting the measurement procedures, it is recommended to decontaminate all used materials by immersing them in a 10% HNO_3 (v/v) bath for 24 h, followed by rinsing with Milli-Q water.
2. It is necessary before starting the measurement to perform a calibration of the ICP-MS spectrometer using a multielement mixture of metallic standards from Agilent Technologies, Japan.
3. The calibration curve should cover a range of concentrations that encompasses the expected levels of P and Cd in the lettuce samples.
4. It is important to ensure that the calibration curve for the elements demonstrates good linearity across the entire concentration range, with coefficients of determination (R^2) higher than 0.90.
5. Regular cleaning and maintenance of the ICP-MS instrument are essential to ensure accurate and reliable results.
6. Sample preparation is a critical step in ICP-MS analysis. It involves homogenizing the lettuce samples to ensure representative subsamples and minimizing variations within the batch.
7. The digestion process is necessary to break down complex matrices and release the analytes of interest. Microwave digestion using appropriate reagents, such as HNO_3, may be employed to achieve efficient sample digestion.
8. Instrument stability and drift should be monitored throughout the analysis. Regular analysis of blank samples and reference materials can help identify any instrumental drift or background interferences.

9. Method validation is essential to assess the accuracy, precision, and reliability of the analytical method. This involves analyzing CRMs and comparing the obtained results with the certified values.
10. QC measures should be implemented to ensure the accuracy and precision of the analytical results. This includes the use of standard reference materials, matrix-matched calibration standards, and blank samples to identify and correct any potential sources of error.
11. The reporting of results should include both the measured concentrations of P and Cd in the lettuce samples as well as the associated uncertainties or confidence intervals to provide a comprehensive representation of the data.
12. When interpreting the analytical results, it is crucial to consider the acceptable limits set by regulatory agencies and the potential health risks associated with the observed levels of trace elements in the food samples.

9 Conclusion

In conclusion, this protocol provides a detailed method for the use of ICP-MS to assess the levels of P and Cd in lettuce samples. This method can be used to monitor the levels of these elements in lettuce and other food samples for safety and nutritional purposes. QC measures should be implemented throughout the analysis to ensure the accuracy and precision of the results.

References

1. Singh RL, Mondal S (2019) Chapter 1—Current issues in food safety with reference to human health. In: Singh RL, Sukanta BT (eds) Food safety and human health mondal. Academic Press, pp 1–14. https://doi.org/10.1016/B978-0-12-816333-7.00001-1
2. Almaghrabi MA (2023) Assessing public interest, risk perceptions, and awareness of food safety in Saudi Arabia: a cross-sectional study. Food Control 151:109810. https://doi.org/10.1016/j.foodcont.2023.109810
3. El-Kady AA, Abdel-Wahhab MA (2018) Occurrence of trace metals in foodstuffs and their health impact. Trends Food Sci Technol 75:36–45. https://doi.org/10.1016/j.tifs.2018.03.001
4. Rzymski P, Budzulak J, Niedzielski P, Klimaszyk P, Proch J, Kozak L, Poniedziałek B (2019) Essential and toxic elements in commercial microalgal food supplements. J Appl Phycol 31:3567–3579. https://doi.org/10.1007/s10811-018-1681-1
5. Kim MJ, Moon Y, Tou JC, Mou B, Waterland NL (2016) Nutritional value, bioactive compounds and health benefits of lettuce (Lactuca sativa L.). J Food Compos Anal 49:19–34. https://doi.org/10.1016/j.jfca.2016.03.004
6. Shi M, Gu J, Wu H, Rauf A, Emran TB, Khan Z, Mitra S et al (2022) Phytochemicals, nutrition, metabolism, bioavailability, and health benefits in lettuce—a comprehensive review. Antioxidants 11. https://doi.org/10.3390/antiox11061158
7. Song J, Huang H, Hao Y, Song S, Zhang Y, Wei S, Liu H (2020) Nutritional quality, mineral and antioxidant content in lettuce affected by interaction of light intensity and nutrient solution concentration. Sci Rep 10:2796. https://doi.org/10.1038/s41598-020-59574-3

8. Martín-Diana AB, Rico D, Frías JM, Barat JM, Henehan GTM, Barry-Ryan C (2007) Calcium for extending the shelf life of fresh whole and minimally processed fruits and vegetables: a review. Trends Food Sci Technol 18:210–218. https://doi.org/10.1016/j.tifs.2006.11.027
9. Zeb A, Liu W, Meng L, Lian J, Wang Q, Lian Y, Chen C, Wu J (2022) Effects of polyester microfibers (PMFs) and cadmium on lettuce (Lactuca sativa) and the rhizospheric microbial communities: a study involving physio-biochemical properties and metabolomic profiles. J Hazard Mater 424:127405. https://doi.org/10.1016/j.jhazmat.2021.127405
10. Neocleous D, Savvas D (2019) The effects of phosphorus supply limitation on photosynthesis, biomass production, nutritional quality, and mineral nutrition in lettuce grown in a recirculating nutrient solution. Sci Hortic 252:379–387. https://doi.org/10.1016/j.scienta.2019.04.007
11. do Nascimento da Silva E, Heerdt G, Cidade M, Pereira CD, Morgon NH, Cadore S (2015) Use of in vitro digestion method and theoretical calculations to evaluate the bioaccessibility of Al, cd, Fe and Zn in lettuce and cole by inductively coupled plasma mass spectrometry. Microchem J 119:152–158. https://doi.org/10.1016/j.microc.2014.12.002
12. Serna J, Bergwitz C (2020) Importance of dietary phosphorus for bone metabolism and healthy aging. Nutrients 12. https://doi.org/10.3390/nu12103001
13. Bird RP, Michael Eskin NA (2021) Chapter two—the emerging role of phosphorus in human health. The latest research and development of minerals in human nutrition. In: N. A., Michael BT (eds) Advances in food and nutrition research Eskin, vol 96. Academic Press, pp 27–88. https://doi.org/10.1016/bs.afnr.2021.02.001
14. Eltohamy KM, Liu C, Khan S, Niyungeko C, Jin Y, Hosseini SH, Li F, Liang X (2021) An internet-based smart irrigation approach for limiting phosphorus release from organic fertilizer-amended paddy soil. *Journal of Cleaner Production* 293. Elsevier Ltd: 126254. https://doi.org/10.1016/j.jclepro.2021.126254
15. Calvo MS, Uribarri J (2013) Public health impact of dietary phosphorus excess on bone and cardiovascular health in the general population 1,2,3. Am J Clin Nutr 98:6–15. https://doi.org/10.3945/ajcn.112.053934
16. Uribarri J, Calvo MS (2017) Chapter 35—molecular mechanisms of adverse health effects associated with excess phosphorus intake. In: James FBT (ed) Molecular Collins genetic, and nutritional aspects of major and trace minerals. Academic Press, Boston, pp 429–436. https://doi.org/10.1016/B978-0-12-802168-2.00035-X
17. Schaefer HR, Dennis S, Fitzpatrick S (2020) Cadmium: Mitigation strategies to reduce dietary exposure. *Journal of Food Science* 85. John Wiley & Sons, Ltd:260–267. https://doi.org/10.1111/1750-3841.14997
18. Soisungwan S, Garrett Scott H, Ann SM, Sens Donald A (2010) Cadmium, environmental exposure, and health outcomes. *Environmental Health Perspectives* 118:182–190. https://doi.org/10.1289/ehp.0901234
19. Genchi G, Sinicropi MS, Lauria G, Carocci A, Catalano A (2020) The effects of cadmium toxicity. Int J Environ Res Public Health 17. https://doi.org/10.3390/ijerph17113782
20. Eltohamy KM, Milham PJ, Gouda M, Menezes-Blackburn D, Khan S, Liu B, Jin J, Ye Y, Liang X (2023) Size and composition of colloidal phosphorus across agricultural soils amended with biochar, manure and biogas slurry. Carbon Res 2:16. https://doi.org/10.1007/s44246-023-00048-2
21. He P, Lu Y, Liang Y, Chen B, Wu M, Li S, He G, Jin T (2013) Exposure assessment of dietary cadmium: findings from shanghainese over 40 years, China. BMC Public Health 13: 590. https://doi.org/10.1186/1471-2458-13-590
22. Palmer SC, Hayen A, Macaskill P, Pellegrini F, Craig JC, Elder GJ, Strippoli GFM (2011) Serum levels of phosphorus, parathyroid hormone, and calcium and risks of death and cardiovascular disease in individuals with chronic kidney disease: a systematic review and meta-analysis. JAMA 305:1119–1127. https://doi.org/10.1001/jama.2011.308
23. Da J, Xie X, Wolf M, Disthabanchong S, Wang J, Zha Y, Lv J, Zhang L, Wang H (2015) Serum phosphorus and progression of CKD and mortality: a meta-analysis of cohort studies. Am J Kidney Dis 66. Elsevier:258–265. https://doi.org/10.1053/j.ajkd.2015.01.009
24. Balcaen L, Bolea-Fernandez E, Resano M, Vanhaecke F (2015) Inductively coupled plasma—tandem mass spectrometry (ICP-MS/MS): a powerful and universal tool for the interference-free determination of (ultra)trace elements—a tutorial review. Anal Chim Acta 894:7–19. https://doi.org/10.1016/j.aca.2015.08.053
25. Ammann AA (2007) Inductively coupled plasma mass spectrometry (ICP MS): a versatile

tool. *Journal of Mass Spectrometry* 42. John Wiley & Sons, Ltd:419–427. https://doi.org/10.1002/jms.1206

26. Nardi EP, Evangelista FS, Tormen L, Saint'-Pierre TD, Curtius AJ, de Souza SS, Barbosa F (2009) The use of inductively coupled plasma mass spectrometry (ICP-MS) for the determination of toxic and essential elements in different types of food samples. Food Chem 112:727–732. https://doi.org/10.1016/j.foodchem.2008.06.010
27. Mazarakioti EC, Zotos A, Thomatou A-A, Kontogeorgos A, Patakas A, Ladavos A (2022) Inductively Coupled Plasma-Mass Spectrometry (ICP-MS), a Useful Tool in Authenticity of Agricultural Products' and Foods' Origin. Foods 11:3705. https://doi.org/10.3390/foods11223705
28. Krzciuk K (2016) Intelligent analysis of samples by semiquantitative inductively coupled plasma-mass spectrometry (ICP-MS) technique: a review. *Critical Reviews in Analytical Chemistry* 46. Taylor & Francis:284–290. https://doi.org/10.1080/10408347.2015.1053106
29. Filipiak-Szok A, Kurzawa M, Szłyk E (2015) Determination of toxic metals by ICP-MS in Asiatic and European medicinal plants and dietary supplements. J Trace Elem Med Biol 30: 54–58. https://doi.org/10.1016/j.jtemb.2014.10.008
30. Chaves ES, dos Santos EJ, Araujo RGO, Oliveira JV, Frescura VLA, Curtius AJ (2010) Metals and phosphorus determination in vegetable seeds used in the production of biodiesel by ICP OES and ICP-MS. Microchem J 96: 71–76. https://doi.org/10.1016/j.microc.2010.01.021
31. Lee J, Park Y-S, Lee DY (2023) Fast and green microwave-assisted digestion with diluted nitric acid and hydrogen peroxide and subsequent determination of elemental composition in brown and white rice by ICP-MS and ICP-OES. LWT 173:114351. https://doi.org/10.1016/j.lwt.2022.114351
32. El Hosry L, Sok N, Richa R, Al Mashtoub L, Cayot P, Bou-Maroun E (2023) Sample preparation and analytical techniques in the determination of trace elements in food: a review. Food Secur 12. https://doi.org/10.3390/foods12040895

Chapter 13

Detecting Apple Valsa Canker (AVC) at an Early Stage Using Micro-SERS Combined with Chemical Imaging Analysis

Shiyan Fang, Yanru Zhao, Yan Wang, Junmeng Li, Fengle Zhu, and Keqiang Yu

Abstract

Surface-enhanced Raman scattering (SERS) spectroscopy with simplified detection procedures and improved detection efficiency is a potential method for plant disease detection. This chapter details how to detect Apple Valsa canker (AVC) with early incubation using micro-SERS combined with chemical imaging analysis. Chemometrics was used to eliminate the interference of baseline shifts. Machine learning methods were utilized to establish discriminative models to detect of the AVC disease stage. Further, chemical distribution imaging was successfully applied to the analysis of disease dynamics.

Key words Apple Valsa canker, Early detection, Surface-enhanced Raman scattering, Chemical imaging, Machine learning

1 Introduction

Apple Valsa canker (AVC), is a severe apple tree disease resulting in serious economic losses in Southeast Asia and China [1]. The fungal pathogen mainly infected the subcutaneous phloem through the wounded bark tissue at the initial infected stage. After infection, fungus hypha colonized the phloem tissues, leading to severe tissue cell death [2]. However, when visible symptoms appear, it is challenging to prevent AVC from spreading throughout the orchard by conventional treating methods such as spraying fungicides, manually removing the diseased areas, and pruning the dead branches. Unfortunately, there were no adequate methods for AVC treatment due to the complicated pathogenic mechanism so far. Thus, early

The original version of the chapter has been revised. A correction to this chapter can be found at https://doi.org/10.1007/978-1-0716-3938-2_19

Mostafa Gouda et al. (eds.), *Plant Chemical Compositions and Bioactivities*, Methods and Protocols in Food Science,
https://doi.org/10.1007/978-1-0716-3938-2_13,

detection of the infected trees is necessary to prevent the rapid development of the disease in orchards.

Various molecular biology methods, including enzyme-linked immunosorbent assay (ELISA) and polymerase chain reaction (PCR), were developed for the isolation and identification of pathogen [3, 4]. ELISA kits have been widely utilized due to the low cost, but are ineffective for detecting symptomless tissue [5], while PCR is an effective detection method. Zang et al. (2012) developed a nested PCR assay to detect the presence of *Valsa mali* in apple trees and achieved an accuracy of 64.7% [6]. However, DNA deriving from the woody plant tissues contained PCR-inhibiting compounds and could affect the accuracy of PCR reaction [7]. What's worse, a well-equipped laboratory and experienced personnel are also required, which was not feasible for on-site detection using the PCR [8]. Therefore, it is of great significance to develop a fast, nondestructive, and economical method for accurate detection of AVC.

Raman spectroscopy (RS) is a noninvasive, rapid, and high-throughput spectroscopic technique [9–11]. Raman shift is only related to the vibration frequency of the molecular functional group, but not to the incident light. Therefore, the Raman "fingerprint" of each sample is unique [12]. Significantly, RS could provide essential information related to the biochemical composition of the tree tissue cell, such as protein, polysaccharide, and lipid. Neither symptomatic nor asymptomatic trees, but these biochemical compositions are significantly different between diseased and healthy tissue. These composition changes can be reflected in Raman shifts or intensity changes of specific Raman bands assigned to those molecules. Therefore, RS provides an accessible way to identify subtle changes in the molecular compounds, which offers theoretical evidence for detecting tree diseases. Vallejo et al. (2016) investigated the application of RS combined with statistical analysis for detecting citrus Huanglongbing (HLB) infection in the field, and a good result was obtained with an overall classification accuracy of about 89.2% [13]. Sanchez et al. (2019b) readily distinguished between healthy and early-HLB citrus trees using a handheld Raman system and achieved an accuracy of 94% [14]. In their following study, Sanchez et al. (2019a) demonstrated that utilizing a handheld Raman spectrometer combined with chemometric analyses enabled the detection and identification of the secondary disease on HLB-infected orange trees [15]. Those researches indicated that the RS technique combined with chemometrics methods could detect diseased trees.

Surface-enhanced Raman scattering (SERS), based on the improvement of traditional Raman spectroscopy, uses certain metallic nanosubstrates such as gold or silver nanoparticles (AgNPs) to enhance signals under low laser powers, which maximizes fluorescence suppression. Meanwhile, the Raman system combined with the microimaging technology allows for scanning micron-scale Raman collection points (e.g., one-micron pixel) [16], which offers chemical information on the constituents at a high spatial resolution in situ. Qin et al. (2011) developed a Raman chemical imaging system to visualize the internal distribution of lycopene in postharvest tomatoes and established a Raman chemical image to visualize the spatial distribution of lycopene at different stages of maturity [17]. Yang et al. (2018) used a Raman imaging system to detect the spatial distribution of chemical components in maize seeds [18]. These studies manifested that Raman chemical imaging has great potential in the visualizing of plant tissue components. Therefore, this study aimed to develop a fast, noninvasive, and in situ diagnosis method for detecting AVC at early infection stages using SERS combined with microimaging technology.

The main objectives are to:

1. Optimize experimental conditions (i.e., laser intensity and exposure time) for obtaining valid SERS microimaging data, including synthesis and SERS AgNPs characterization;
2. Establish optimal discriminative models for detecting AVC in early infection stages based on machine learning algorithms; and
3. Generate microdistribution maps of cellulose and lignin at the disease–health junction of the tree phloem tissues to reveal the dynamic development characteristics of the disease.

2 Materials

2.1 Materials Preparation

1. Fungus *Valsa mali* from College of Plant Protection of Northwest A&F University (Yangling, Shannxi, China) (*see* **Note 1**).
2. Potato dextrose agar (PDA) medium from College of Plant Protection of Northwest A&F University (Yangling, Shannxi, China).
3. Two-year-old apple branches (*Malus domestica* cv. Fuji) were collected from the Economic Tree Garden of Northwest A&F University (Yangling, Shannxi, China).
4. The selected branches were pruned into 15 cm segments, and the surface of the branches was disinfected with 75% alcohol for 15 min (*see* **Note 2**).
5. The ends of the branches were sealed with a wet skimmed cotton to keep them fresh, followed by punching holes in the branches with a hole puncher (hole diameter 5 mm).
6. The activated *Valsa mali* fungus was inoculated on the wounds of apple branches with two points on each branch (*see* **Notes 3, 4**).
7. AgNPs were synthesized by using the Lee–Meisel method [19] (*see* **Note 5**).

2.2 Equipment and Parameters

1. Tecnai G2 transmission electron microscopy (FEI Inc., Hillsboro, OR, USA).
2. Lambda 35 Spectrophotometer (PerkinElmer Inc., Waltham, MA, USA).
3. DXR3xi Raman microimaging spectrometer (Thermo Fisher Scientific Inc., Waltham, MA, USA).
4. Each sample placed on a glass slide was dripped with the AgNPs.
5. The excitation wavelength was 785 nm; the collected spectral range was 300–3000 shift/cm; the laser intensity was 2.6 mW; the exposure time was 0.00285 s (350 Hz); and the number of scanning was 40.
6. The samples were scanned point by point in 2 μm steps.

2.3 SERS Spectroscopy Acquisition

1. Branches were removed from the incubator, and the inoculation points on the phloem were scraped with a knife as the samples (*see* **Note 2**).
2. Each sample placed on a glass slide was dripped with the AgNPs. Then, each sample was placed on the automatic stage and aligned with a Raman laser.
3. Before starting the Raman measurements, the calibration procedure that came with the instrument was executed automatically.
4. The data acquisition software OMNICxi v1.6 was used to adjust the acquisition parameters.
5. The number of scanning was 40.

2.4 Spectral Data Processing

1. These three algorithms include the multiple spectral baseline correction (MSBC) [20], the asymmetric least squares (AsLS) [21], and the adaptive iterative reweighted penalized least squares (air-PLS) [22] and were used to remove background noises and baselines.
2. Principal component analysis (PCA) [23] was used to reduce the original high-dimensional variable space.
3. Competitive adaptive reweighted sampling (CARS) [24] and random frog (RFrog) [25] algorithms were combined to select the optimal variables.
4. BP artificial neural network (BP-ANN) [26], random forest (RForest) [27], the least squares support vector machine (LS-SVM) [28], and extreme learning machine (ELM) [29] were employed to establish the classification models.

2.5 Software

1. MATLAB R2018b (The MathWorks, Natick, MA, USA).
2. Computer with an Intel Core i5-9400F CPU, 16GB RAM, and a Windows 10 operating system.

2.6 Key Steps

Figure 1 demonstrates key steps for detecting AVC at an early stage based on SERS spectroscopy combined with chemical imaging analysis.

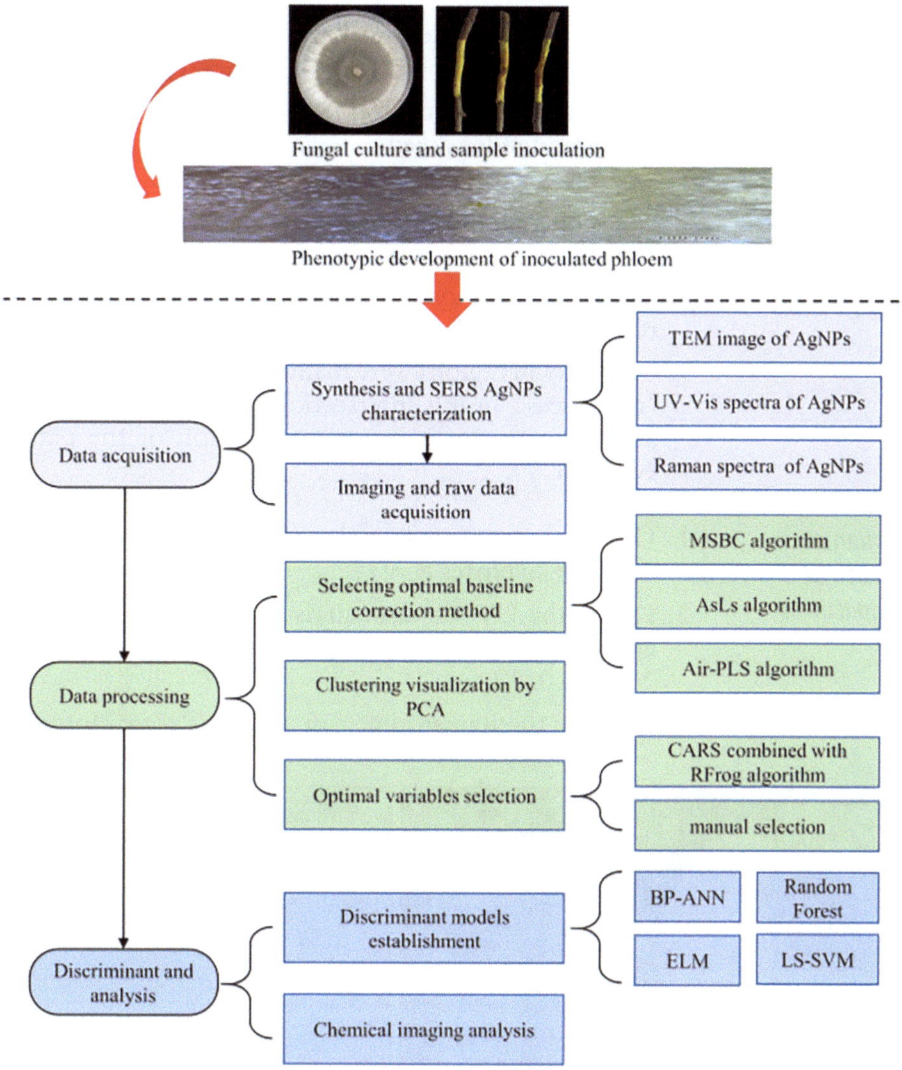

Fig. 1 Key steps for detecting AVC based on SERS combined with chemical imaging analysis

3 Methods and Analysis

3.1 Phenotypic Development of Samples

1. The strains of the fungus *Valsa mali* on the PDA medium in Fig. 2a.
2. The dynamic process of the diseased phloem in the first 7 days in Fig. 2b.
3. The disease–health intersection of the diseased phloem was presented using optical microscopy in Fig. 2c.
4. The bark surface of inoculated branch samples showed no visible symptoms during the first 7 days.
5. The phloem inside the bark appeared with early infection symptoms.
6. The healthy phloem (the first 3 days) had a smooth surface and displayed tender green.
7. The diseased phloem became rough and showed pale brown when the symptoms of mild infection were visible on the 5th day.
8. The diseased phloem appeared dark brown, and the tissue was rotten on the 7th day.
9. The phloem regions were manually labeled as healthy, disease-1 (the disease–health intersection), and disease-2 (late-disease) according to the infection progression of the pathogen (*see* **Notes 6**, **7**).

3.2 Detection of Characteristics of Silver Nanoparticles

1. Obtain the transmission electron microscopy (TEM) image of AgNPs (*see* **Notes 8**, **9**).
2. Detect the UV-Vis characteristic absorption peak (at 410 nm) corresponding to the single plasmon resonance mode (*see* **Notes 10**, **11**).
3. Measure the Raman spectrum of AgNPs (*see* **Note 10**).

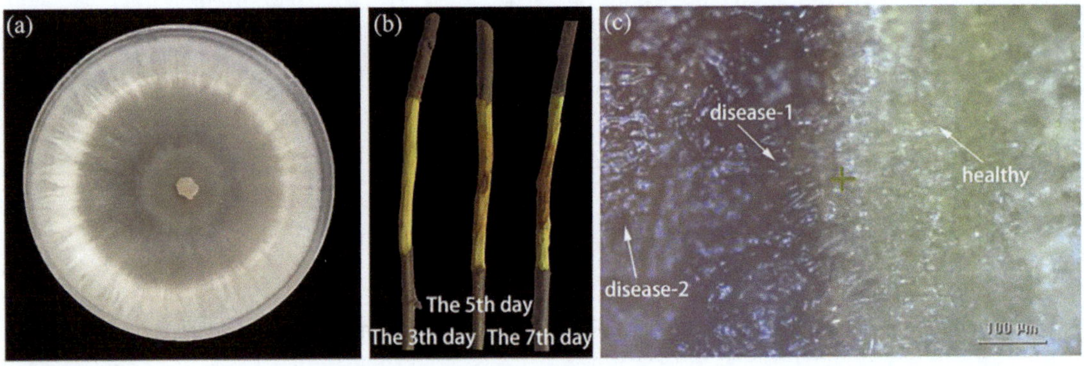

Fig. 2 Phenotypic development of healthy and inoculated branch. (**a**) The strains of the fungus *Valsa mali* on PDA medium. (**b**) The dynamic process of the diseased phloem in the first 7 days. (**c**) Optical micrograph of the disease–health junction

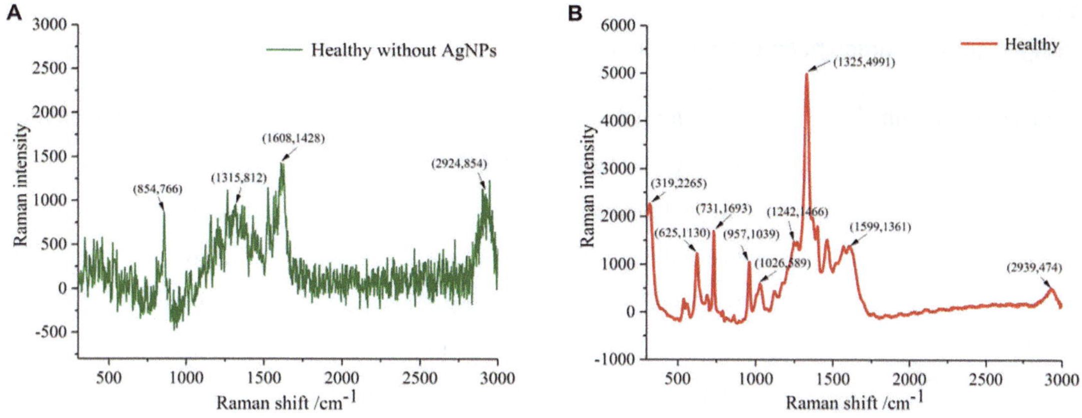

Fig. 3 The spectrum of healthy tissue samples, with and without AgNPs, respectively. (**a**) Raman spectra peaks of healthy samples without AgNPs did not appear. (**b**) The SERS characteristic peaks of healthy samples were obvious, which further proved that AgNPs was effective

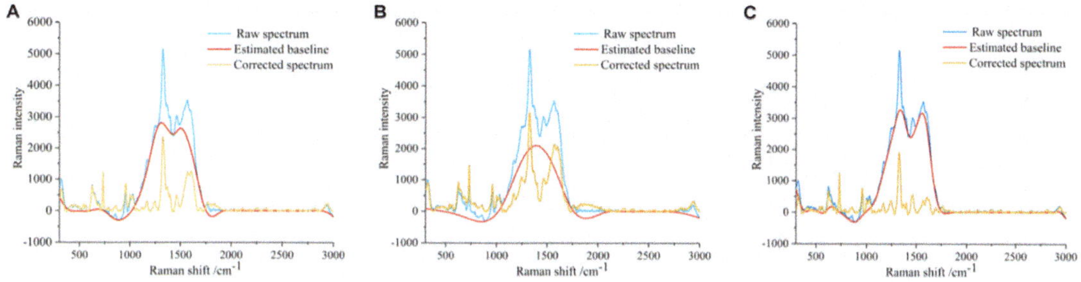

Fig. 4 Spectral baseline correction. (**a**) Baseline correction using MSBC. (**b**) Baseline correction using AsLS. (**c**) Baseline correction using air-PLS. The blue line represents the original spectrum, the red line represents the estimated baseline, and the yellow line represents the corrected spectrum

4. Drop the AgNPs on the tested branches and detect the SERS of the healthy samples.
5. Figure 3 shows the SERS characteristics peaks of healthy samples were obvious (*see* **Note 12**).

3.3 Preprocessing

1. MSBC, AsLS, and air-PLS algorithms were adopted to eliminate the disturbances of the baseline offset (Fig. 4).
2. For the MSBC algorithm, the parameters were set to $\lambda = 150$, $\mu = 8 \times 10^7$, and $\rho = 0$.
3. For AsLS algorithm, the parameters were set to $\lambda = 5000$ and $\rho = 0.0001$.
4. For the air-PLS algorithm, the parameters were set to $\lambda = 150$ and $\rho = 0.01$.
5. The curved baselines were well-fitted and subtracted by the three algorithms.

Table 1
Assignment of characteristic wavenumbers

Wavenumber (cm^{-1})	Assignment	Biological components
319	C–C–C or C–O–C skeletal bending	Cellulose
625	Skeletal bending	Lignin
731	Skeletal bending	Lignin
957	C–C or C–O stretching vibration	Cellulose
1026	C–C or C–O stretching vibration	Cellulose
1165	H–C–C or H–C–O skeletal bending	Cellulose
1242	C=O stretching vibration	Cellulose
1325	C–H bending vibration	Cellulose
1599	C–C aromatic ring	Lignin
2939	C–H asymmetric stretching vibration	Lignin

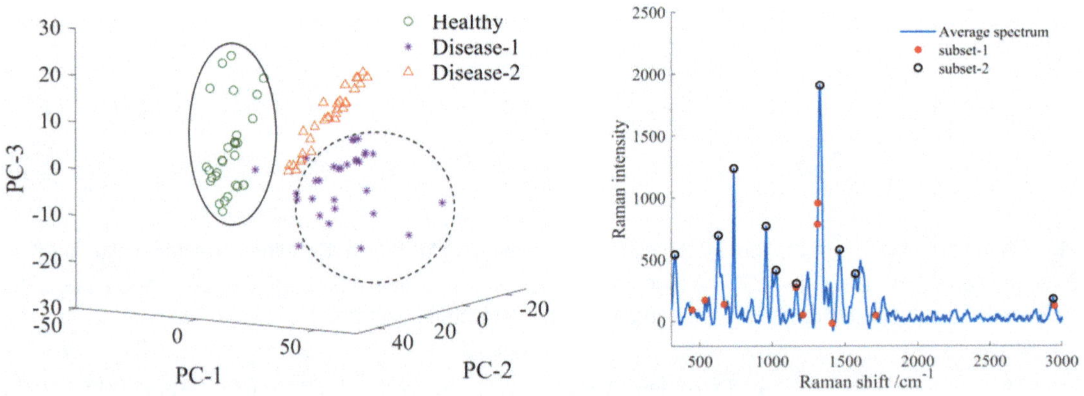

Fig. 5 Feature variables selection. (**a**) Score plots of first three PCs from PCA on spectral data of the three types of samples. (**b**) The characteristic variables for early disease detection of the AVC disease. Two strategies were used to select characteristic variables: algorithm selection (subset-1) and manual selection (subset-2)

3.4 Characteristic Wavenumbers Analysis

1. Observe the characteristic wavenumbers.
2. Table 1 shows the assignment of characteristics wavenumbers.

3.5 Feature Variables Selection

1. PCA was performed on the raw spectra of the total sample set to reduce the high-dimensional data (Fig. 5).
2. The first three principal components were displayed in score scatters plot.
3. Algorithm selection (CARS combined with RFrog) and manual selection were conducted to extract feature variables.

4. PC1, PC2, and PC3 provided 51.74, 15.01, and 11.56% of the variations among samples, respectively.
5. The cumulative contribution of the first three PCs achieved 78.31%.
6. Important variables were extracted from the total 1401 spectral variables.

3.6 Discriminant Models Establishment

1. Set the labels for healthy, disease-1, and disease-2 of 1, 2, and 3, respectively.
2. SERS spectral data were divided into a calibration set and a prediction set at the ratio of 3:1.
3. BPANN, ELM, RForest, and LS-SVM models were established using four variable matrices (x) to classify the healthy, disease-1, and disease-2 samples.
4. Set the parameters of the BP-ANN model: the learning rate was set uniformly to 0.1, and the number of neurons in the hidden layers were 10, 3, 3, and 10, respectively.
5. Set the parameters of the ELM model: the number of neurons in the hidden layer was determined by comparing the performances of the ELM model using different numbers of neurons from 1 to 100 with a step of 1 (the ELM with 34 neurons was selected as the optimal model).
6. Set the parameters of the RForest model: the number of decision trees was determined by comparing the model performances using different numbers of decision trees from 1 to 500 with a step of 1 (the RForest with 100 decision trees was selected as the optimal model).
7. Set the parameters of the LS-SVM model: RBF as the kernel function, optimal penalty coefficient, (c) and the kernel function parameter gamma (g) were obtained by a grid search procedure, the best-c was 379, and the best-g was 45.

The classification accuracy and running time are listed in Table 2.

3.7 Chemical Imaging Analysis

1. The SERS micro-spectral image data cube of each phloem sample was processed by the air-PLS algorithm
2. The processed micro-spectral cube in a pixel-wise manner generated chemical distribution images.
3. The symmetric tensile vibration at 1600 cm^{-1} in lignin was identified as the characteristic peak of lignin components.
4. The bands at 300–550 cm^{-1} were contributed by cellulose.
5. Figure 6 was constructed based on the cellulose signature peak at 300–550 cm^{-1} and lignin signature peak at 1600 cm^{-1}.

Table 2
The discriminant accuracy of the models

Models	Discriminant accuracy (%) Full SERS spectra	Subset-1	Subset-2	Predicted fluorescence baselines	Running time (s)
BP-ANN	86.22	93.17	94.42	91.80	0.28
ELM	92.36	85.35	88.93	95.39	0.01
RForest	98.46	96.67	95.87	99.57	0.15
LS-SVM	98.86	94.49	95.48	98.04	0.91

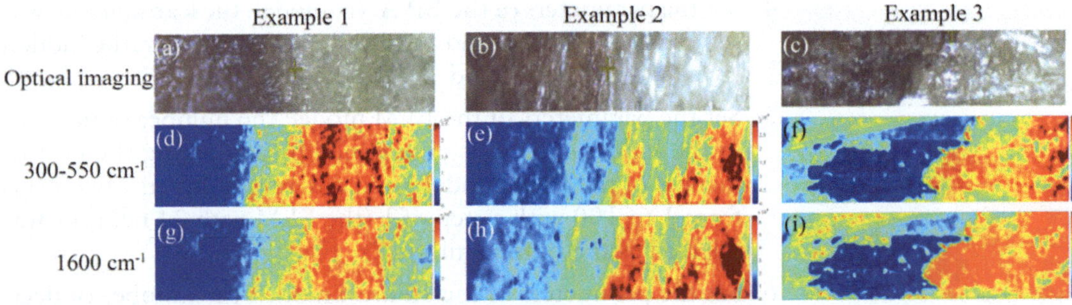

Fig. 6 Chemical distribution images of phloem samples based on SERS spectroscopy. (**a–c**) The optical micrographs of the phloem tissues taken from the branches of three trees. (**d–f**) The chemical imaging results based on the cellulose signature peak at 300–550 cm^{-1}. (**g–i**) The chemical imaging results based on the lignin signature peak at 1600 cm^{-1}

6. The redder-colored the pixels, the stronger the spectral signals of the chemical component in Fig. 6.
7. The bluer-colored the pixels, the weaker the spectral signals in Fig. 6.
8. The SERS signal at the healthy tissue exhibited high intensity with red, bright yellow, and green pixel colors.
9. The diseased phloem tissue exhibited low intensity with blue and green pixel colors, and the disease–health junction exhibited green pixel colors.
10. These differences in SERS imaging of different regions can be attributed to differences in cell wall components.
11. Cellulose and lignin in the cell walls of infected tissues reduced significantly.

3.8 Conclusion

1. Raman microimaging can visualize the intensity and distribution of components of the cell walls in situ through cytological observations.
2. This rapid and noninvasive chemical imaging strategy is superior to the other methods.
3. SERS spectroscopy combined with chemometric methods was applied for early detection of the AVC disease.

4 Notes

1. The fungus *Valsa mali* need to be stored at −80 °C in an ultra-low temperature refrigerator [6].
2. The surface of the branches should be disinfected with 75% alcohol at least for 15 min [1].
3. The activated *Valsa mali* fungus was inoculated on the wounds of apple branches with two points on each branch.
4. After inoculation, the branches were transferred to a 25 °C incubator for further incubation [6].
5. $AgNO_3$ (36 mg) was dissolved in 200 mL of ultrapure water and boiled quickly [19].
6. A solution of 1 wt.% trisodium citrate (6 mL) was charged to the reaction solution and was held on boiling for 25 min accompanied by stirring at 200 rpm [19].
7. Branches were removed from the incubator for capturing Raman images [19].
8. Before acquiring the SERS spectrum, inoculation points on the phloem were scraped with a knife as the samples.
9. Each sample placed on a glass slide was dripped with the AgNPs.
10. Selecting the optimal pretreatment method was necessary to improve the spectral quality [22].
11. Construct relationships between variables and properties of interest are essential to make classification modes [4].
12. It is necessary to remove the skin before observing the symptoms changes of the infected tissue.

References

1. Wang XL, Wei JL, Huang LL, Kang ZS (2011) Re-evaluation of pathogens causing Valsa canker on apple in China. Mycologia 103:317–324. https://doi.org/10.3852/09-165
2. Suzaki K (2008) Population structure of *Valsa ceratosperma*, causal fungus of Valsa canker, in apple and pear orchards. J Gen Plant Pathol 74: 128–132. https://doi.org/10.1007/s10327-008-0078-4
3. Liu M, Elisa M, Julie TC, Julie C, Sylvia KW, Raymond T et al (2015) Detection and identification of selected cereal rust pathogens by

TaqMan® real-time PCR. Can J Plant Pathol 37:92–105. https://doi.org/10.1080/07060661.2014.999123

4. Golhani K, Siva KB, Ganesan V, Biswajeet P (2018) A review of neural networks in plant disease detection using hyperspectral data. Inf Process Agric 5:354–371. https://doi.org/10.1016/j.inpa.2018.05.002
5. Fang Y, Ramasamy PR (2015) Current and prospective methods for plant disease detection. Biosensors 5:537–561. https://doi.org/10.3390/bios5030537
6. Zang R, Yin ZY, Ke XW, Wang XJ, Li ZL, Kang ZS et al (2012) A nested PCR assay for detecting Valsa mali var. mali in different tissues of apple trees. Plant Dis 96:1645–1652. https://doi.org/10.1094/PDIS-05-11-0387-RE
7. Martinelli F, Riccardo S, Salvatore D, Stefano P, Giuseppe S, Paolo R et al (2015) Advanced methods of plant disease detection. a review. Agron Sustain Dev 35:1–25
8. Okiro LA, Matthew AT, Steven GN, Christine DS, Monica LP (2019) Comparative evaluation of LAMP, qPCR, conventional PCR, and ELISA to detect Ralstonia solanacearum in Kenyan potato fields. Plant Dis 103:959–965. https://doi.org/10.1094/PDIS-03-18-0489-RE
9. Farber C, Rebecca B, Li P, Charles R, Dmitry K (2020) Non-invasive characterization of single-, double- and triple-viral diseases of wheat with a hand-held Raman spectrometer. Front Plant Sci 11:1300. https://doi.org/10.3389/fpls.2020.01300
10. Huang CH, Gajendra PS, Su HP, Chua NH, Rajeev JR, Bong SP (2020) Early diagnosis and management of nitrogen deficiency in plants utilizing Raman spectroscopy. Front Plant Sci 11:663. https://doi.org/10.3389/fpls.2020.00663
11. Zhao YR, Fang SY, Ye YK, Yu KQ (2021) Chemometric development using portable molecular vibrational spectrometers for rapid evaluation of AVC (*Valsa mali* Miyabe et Yamada) infection of apple trees. Vib Spectrosc 114:103231. https://doi.org/10.1016/j.vibspec.2021.103231
12. Fang SY, Cui RY, Wang Y, Zhao YR, Yu KQ, Jiang A (2021) Application of multiple spectral systems for the tree disease detection: a review. Appl Spectrosc Rev. https://doi.org/10.1080/05704928.2021.1930552
13. Vallejo P, Moises R, Maria GGM, Miguel GRE, Francisco JG, Hugo RNC et al (2016) Raman spectroscopy an option for the early detection of citrus huanglongbing. Appl Spectrosc 70:829–839. https://doi.org/10.1177/0003702816638229
14. Sanchez L, Shankar P, Xing ZL, Kranthi M, Dmitry K (2019a) Rapid and noninvasive diagnostics of Huanglongbing and nutrient deficits on citrus trees with a handheld Raman spectrometer. Anal Bioanal Chem 411:3125–3133. https://doi.org/10.1007/s00216-019-01776-4
15. Sanchez L, Shankar P, Mike I, Kranthi M, Dmitry K (2019b) Detection and identification of canker and blight on orange trees using a handheld Raman spectrometer. J Raman Spectrosc 50:1875–1880. https://doi.org/10.1002/jrs.5741
16. Li XL, Sha JJ, Chu BQ, Wei YZ, Huang WH, Zhou H et al (2019) Quantitative visualization of intracellular lipids concentration in a microalgae cell based on Raman micro-spectroscopy coupled with chemometrics. Sensor Actuat B Chem 292:7–15. https://doi.org/10.1016/j.snb.2019.04.048
17. Qin JW, Chao KL, Moon SK (2011) Investigation of Raman chemical imaging for detection of lycopene changes in tomatoes during postharvest ripening. J Food Eng 107:277–288. https://doi.org/10.1016/j.jfoodeng.2011.07.021
18. Yang GY, Wang QY, Liu C, Wang XB, Fan SX, Huang WQ (2018) Rapid and visual detection of the main chemical compositions in maize seeds based on Raman hyperspectral imaging. Spectrochim Acta A Mol Biomol Spectrosc 200:186–194. https://doi.org/10.1016/j.saa.2018.04.026
19. Lee PC, Meisel D (1982) Adsorption and surface-enhanced Raman of dyes on silver and gold sols. J Phys Chem 86(17):3391–3395. https://doi.org/10.1021/j100214a025
20. Peng JT, Peng SL, Jiang A, Wei JP, Li CW, Tan J (2010) Asymmetric least squares for multiple spectra baseline correction. Anal Chim Acta 683:63–68. https://doi.org/10.1016/j.aca.2010.08.033
21. Eilers PHC (2003) A perfect smoother. Anal Chem 75:3631–3636. https://doi.org/10.1021/ac034173t
22. Baek SJ, Aaron P, Young JA, Jaebum C (2015) Baseline correction using asymmetrically reweighted penalized least squares smoothing. Analyst 140:250–257. https://doi.org/10.1039/c4an01061b
23. Dong CW, Yang Y, Zhang JQ, Zhu HK, Liu F (2014) Detection of thrips defect on green-peel citrus using hyperspectral imaging technology combining PCA and B-spline lighting correction method. J Integr Agric 13:2229–2235. https://doi.org/10.1016/s2095-3119(13)60671-1

24. Li QQ, Yue H, Song XZ, Zhang JX, Min SG (2019) Moving window smoothing on the ensemble of competitive adaptive reweighted sampling algorithm. Spectrochim Acta A Mol Biomol Spectrosc 214:129–138. https://doi.org/10.1016/j.saa.2019.02.023
25. Li HD, Qing SX, Liang YZ (2012) Random frog: an efficient reversible jump Markov Chain Monte Carlo-like approach for variable selection with applications to gene selection and disease classification. Anal Chim Acta 740:20–26. https://doi.org/10.1016/j.aca.2012.06.031
26. Zhang L, Wang FL, Sun T, Xu B (2018) A constrained optimization method based on BP neural network. Neural Comput Appl 29: 413–421. https://doi.org/10.1007/s00521-016-2455-9
27. Lussier F, Vincent T, Benjamin C, Gregory QW, Jean FM (2020) Deep learning and artificial intelligence methods for Raman and surface-enhanced Raman scattering. TrAC Trend Anal Chem 124:115796
28. Lucay FA, Luis AC, Edelmira DG (2020) An LS-SVM classifier based methodology for avoiding unwanted responses in processes under uncertainties. Comput Chem Eng 138: 106860. https://doi.org/10.1016/j.compchemeng.2020.106860
29. Qiu SS, Gao LP, Wang J (2015) Classification and regression of ELM, LVQ and SVM for E-nose data of strawberry juice. J Food Eng 144:77–85. https://doi.org/10.1016/j.jfoodeng.2014.07.015

Chapter 14

Gas Chromatography for Unsaturated Fatty Acid Profiles of Baobab Plant (*Adansonia digitata*)

Musa Abubakar Tadda, Abubakar Shitu, Umar Abdulbaki Danhassan, Hassan Idris Abdu, Pei Jinjin, and Songming Zhu

Abstract

Gas chromatography (GC) has been a powerful tool for the quantitative and structural analysis of various plant species' lipids. Unsaturated fatty acids (UFAs) are vital for human nutrition and health and should be a major part of any diet. Herein, a protocol to determine UFAs profiles of the baobab plant (*Adansonia digitata*) by gas chromatography (GC) detection has been described. The two most important parts were considered from the baobab plant: the pulp and the seeds. First, the baobab whole ripe fruit is dried under a shed and broken to separate the pulp from the seeds and fiber. Both the pulp and the seeds are then ground into fine powder. The powdered samples are then dissolved in ultrapure water and prepared for the GC analysis following standard operation procedures. The procedure described in this chapter is designed for 100 mg of the pulp and seed per sample to establish the UFAs profiles of the baobab plant for the determination of its relative concentration and purity.

Key words Baobab plant, Baobab fruit-pulp, Baobab seeds, Gas chromatography, Unsaturated fatty acids (UFAs)

1 Introduction

Unsaturated fatty acids (UFAs) are a class of lipids that play a crucial role in human nutrition and physiology [1]. They are integral components of cell membranes and serve as energy storage molecules [2]. Historically, UFAs have been a subject of scientific investigation, leading to a deeper understanding of their significance for human health. The identification of key fatty acids, including the omega-3 fatty acid, alpha-linolenic acid, and the omega-6 fatty acid linoleic acid, showed how important UFAs were to sustaining health [3]. Increased curiosity on dietary sources of UFAs and their possible health advantages resulted from these discoveries [4]. Additionally, UFAs have a wide range of applications, primarily in the food, pharmaceutical, and cosmetic industries. Their

Mostafa Gouda et al. (eds.), *Plant Chemical Compositions and Bioactivities*, Methods and Protocols in Food Science, https://doi.org/10.1007/978-1-0716-3938-2_14,

utilization is diverse, owing to their unique chemical properties and health benefits [5].

Baobab (*Adansonia digitata*) is a magnificent naturally occurring and longest lived tree majorly found in the arid and semiarid areas of sub-Saharan Africa (SSA) and Madagascar, with most of its parts (fruit pulp, seeds, leaves, and fibrous filaments) considered to have over 300 different applications such as in medicine, food, building materials, and so on [6, 7]. Interestingly, previous literature has found many beneficial compounds and essential nutrients, especially in the fruit pulp, such as carbohydrates, vitamin C, phosphorus, calcium, potassium, lipids, fibers, and proteins, which can serve as additives and appetizers in food [6].

For the baobab seeds, substantial amounts of sodium, iron, phosphorus, zinc, manganese, and magnesium and significant amounts of thiamine, lysine, and so on were all discovered. It has been shown that baobab seeds contain around 32% oil, and the extracted oil from the seeds can be utilized in manufacturing cosmetics and other antioxidant additives. Thus, considering the benefits mentioned above, these two special parts of the baobab plant, the pulp and the seeds, can be further investigated, targeting enhancing human health when used as food and providing the needed nutrients to plants for proper growth [8]. Besides, the pulp and seeds can be consumed directly. Roasted seeds are often utilized as coffee substitute and as food ingredient in porridges, and beverages by the local communities [9]. Moreover, the seeds could also be used as additives or as filler fiber to produce carbon sources for denitrifying wastewater, as it was proven to enhance the biodegradation of synthetic plastic materials such as poly(butylene succinate) (PBS) [10, 11].

Previously, the proximate, mineral content, and amino acid compositions of baobab seeds and pulp were investigated by Osman [12]. For example, it was reported in the literature that the major unsaturated fatty acids found in the baobab seeds were oleic acid (35.8%) and linoleic acid (30.7%), while palmitic acid (24.2%) was the major saturated acid found [12]. However, in another study conducted by Abeer et al. [13], linoleic acid (30.63%) was the highest unsaturated fatty acid found though with very similar amount as reported by Osman [12], while oleic acid (23.34%) comes second. Table 1 contains the details of the composition, structure, and other relevant parameter as reported previously by Abeer et al. [13]. In another study, baobab seed was discovered to have contained monounsaturated fatty acids (MUFA: 32–38%) and polyunsaturated fatty acids (PUFA: 22–26%), which were found to regulate body metabolism as well as reducing cholesterol levels [9, 14]. However, this majestic food source's unsaturated fatty acid profiles have not been well researched, especially using GC analysis. The baobab plant can be considered a non-model food source, as highly limited genomics,

Table 1
Unsaturated fatty acid composition of baobab seed oil [13]

Fatty acid	Formula	Systematic name	Structure	Composition (%)
Linoleic acid	$C_{18}H_{32}O_2$	9,12-Octadecadienoic acid	C18:2	30.63
Oleic acid	$C_{16}H_{34}O_2$	9-Octadecenoic acid	C16:1	23.34
Malvalic acid	$C_{18}H_{32}O_2$	Methy-l2-propene-1-butyl-cyclo heptanoic acid	C18:1	5.52
Cis-10-nonadecenoic acid	$C_{19}H_{36}O_2$	Cis-10-nonadecenoic acid	C19:1	2.67
Sterculic acid	$C_{19}H_{34}O_2$	Methy-l2-butyl-cyclopropane-1-octanoic acid	C19:1	1.61
Elaidic acid	$C_{18}H_{34}O_2$	9-Octadecenoic acid (E)	18:1	0.94
Cis-10-heptadecenoic acid	$C_{17}H_{32}O_2$	Cis-10-heptadecenoic acid	C17:1	0.48
8,11-Octadecadienoic acid	$C_{18}H_{32}O_2$	8,11-Octadecadienoic acid	C18:2	0.38
Gondoic acid	$C_{20}H_{38}O_2$	Cis-11-eicosenoic acid	C20:1	0.37
Palmitoleic acid	$C_{16}H_{30}O_2$	9-Hexadecenoic acid	C16:1	0.36
Terephthalic acid	$C_9H_8O_4$	1,4-Benzenedicarboxylic acid	C9:3	0.10
7,10-Hexadecadienoic acid	$C_{16}H_{28}O_2$	7,10-Hexadecadienoic acid	C16:2	0.02

transcriptomics, and proteomics data were available, which may lead to error or lower quality identification rates of its profiles. Therefore, this chapter aims to detail the protocols for establishing the profile of unsaturated fatty acid laden in the baobab seeds and pulp.

2 Materials

The materials comprised the baobab plant parts, the tools, devices, and equipment needed for the GC analysis of the unsaturated fatty acid profiles from the baobab plant. Ultrapure water and analytical-grade reagents will be used to prepare the solution unless indicated otherwise. Also, the prepared solutions and the reagent will be stored at room temperature unless indicated otherwise. For standards prepared, they should be stored at −20 °C. The waste materials are to be carefully disposed of following the standard laboratory regulations on disposal.

Fig. 1 Baobab plant parts: (**a**) dried whole fruit; (**b**) dried seeds; (**c**) dried fruit pulp

2.1 Baobab Plant Materials

Baobab seeds and pulp (Fig. 1).

2.2 Materials for Unsaturated Fatty Acid Derivatization

The following materials are needed for derivatizing the baobab seeds and pulp to make them suitable for the GC analysis:

1. Fume hood.
2. GC vials with Teflon-lined caps.
3. Teflon type of lined screw caps.
4. Nitrogen gas equipped sample evaporator.
5. Water bath with time and temperature control.
6. Micropipette plastic tips.
7. 12 × 100 mm glass tubes.
8. 8 mL glass vials with threads.
9. 150 mm Pasteur glass pipettes.
10. Methanolic HCl.
11. Centrifuge machine.
12. Dimethyl disulfide (DMDS).
13. HPLC-grade methanol.
14. HPLC-grade chloroform.
15. HPLC-grade water.
16. Iodine (crystalline, I_2).
17. Diethyl ether.
18. Conical flask.
19. Anhydrous pyridine.
20. Deionized water.

21. *N*-methyl-*N*-(trimethylsilyl)-trifluoroacetamide (MSTFA).
22. Sodium thiosulfate ($Na_2S_2O_3$) solution (10%, w/v) in H_2O (to be freshly prepared when before the analysis).
23. Sodium methylate solution (basic catalysis).

2.3 Gas Chromatography Details and Settings

1. Gas chromatograph (Agilent Technologies 7890A) equipped with mass detector (MS).
2. GC column for unsaturated fatty acid methyl esters (FAMEs) analysis. Nonpolar capillary type—100% dimethyl-polysiloxane (30 m × 0.25 mm, and 0.25 μm film thickness).
3. GC autosampler vials with inserts (2 mL) and Teflon-lined caps.
4. The injector should be splitless at 250 °C injection temperature and 1 μL/L injection volume.
5. Helium gas should be the carrier gas at 1 mL/min flow rate.
6. Temperature setting: 60 °C for 3 min, then 240 °C at 3 °C/min and held for 10 min.
7. The total run time is 93 min.

3 Methods

First, HPLC-analysis grade solvents should be used to rinse all glassware containers and tubes using organic chloroform–methanol (2:1, v/v) before use. Carry out all procedures at room temperature unless otherwise specified. To extract the oil from the baobab seeds, the method followed by Abeer et al. [13] was adapted with some modifications. The extracted oil from the baobab seeds was converted into fatty acid methyl ester through derivatization [15] to make them volatile (turn into a gas) for possible detection. A 2 M methanolic potassium hydroxide (KOH-solution) solution is then prepared. Measure 2 mL of the oil sample into a test tube and dissolve it in 10 mL of hexane. Then, add 1 mL of KOH-solution to the same test tube and vortex for 15 min. Collect the hexane phase, wash it twice with 4 mL of deionized water, and then dry it over anhydrous sodium sulphate. Then analyze the samples using a GC device with details of its settings described in Subheading 2.3.

For the unsaturated fatty acids from the baobab pulp, 5 mL methanol (100%) was added to 100 mg of the finely powdered pulp and then sonicated for 30 min at 53 kHz, followed by centrifugation for 10 min at 12,000 rpm. The sample preparation and measurements were conducted in triplicates under the same conditions. After that, 100 mL of the extract of methanol was aliquoted into cap-tight vials and allowed to completely dry up under a nitrogen gas stream. The dried methanol extract was then mixed with

150 mL of *N*-methyl-*N*-(trimethylsilyl)-trifluoroacetamide (MSTFA), which had been previously diluted (1:1%) with anhydrous pyridine. The mixture was then incubated for 45 min at 60 ° C before being analyzed using a GC apparatus. Hence, the mono- and polyunsaturated fatty acid compositions are summed up as the total unsaturated fatty acids.

4 Notes

1. Liquid nitrogen helps make the samples brittle to facilitate grounding them into powder and protect against potential protein degradation of the samples before analysis.
2. Usage of blank samples is recommended to identify potential contaminations of the samples.
3. It is recommended to use a 0.22 μm membrane filter to filter the samples before use further, thus reducing the chances of GC column clogging.

Acknowledgments

The authors are very grateful to Hangzhou Xiaoshan East-Sea Aquaculture Co. Ltd. for providing the funding and other necessities. We are also grateful to the Zhejiang University for the laboratory resources.

References

1. Lee H, Park WJ (2014) Unsaturated fatty acids, desaturases, and human health. J Med Food 17:189–197. https://doi.org/10.1089/jmf.2013.2917
2. Wu Y, Zhang W, Wang Y, Lu Q, Zhou J, Chen Z, Yang Z, Loor JJ (2023) Molecular mechanisms in the miR-33a/LPPR4 pathway regulating unsaturated fatty acid synthesis in bovine mammary epithelial cells. Anim Biotechnol 34:2636–2648. https://doi.org/10.1080/10495398.2022.2111308
3. Saini RK, Keum Y (2018) Omega-3 and omega-6 polyunsaturated fatty acids: dietary sources, metabolism, and significance—a review. Life Sci 203:255–267. https://doi.org/10.1016/j.lfs.2018.04.049
4. Gibson RA, Muhlhausler B, Makrides M (2011) Conversion of linoleic acid and alpha-linolenic acid to long-chain polyunsaturated fatty acids (LCPUFAs), with a focus on pregnancy, lactation and the first 2 years of life. Matern Child Nutr 7:17–26. https://doi.org/10.1111/j.1740-8709.2011.00299.x
5. Pehlivanoglu H, Ozulku G, Yildirim RM, Demirci M, Toker OS, Sagdic O (2018) Investigating the usage of unsaturated fatty acid-rich and low-calorie oleogels as a shortening mimetics in cake. J Food Process Preserv 42: e13621. https://doi.org/10.1111/jfpp.13621
6. Rahul J, Jain MK, Singh SP, Kamal RK, Anuradha AN, Gupta AK, Mrityunjay SK (2015) Adansonia digitata L. (baobab): a review of traditional information and taxonomic description, Asian Pac. J Trop Biomed 5:79–84. https://doi.org/10.1016/S2221-1691(15)30174-X
7. Kaimba GK, Mithöfer D, Muendo KM (2021) Commercialization of underutilized fruits: baobab pulp supply response to price and non-price incentives in Kenya. Food Policy 99:101980. https://doi.org/10.1016/j.foodpol.2020.101980
8. Ansari AA, Singh S (2017) Essential Plant Nutrients, 1st edn. Springer International

Publishing, Cham. https://doi.org/10.1007/978-3-319-58841-4

9. Muthai UK, Indieka AS, Muchugi A, Karori SM, Mng'omba S, Ky-Dembele C, Jamnadass R (2019) Quantitative variation of fatty acid composition in seed oil from baobab (Adansonia digitata L.) wild populations in sub-Sahara Africa. South Afr J Bot 123:1–8. https://doi.org/10.1016/j.sajb.2019.01.026
10. Tadda MA, Gouda M, Shitu A, Yu Q, Zhao X, Ying L, Zhu S, Liu D (2023) Baobab fruit powder promotes denitrifiers' abundance and improves poly(butylene succinate) biodegradation for a greener environment. J Environ Chem Eng 11:109654. https://doi.org/10.1016/j.jece.2023.109654
11. Tadda MA, Altaf R, Gouda M, Rout PR, Shitu A, Ye Z, Zhu S, Liu D (2021) Impact of saddle-chips biocarrier on treating mariculture wastewater by moving bed biofilm reactor (MBBR): mechanism and kinetic study. J Environ Chem Eng 9:106710. https://doi.org/10.1016/j.jece.2021.106710
12. Osman MA (2004) Chemical and nutrient analysis of baobab (Adansonia digitata) fruit and seed protein solubility. Plant Foods Hum Nutr 59:29–33. https://doi.org/10.1007/s11130-004-0034-1
13. Abeer AI, Azhari HN, Mahmoud MA, Ibrahim YE, Omer AOI (2020) Physicochemical properties and fatty acids composition of Sudanese baobab (Adansonia Digitata L.) Seed oil. Int J Pharma Bio Sci 11:34–42. https://doi.org/10.22376/ijpbs.2020.11.1.p34-42
14. Offiah VO, Falade KO (2023) Potentials of baobab in food systems. Appl Food Res 3:100299. https://doi.org/10.1016/j.afres.2023.100299
15. Bartels D (2020) Plant lipids. Blackwell. https://doi.org/10.1201/9780367813123

Chapter 15

Antioxidant Bioactivity of Single-Cell Microalgae Assessment by Electrochemical Sensors

Mostafa Gouda and Xiaoli Li

Abstract

One of the most significant developed technologies is the electrochemical sensors to determine the chemical structures of biological cells and their bioactivities. Besides, the new electrochemical techniques for visualizing plant cellular chemical compositions could significantly open the way toward advanced analytical technologies. For instance, electrochemical sensors based on nanogold will be used to identify the antioxidant activity. Therefore, this chapter focuses on the advanced electrochemical technologies for tracking the composition changes of microalgae single cells. Also, a few key configurations of the electrochemical sensors and their different wave patterns applications in biomedical and microfluidic media progress will be discussed..

Key words Electrochemical sensors, Quartz crystal microbalance, Emerging technologies, Piezoelectric materials, Electrochemical devices

1 Introduction

Microalgae are considered as potential new sources of antioxidants because of their composition and phytochemical content [1, 2]. The nutritional composition varies tremendously among microalgal species Important nutritional components to consider are phytochemicals, vitamins, and mineral contents, all of which are known to positively impact human health. Also, there are only a few microalgae that are generally recognized as safe (GRAS) by the FDA [3]. These algae include *Arthrospira platensis*, *Chlorella vulgaris*, *Phaeodactylum tricornutum*, and *Scenedesmus obliquus*. They are commonly consumed in the United States, Canada, China, EU, India, and Japan [4]. Also, microalgae phytochemicals scavenge a wide range of water-soluble and lipid-soluble radicals. Several studies reported that microalgae extracts have potent radical scavenging activity against 2,2-diphenyl-1-picrylhydrazyl (DPPH), hydroxyl radical, methyl radical, and nitric oxide. On the other hand,

Mostafa Gouda et al. (eds.), *Plant Chemical Compositions and Bioactivities*, Methods and Protocols in Food Science, https://doi.org/10.1007/978-1-0716-3938-2_15,

scientists are working hard to find fast and nondestructive methods to explore the scientific essence of phytochemicals' antioxidant principles, especially phenolic and flavonoid compounds [5]. These phytochemicals increase the different biological media antioxidant, anticancer, antimicrobial, and other physicochemical and biological efficiencies [6–10].

The development of electrochemical sensors is a very important scientific and technical issue that could solve the negative problems of the traditional analysis such as the fluorescence of the phytochemicals' intervention in the studies of the spectroscopic techniques. Currently, chemical and chromatographic methods are used for the evaluation and analysis of microalgae antioxidants. However, these methods are toxic and time-consuming, and they cannot obtain detailed information about the single cell without affecting its structure [11]. On the other hand, the microalgae electrochemical charge is a very important parameter for maximizing the microalgae process such as biomass harvesting. Zeta potential is a measure of the surface electrical charges of microalgae across its surface solid phase and the liquid media. In addition, it could be used as a reference method for the characterization of its electronegativity [12–14]. Electrochemical sensors are widely applied in analytical systems that monitor the biological and chemical signaling devices. This technology is based on electrochemical sensation, which has become one of the important emerging technologies. In our previous study, we discussed the electrochemical characteristics of microalgae and their influence on surface electric charge [15]. That study used single-cell electrochemical current and other physicochemical methods to analyze four microalgae species and their intracellular phenolic and flavonoid contents. The results showed promising approach in terms of higher phenolics, flavonoids, and antioxidant activities correlations. Moreover, electrochemical devices have been commercially distributed in many fields for several decades. Several of the emerging applications for electrochemical devices as sensors include biotechnological and biomedical applications (such as tracking the intra-electrochemical pulses by the single-cell chemical components). That is due to these sensors' sensitivity and intrinsic reliability.

In addition, single-cell electrochemical current by microelectrode has emerged as an increasingly important technique for fundamental studies of single-cell activities and functionality [16, 17]. It is known that the combination of biomolecules, especially antioxidants, with nanoparticles creates interesting features for the development of nanosensors [18]. The most straightforward pathway is using metal nanoparticles such as gold or silver for the detection process [19]. The critical dimension of these kinds of microelectrodes is so small, with nano-diameters, that they can be used to detect the cellular biomolecules activities. Also, gold nanoprobe biosensors can be used in micro-space electrochemistry and its pulled end can be precisely controlled down to 50 nm of attachment with the single cell [20, 21]. Recently, the nanometer-sized electrodes were used to record the intracellular reactive oxidative markers such as hydrogen peroxide and nitrogen monoxide [18]. However, specific care is needed to maintain a very small current response (10^{-10} amp) relation with the actual biochemistry of cells. Therefore, continuous development of intracellular electrochemical detection and its relationship with the chemical invasive methods and the other noninvasive methods such as micro-Raman spectroscopy should be more studied to solidify the electrochemical analysis and give spatial information on the detection [22]. Therefore, more reliable information should be obtained by correlating it with the chemical and chemometric analyses for verification of the bands' relationship with the standard chemical methods, where the concentration of a complex compound is determined by the synergistic combination of multiple bands [23, 24].

Thus, the aim of this chapter is to introduce a novel concept of living microalgae electrochemical sensation for micro-biochemical molecules detection. It presents a microalgae cellular biosystems novel method for quickly evaluate the single-cell antioxidant activity based on single-cell current to evaluate the antioxidant and electrochemical potentials of different microalgae species. Our results suggest that gold nanoprobe can be used as a nanodevice for extracting molecular information from the microalgae microenvironment. Besides, the microlevel study of single-cell phytochemicals can better explain their exact functional effects compared to the methods of macro-analysis that use millions of cells, which reduces their sensitivity. This chapter provides a microlevel understanding of single-cell phytochemicals and their functional impacts. Also, it provides the latest achievements related to the design, fabrication, modeling, testing, characterization, and advanced research trends in electrochemical sensor technology as efficient analytical methods.

2 Materials

2.1 Microalgae Species That Could Be Obtained from Shanghai Biotech. Co., Ltd. (Shanghai, China)

1. *Arthrospira platensis*.
2. *Chlorella vulgaris*.
3. *Phaeodactylum tricornutum*.
4. *Scenedesmus obliquus*.

2.2 Chemicals for Culture

1. Sodium nitrate ($NaNO_3$).
2. Calcium chloride ($CaCl_2$).
3. Magnesium sulfate ($MgSO_4$).

4. Dipotassium phosphate (K_2HPO_4).
5. Sodium chloride (NaCl).
6. Citric acid.
7. Vitamin B_{12}.
8. Deionized water.
9. Ethanol (70%) (Sinopharm Chemical, China).

2.3 Instruments

1. Undercooling centrifuge (Thermo, USA) at 5000 × *g*.
2. Ultrasonic bath (KQ-500B, 42 kHz, 100 W; China).
3. Syringe filters (nylon membrane, pore size 0.45 μm).
4. Fridge at 4 °C.

3 Methods

3.1 Sample Preparation

1. Microalgae species will be cultivated using controlled photobioreactors and standard culture media, which mainly contained (H_2O, $NaNO_3$, $CaCl_2$, $MgSO_4$, K_2HPO_4, NaCl, citric acid, and vitamin B_{12}).
2. The four microalgae species will be isolated from the culture media by centrifuge (Thermo, USA) at 5000 × *g* for 5 min under 4 °C.
3. After that, the fresh biomass will be washed 6 times using deionized water.

3.2 Single-Cell Electrochemical Characterization by Gold Microelectrode

1. Current levels of the single live microalgae cells will be measured by using gold-coated microelectrode nanoprobes designed at Rayme biotechnology labs (Rayme Biotechnology Co. Ltd., Wuxi, China) (Fig. 1).
2. Optical fiber coated with thickness of 100 nm from gold nanoparticles, and the size of its active site pulled end will be 297.7 nm.
3. The counts of current (mV) will be constantly detected and calculated for a 2000-s period.
4. Electrochemical measurements will be carried out using Rayme electrochemical station mainly including an inverted microscope (IX-71, Olympus Corp., Beijing, China) and a nano manipulator (0.5 μm precision) (Rayme Biotechnology Co. Ltd., Wuxi, China).
5. Ag/AgCl electrode calibrated with saturated KCl will be applied as the reference electrode.

The amperometry curve will be acquired at an applied potential as follows:

(a)

(Microelectrical probe)

(b)

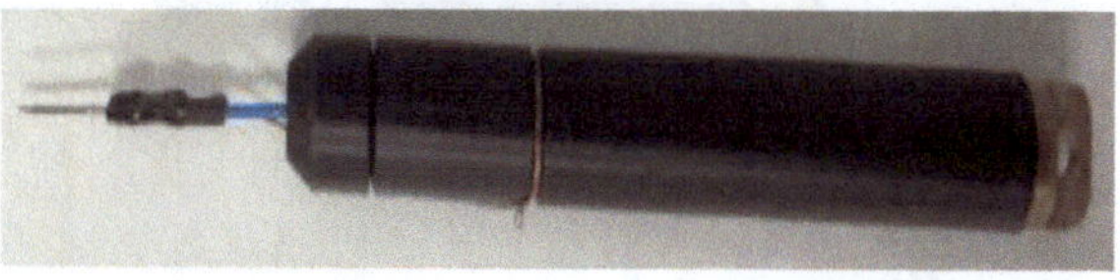

(Microelectrical probe into the fixed holder)

(c)

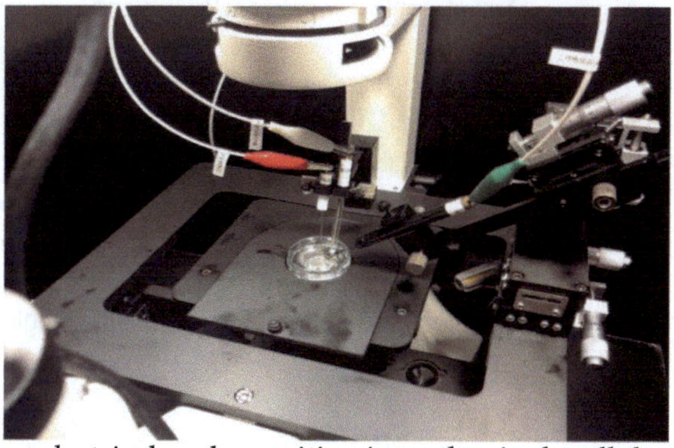

(Microelectrical probe position intro the single cell device)

Fig. 1 Used gold microelectrode accessories for microalgae single-cell antioxidants method

(a) The gold-coated nanoprobe and Ag/AgCl reference electrode will be immersed into electrodeposition solution containing 2.5 mM $FeCl_3$, 0.1 M HCl, 0.1 M KCl, and 2.5 mM $K_3[Fe(CN)_6]$ (Macklin, China) to assist the probe to attach the microalgae surface area and make an electrochemical salt bridge (Fig. 2).

(b) Amperometric curve will be set to the initial voltage of 0.4 V for 100 s.

(c) After deposition, the two electrodes will be immersed in the calibration solution containing 0.1 M KCl and 0.1 M HCl, and the electric probe will be scanned by cyclic voltammetry (CV) for 4 laps at −0.2 to 0.6 V and 50 mV/s speed (Fig. 3).

(d) Then, they will be performed in the microalgae aquatic solution and the gold nanoprobe will be attached to the single cells for the electrochemical measurement.

(e) The software will be set on the "Amperometric curve" with the following condition: initial voltage (0.6 V), run time (2000 s), and sensitivity (10^{-9} amp/V).

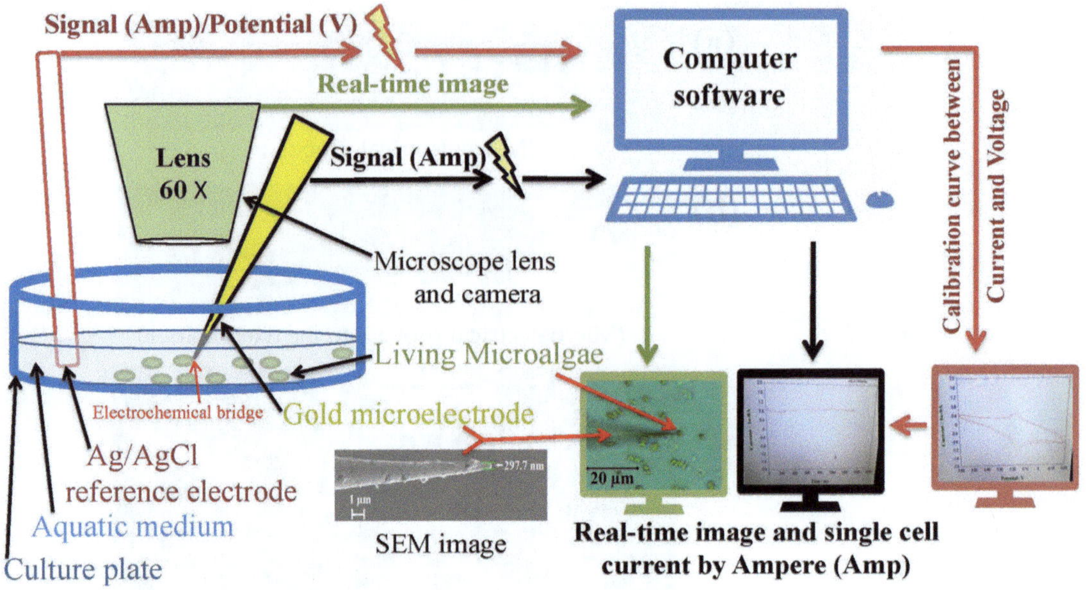

Fig. 2 Fabricated gold microelectrode method for microalgae single-cell antioxidants. (Copyright permission no.: 5750550737759)

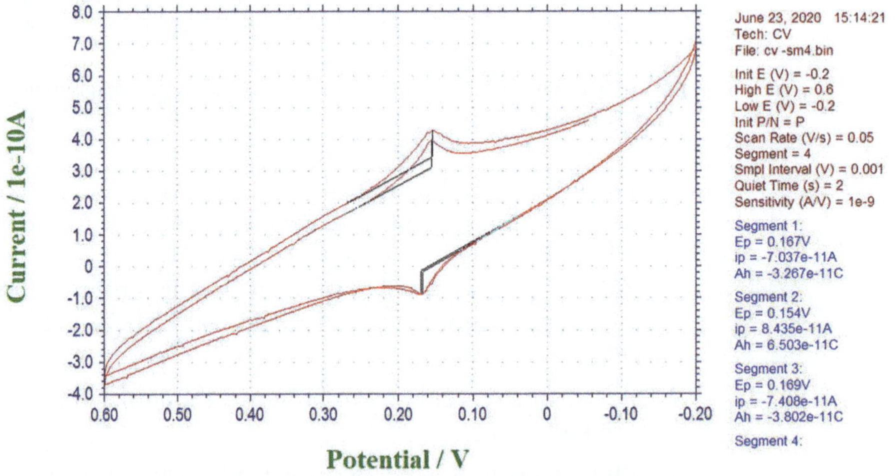

Fig. 3 Cyclic voltammetry (CV) for 4 laps at −0.2 to 0.6 V and 50 mV/s speed image

(f) Fifteen replicates will be measured from each microalgae species and the same electrode will be applied for the measurements of the four species and consider it as one replicate.

The zeta potential (ZP) of samples will be measured by using Zetasizer (Nano-ZS instrument, Malvern Instruments, Worcestershire, UK):

(a) Fresh microalgae samples will be diluted and adjusted with PBS.

(b) Samples will be injected into the apparatus, and the averages of three measurements will be reported as ZP and ZS at 25 °C.

3.3 Phytochemical Extraction

1. Ultrasound-assisted method will be used with ethanol (70%) (Sinopharm Chemical, China) as a solvent.
2. 10 mL of ethanol (70%) will be added to 0.5 g of freeze-dried biomass (DB).
3. Perform for 2 h using an ultrasonic bath (KQ-500B, 42 kHz, 100 W; China) with controlling the temperature of the water bath so that it did not exceed 25 °C by using ice cubes.
4. The liquid will be separated from solids by centrifugation at 3000 × *g* and then dried.
5. The extracted yields will be weighted and calculated, then the dry extracts will be dissolved again to give a concentration of 10 mg/mL.
6. Filter through using syringe filters (nylon membrane, pore size 0.45 μm) and store in vials in the fridge at 4 °C until analysis.

3.3.1 Antioxidant by DPPH, ABTS, and Nitric Oxide (NO)

1. Antioxidant activity by DPPH radical (Aladdin Chem., China) will be measured.
2. 20 μL of the sample (0.25–10 mg/mL extract) will be mixed with 200 μL of methanolic solution of DPPH (0.04 mg/mL) and left for 30 min in the dark.
3. The Abs will be measured at 517 nm wavelength using EPOCH2 microplate reader (BioTek, Inc., USA).
4. GAL and QUR will be used as standards [25].
5. The percentage of free radical scavenging will be calculated using the formula:

$$\text{INH\%} = [(\text{Sample (Abs)} - \text{Negative control (Abs)} \\ /\text{Negative control (Abs)})] \times 100.$$

The EC_{50} values will be calculated by using a regression equation between the concentration and the antioxidant percentage of each sample. Results will be expressed in INH%/DB, EC_{50}/DB, μg GAL/gDB, and μg QUR/gDB.

6. The antiradical power will be measured using ABTS radical scavenging method.
7. A stock solution of 2.2′-azino-bis-3-ethylbenzothiazoline-6-sulphonic acid (ABTS) (Macklin, China) will be prepared in water (7 mM).

8. ABTS stock solution will be mixed with potassium persulphate solution (2.6 mM) (1:1) and kept in the dark at room temperature for 12–16 h before use.
9. 280 μL of ABTS will be mixed with 20 μL of each sample (0.1–10 mg/mL extract).
10. QUR and trolox (T; TargetMol, China) will be used as standards. Reaction mixtures are incubated for 2 h in the dark.
11. The Abs will be measured at 734 nm.
12. The Inhibition (INH) % and EC_{50} will be measured using the same formula of DPPH method.
13. Results will be expressed as INH%/DB, EC_{50}/DB, μg QUR/gDB, and μg T/gDB.
14. 80 μL of the extract will be added to 20 μL of sodium nitroprusside solution (25 mM) for scavenging effects on nitric oxide (NO).
15. Incubate at 37 °C for 150 min.
16. A volume of 60 μL of Griess reagent (1% sulfanilamide in 5% H_3PO_4 and 0.1% naphthyl ethylenediamine dihydrochloride) will be mixed with the prepared solution in **step 14**.
17. The absorbance will be read at 570 nm, and sodium nitrite salt will be treated in the same way with Griess reagent, which will be used as a standard for sodium nitrite equivalent% result values [26].

4 Data Analysis

1. Before spectra band analysis, the experimental data will be first processed to eliminate high-frequency noise and to correct the baseline background by using Origin 8 (Massachusetts, USA).
2. Then, SPSS 16.0 (IBM, USA) and Matlab R2017b (MathWorks, USA) will be used for partial least square (PLS) model for processing the principal component analysis (PCA).
3. Characteristic spectra bands PCA with full cross-validation will be used to examine general trends in the dataset.
4. PCA scores plots and heatmaps will be used to visualize any clustering of the samples, and VIP and loadings plots will be used to determine which spectral region most contributed to the variance in the dataset.
5. Average values ± standard deviation (SD) are tabled.
6. Analysis of variance (ANOVA) and correlation efficient (r^2) will be applied among all the measurements, and $P \leq 0.05$ will be considered statistically significant or correlate and $P \leq 0.01$ will be considered as highly significant.

7. Also, the correlation figures will be made by Matlab software between wavenumber abundance (current, intensity, or transmittance) of the different spectral methods and the chemical composition values of the same group replicates.
8. Duncan, least significant differences (LSD), and Pearson tests have been calculated to measure the significance among the tested groups and properties.
9. Also, the fabricated method reliability and intraclass correlation coefficient (ICC) will be measured by Cronbach's Alpha (CA) model using SPSS 16.0.
10. The repeatability will be measured by using Gage repeatability and reproducibility (GR&R) model, and the stability of the used microelectrode will be measured by using Chow test for the system's stability by using Minitab 18.0 (Minitab HK Ltd., Hong Kong) and NumXLUI (NumXL, USA).

5 Notes

1. The volumes of media to prepare should be less than 1 mL, depending on the number of cells to be analyzed in each plate.
2. This protocol can be easily adapted to other single-cell current systems, depending on the objective of the target ROS released chemicals of the cells, and it might even be used to measure the ROS of animal cells.
3. The instrument to select the probe could use silver and gold nanoprobes or it could be modified to use the optical probe for the photosynthetic potential reaction.
4. Given the photosensitivity of some phenolic compounds, all solution preparation should be carried out with protection from light.
5. Depending on the composition and concentration of the original extract, it is important not to use the nanoprobe for more than 5 times based on its efficiency. Also, calibration procedure should be proceeded frequently to ensure the sensitivity of the used probe and to verify the detector signal.

References

1. Li X et al (2023) 13C-metabolic flux analysis of lipid accumulation in the green microalgae Tetradesmus obliquus under nitrogen deficiency stress. Bioresour Technol 388:129740
2. Zhan Z, Gouda M, Li X (2023) Raman-stable isotope technology for tracking single-cell plant metabolism. Trends Plant Sci 28(9): 1081–1082
3. Gouda M et al (2021) Physicochemical impact of bioactive terpenes on the microalgae biomass structural characteristics. Bioresour Technol 334:125232
4. Gouda M et al (2022) Microalgae bioactive carbohydrates as a novel sustainable and eco-friendly source of prebiotics: emerging health functionality and recent technologies

for extraction and detection. Front Nutr 9: 806692

5. Yu W et al (2018) The Effects of Radix Astragali water abstract on energy metabolism in rat yang-deficiency cold syndrome model through PPAR signaling pathway. Evid Based Complement Alternat Med 2018:9194362
6. Ahmed FE et al (2017) Role of melt curve analysis in interpretation of nutrigenomics' microRNA expression data. Cancer Genomics Proteomics 14(6):469–481
7. Fouad MT et al (2015) In-vitro antioxidant and antimicrobial activities of selected fruit and vegetable juices and fermented dairy products commonly consumed in Egypt. Res J Pharm Biol Chem Sci 6(2):541–550
8. Gouda M et al (2016) Three week dietary intervention using apricots, pomegranate juice or/and fermented sour sobya and impact on biomarkers of antioxidative activity, oxidative stress and erythrocytic glutathione transferase activity among adults. Nutr J 15(1):52
9. Gouda M et al (2017) Effects of four natural antioxidant phenyl terpenes on emulsifying and rheological properties of egg yolk. LWT Food Sci Technol 83:59–67
10. Gouda M et al (2018) Influence of bio-active terpenes on the characteristics and functional properties of egg yolk. Food Hydrocoll 80: 222–230
11. Li X et al (2019) Quantitative visualization of intracellular lipids concentration in a microalgae cell based on Raman micro-spectroscopy coupled with chemometrics. Sensors Actuators B Chem 292:7–15
12. Pandey A et al (2019) Experimental studies on zeta potential of flocculants for harvesting of algae. J Environ Manag 231:562–569
13. Lananan F et al (2016) Optimization of biomass harvesting of microalgae, Chlorella sp. utilizing auto-flocculating microalgae, Ankistrodesmus sp. as bio-flocculant. Int Biodeterior Biodegradation 113:391–396
14. Emami Moghaddam SA et al (2018) Potential of zeolite and algae in biomass immobilization. Biomed Res Int 2018:6563196
15. Gouda M et al (2021) Detection of microalgae single-cell antioxidant and electrochemical potentials by gold microelectrode and Raman micro-spectroscopy combined with chemometrics. Sensors Actuators B Chem 329: 129229
16. Adams KL, Puchades M, Ewing AG (2008) In vitro electrochemistry of biological systems. Annu Rev Anal Chem (Palo Alto, Calif) 1:329
17. Roxby DN et al (2020) Microalgae living sensor for metal ion detection with nanocavity-enhanced photoelectrochemistry. Biosens Bioelectron 165:112420
18. Liu H et al (2018) A multiparameter pH-sensitive nanodevice based on plasmonic nanopores. Adv Funct Mater 28(1):1703847
19. Tan SF, Anand U, Mirsaidov U (2017) Interactions and attachment pathways between functionalized gold nanorods. ACS Nano 11(2):1633–1640
20. Liu S et al (2012) Studies of ionic current rectification using polyethyleneimines coated glass nanopipettes. Anal Chem 84(13): 5565–5573
21. Fang W et al (2015) Portable SERS-enabled micropipettes for microarea sampling and reliably quantitative detection of surface organic residues. Anal Chem 87(18):9217–9224
22. He RQ et al (2016) Electrochemical visualization of intracellular hydrogen peroxide at single cells. Anal Chem 88(4):2006–2009
23. Heraud P et al (2006) Effects of pre-processing of Raman spectra on in vivo classification of nutrient status of microalgal cells. J Chemom 20(5):193–197
24. Korzeniowska K, Łęska B, Wieczorek PP (2020) Isolation and determination of phenolic compounds from freshwater Cladophora glomerata. Algal Res 48:101912
25. Messyasz B et al (2017) Valuable natural products from marine and freshwater macroalgae obtained from supercritical fluid extracts. J Appl Phycol 30(1):591–603
26. Liu J-G et al (2011) Antioxidant effects and UVB protective activity of Spirulina (Arthrospira platensis) products fermented with lactic acid bacteria. Process Biochem 46(7): 1405–1410

Chapter 16

The Impact of Rosmarinic Acid on the Oxidative Stress Metabolism: Evaluation Protocol Based on Glutathione-Dependent and -Independent Antioxidant Enzymes

Duygu Aydemir and Nuriye Nuray Ulusu

Abstract

Phytochemicals are plant-based bioactive molecules produced by plants for their regular metabolism and to protect them against pathogens. Phytochemicals are classified into six classes according to their chemical structure: carbohydrates, lipids, phenolics, terpenoids, alkaloids, and other nitrogen-containing compounds. Several studies have reported that phytochemicals exert health-promoting effects such as anticancer, antioxidant, anti-inflammatory, antimicrobial, cardioprotective, anti-allergic, immune-modulatory, and antiaging effects. Among various phytochemicals, rosmarinic acid has been reported as a powerful antioxidant agent due to its ROS-scavenging properties in vitro and in vivo studies. Rosmarinic acid ($C_{18}H_{16}O_8$) can be isolated from herbal balm mint plants, including Rosmarinus officinalis, Prunella vulgaris, and Melissa officinalis, containing two aromatic amino acids, phenylalanine, and tyrosine. Rosmarinic acid is extracted from approximately 162 different plants with different extraction solutions and chromatographic methods, HPLC, capillary electrophoresis, gas chromatography, and other separation techniques including vibration, maceration with continuous stirring, heat reflux, Soxhlet solvent extraction, ultrasound-assisted extraction, and ethanol extraction. Rosmarinic acid is also characterized by HPLC, gas chromatography, GC-MS, mass spectrophotometry, and proton NMR analysis.

Rosmarinic acid is accepted as a potent antioxidant agent, confirmed by several studies; for instance, it alleviates adverse effects of UV radiation by enhancing antioxidant response in the cell and exerts excellent scavenging properties on the free radicals, including per hydroxyl radical· and hydroxyl radical. Oxidative stress metabolism is described by the balance between the antioxidant response and the reactive oxygen species (ROS) production. Reduced antioxidant response or enhanced production of ROS contributes to the increased oxidative stress in the cells and the tissues, contributing to the disease pathogenesis, including diabetes, cancer, metabolic syndrome, endocrine disorders, and autoimmune diseases. In this book chapter, we aim to give detailed information about the health-promoting effects of phytochemicals, especially rosmarinic acid, on the human health associated with oxidative stress metabolism. Isolation, characterization, and in vitro bioactivity evaluation of phytochemicals and rosmarinic acid on oxidative stress metabolism have been explained in detail. In this context, cell culture administration, bioavailability, cell lysate isolation, and specific enzyme activity techniques have been defined over glucose-6-phosphate dehydrogenase (G6PD), 6-phosphogluconate dehydrogenase (6PGD), glutathione reductase (GR), glutathione s-transferase (GST) and glutathione peroxidase (GPx) enzymes, which can be used in the bioactivity evaluation of the phytochemicals on the biological samples associated with the oxidative stress metabolism.

Mostafa Gouda et al. (eds.), *Plant Chemical Compositions and Bioactivities*, Methods and Protocols in Food Science,
https://doi.org/10.1007/978-1-0716-3938-2_16,

Key words Oxidative stress metabolism, Cancer, Phytochemicals, Rosmarinic acid, Bioactivity, Glutathione

1 Introduction

Plants comprise more than 5000 phytochemicals, which are plant-based molecules required for their regular metabolism and to protect against pathogens. Generally, phytochemicals are classified into six classes according to their chemical structure: carbohydrates, lipids, phenolics, terpenoids, alkaloids, and other nitrogen-containing compounds [1]. These molecules can also be classified according to their molecular properties: proanthocyanidins, organosulfur compounds, polyphenols, flavonoids, phenolic acids, carotenoids, lignans, and curcumins [2]. The classification changes according to the researchers' opinions; for instance, the plant's bioactive compounds can be categorized into three groups: in the first group, there were approximately 25,000 types of bioactive compounds classified under terpenes and terpenoids; in the second class, there were about 12,000 types of compounds named alkaloids; and in the final group, 8000 types of compounds were found and named phenolic compounds [3]. The phytochemicals' classification can also be done according to the molecules that have health benefits for humans such as carotenoids, phytosterols, limonoids, polyphenols (flavonoids, isoflavonoids, anthocyanidins), glucosinolates, phytoestrogens, terpenoids, fibers, polysaccharides, and saponins [4]. Polyphenols have many biochemical activities and many roles in human health such as antioxidant, anticarcinogenic, and antimutagenic [5], and protective roles against disease and infections [6].

In general, most of the phytochemicals have antioxidant, antimicrobial, antifungal, antivirus, anticancer, anti-inflammatory, antiaging, immune-modulatory, and hepatoprotective properties [1, 7]. These molecules have crucial roles in detoxifying toxic radicals and pharmaceutical potential that can reduce the risk of diseases [7]. Phytochemicals can be termed phytonutrients positively affecting metabolism, such as protection against oxidative stress, inflammation, glucose intolerance, insulin resistance, hypertension, and dyslipidemia [8]. The differences in chemical structures and bonding arrangements define the basic chemical and physical properties, molecular characteristics, and biological activities of the phytochemicals. Therefore, different structures affect the diversity of the molecular interaction with different molecules and cause different biological activities [9]. Sulforaphane, curcumin, allicin, resveratrol, and other polyphenol molecules that are isolated from grapes have antioxidant properties [8, 10]. The flavonoid polyphenols such as epigallocatechin-3-gallate (EGCG) are found in green tea, and quercetin in apples has antioxidant and health-promoting effects on humans [11]. On the other hand, curcumin and resveratrol are accepted as non-flavonoid polyphenols. Rosmarinic acid or carnosic acid are accepted as phenolic acids and phenolic diterpenes that can be isolated from rosemary. Garlic and broccoli contain thiosulfonate, isothiocyanate and L-sulforaphane. These natural compounds have antioxidant properties and known as organosulfur compounds [8, 10]. The alkaloids have different medicinal properties; for instance, morphine and cocaine can be used as painkillers; quinine for fever; vinblastine and berberine for malignancy; ephedrine for asthma; quinidine for dysrhythmias; piperine for hyperglycemia; reserpine for hypertension; and ciprofloxacin for bacterial infections. On the other hand, piperidine alkaloids have respiratory arrest effects and are known as the "killer of Socrates" [12].

In vitro, plant extracts' enzyme inhibitory or activator effects can be evaluated using pure enzymes or biological sources. Different concentrations of plant extracts can be directly added to the enzymatic reaction mixture, and enzyme activities of interest can be measured spectrophotometrically by using pure enzymes, either purchased or isolated [13]. Enzymes can be purified from different biological sources, or pure enzymes of interest can be bought from companies. On the other hand, healthy and cancer cell lines are treated with increased concentrations of plant extracts to evaluate their impact on different metabolisms, including oxidative stress, aging, inflammation, glucose, pentose phosphate pathway (PPP), and energy in both cytosol and mitochondria. Antiaging, anticancer, antiproliferative, anti-inflammatory, health-promoting, antioxidant, and immune-modulatory effects of plant extracts are evaluated by using different cell lines depending on the aim of the study [14–18]. In this book chapter, we aim to give detailed information about the impact of phytochemicals, especially rosmarinic acid, on the human health associated with oxidative stress metabolism. Moreover, we provide detailed information about the isolation, characterization, and in vitro evaluation of the phytochemicals on oxidative stress metabolism. We explained cell culture administration, bioavailability, cell lysate isolation, and specific enzyme activity techniques that can be used in the bioactivity evaluation of the phytochemicals on the biological samples.

1.1 Rosmarinic Acid and Its Impact on Oxidative Stress Metabolism

Rosmarinic acid ($C_{18}H_{16}O_8$) was first isolated from rosemary [19] containing two aromatic amino acids: phenylalanine and tyrosine [20]. It is one of the well-known phytochemicals, an ester of caffeic acid and 3,4-dihydroxyphenyllactic acid [21, 22]. Rosmarinic acid is isolated from 162 different plants with different extraction solutions and methods, including vibration, maceration with continuous stirring, heat reflux, Soxhlet solvent extraction [19], ultrasound-assisted extraction [23], and ethanol extraction

[24]. Metabolic and enzyme inhibitory effects of rosmarinic acid have been widely investigated in both in vivo and in vitro studies [25–27]. It can be isolated from herbal balm mint plants, such as *Rosmarinus officinalis*, *Prunella vulgaris*, and *Melissa officinalis* [6]. Several studies have reported that rosmarinic acid exerts anti-cancer, antioxidant, anti-inflammatory, antimicrobial, cardioprotective, anti-allergic, immune-modulatory, and antiaging effects. Among indicated health effects, rosmarinic acid is known as a powerful antioxidant agent due to its ROS-scavenging properties [19].

Oxidative stress metabolism is described by the balance between the antioxidant response and the reactive oxygen species (ROS) production. Reduced antioxidant response or enhanced production of ROS contributes to the increased oxidative stress in the cells and the tissues. Antioxidant enzymes, including glucose-6-phosphate dehydrogenase (G6PD), 6-phosphogluconate dehydrogenase (6PGD), glutathione reductase (GR), glutathione s-transferase (GST), and glutathione peroxidase (GPx) are mainly responsible for the regulation of the oxidative stress metabolism [14, 28]. G6PD is the rate-limiting enzyme of the pentose phosphate pathway (PPP), and together with 6PGD, they are responsible for NADPH production. G6PD and 6PGD enzymes are vital for oxidative stress metabolism since they produce NADPH, which the GR enzyme uses to convert GSSG to GSH [14, 17, 29]. Glutathione is the primary antioxidant molecule in the cell and tissues, existing in two forms: reduced glutathione (GSH) and oxidized glutathione (GSSG). GSH is the desired form in cells and tissues since it is responsible for detoxification and antioxidant reactions. GSH is used by GST, where GSSG is converted into GSH by the GR enzyme [28, 30]. The GSH/GSSG ratio is the major indicator of the cell's antioxidant response and is considered a significant biomarker for oxidative stress status [31]. GST enzymes are multifunctional, catalyzing GSH conjugation with xenobiotics, electrophiles, or ROS to form glutathione-conjugates, which can be excreted from the body [32].

Rosmarinic acid ($C_{18}H_{16}O_8$) was first isolated from rosemary [5], containing two aromatic amino acids, phenylalanine, and tyrosine. It can be separated from herbal balm mint plants, such as *Rosmarinus officinalis*, *Prunella vulgaris*, and *Melissa officinalis* [6]. Several studies have reported that rosmarinic acid exerts anti-cancer, antioxidant, anti-inflammatory, antimicrobial, cardioprotective, anti-allergic, immune-modulatory, and antiaging effects. Rosmarinic acid is accepted as a potent antioxidant agent, confirmed by several studies; for instance, it alleviates adverse effects of UV radiation by enhancing antioxidant response in the cell [33]. It exerts excellent scavenging properties on the free radicals, including HOO· and HO· [34]. Oral rosmarinic acid administration treated allergic rhinoconjunctivitis via scavenging of ROS and inhibition of the inflammation, according to the literature [35]. Furthermore, anti-Warburg effects of rosmarinic acid have been reported on gastric cancer in vitro and in vivo, which can result from inhibitory effects of rosmarinic acid on metabolic enzyme activities such as G6PD, 6PGD, GR, and GST at high concentrations [13, 36]. Since rosmarinic acid inhibits PPP and glycolysis enzymes, including G6PD and 6PGD, at higher concentrations, it contributes to the regulation of glucose metabolism and, thus, diabetes [37]. On the other hand, rosmarinic acid also inhibits the amylase enzyme, and amylase enzyme inhibitors have modulatory roles in diabetes [38]. Although rosmarinic acid and other phytochemicals exert health-promoting effects on animals and humans, they can exert adverse effects at higher dosages, such as blocking iron uptake, prooxidative activity, mutagenic, carcinogenic, and genotoxic impacts. Therefore, dose-dependent studies should be widely performed to determine the safest and most effective dosages of phytochemicals, including rosmarinic acid [39].

2 Materials

6-Phosphogluconate (6-PG), glucose-6-phosphate (G6P), nicotinamide adenine dinucleotide phosphate ($NADP^+$), reduced nicotinamide adenine dinucleotide phosphate ($NADPH + H^+$), oxidized glutathione (GSSG), magnesium chloride ($MgCl_2$), cOmplete™ Protease Inhibitor Cocktail, sodium phosphate monobasic and dibasic, tris [Tris (hydroxymethyl) aminomethane], Baker's yeast glutathione reductase (specific activity 1.25 U/mg) (GR), hydrogen peroxide (H_2O_2), sodium azide, and ethylenediaminetetraacetic acid (EDTA) were obtained from Sigma-Aldrich Inc. (St. Louis, MO, USA). All chemicals were obtained with the highest purity and used without further purification. Pierce™ BCA Protein Assay Kit was purchased from Thermo Scientific (USA) [15, 40]. For purification of G6PD and GR enzymes, 2′5′-ADP-Sepharose 4B affinity chromatography and DEAE Sepharose Fast Flow ion exchange column chromatographies were used after the homogenization of the tissues and ultracentrifugation steps [41, 42]. For extraction of rosmarinic acid, methanol (MeOH), $ChCl_3$–MeOH, and EtOAc–MeOH–H_2O are used; for isolation of rosmarinic acid, silica gel chromatography and Sephadex LH-20 chromatography columns were used; and for characterization of the isolated rosmarinic acid, ^1H NMR, ^{13}C NMR, COSY, HMQC, HMBC 500 MHz, CD3OD spectral methods were used [43].

3 Methods

3.1 Extraction and Isolation of Bioactive Compounds

The extraction and isolation of the bioactive compounds and herbal preparations for medicinal usage from specific plants need analytical methodologies and techniques. Special screening assays can do this using one or more chromatographic and non-chromatographic techniques [44]. The extraction of bioactive compounds can be done conventionally with Soxhlet extraction, maceration, hydro-distillation, or nonconventional methods. The nonconventional extraction methods may be classified as ultrasound-assisted extraction, pulse electric field extraction, enzyme-assisted extraction, microwave-assisted and pressurized-assisted extraction methods. Supercritical fluid extraction is one of the methods that can also be used in decaffeinated coffee [3]. In the isolation of phytochemicals from the target plant, the chemical properties of the target bioactive compound are very important, for example, if the bioactive compound is volatile then the best extraction method would be hydrolyzation, supercritical fluid extraction, and shaking maceration [45]. Sequential extraction methods can be used to increase the yield of specific bioactive compounds [46]. Ionic liquid analogs that are also known as deep eutectic solvents (DES) have been used for the extraction of phenolic acids, flavonoids, tanshinone, keratin, tocols, terpenoids, carrageenans, xanthones, isoflavons, α-mangostin, genistin, apigenin, and other bioactive compounds from plants [47]. It is very important to protect the biological activity during the extraction of phytochemicals [48]. Other green extraction techniques apply to bioactive compounds besides DES such as ultrasound-assisted extraction (UAE), supercritical fluid extraction (SFE), and supercritical water extraction (SWE) [49]. The chromatography techniques such as gel filtration, silica gel chromatography, ion-exchange chromatography, preparative liquid chromatography, HPLC, preparative gas chromatography, supercritical fluid chromatography, and countercurrent chromatography methods can be used for obtaining the target substances and separating them from the other biological molecules [50, 51]. These techniques are not limited to extraction and chromatography. During the isolation of desired molecules, other separation techniques also can be used such as microwave-assisted extraction, ultrasonic-assisted extraction, hydrodistillation, steam distillation, precipitation, and/or membrane technology and equipment (ultrafiltration, affinity membranes) [50, 52, 53].

The isolation must be done in a short time to protect their antioxidant, antimicrobial, and other bioactive properties [51]. The identification, isolation, and characterization of the target bioactive phytochemicals and phytomedicines using the best methods is crucial. After isolation, the bioactive compound activity must not be reduced or lost due to environmental and storage factors [54].

The first step is the identification and occurrence of the bioactive compounds for biological activities and potential pharmacological applications [55]. There are various techniques used for the identification of target molecules from the plants such as UV-visible spectroscopy, infrared spectroscopy, nuclear magnetic resonance spectroscopy, mass spectrometry (MS), proton transfer reaction (PTR)-MS, inductively coupled plasma (ICP)-MS, and nuclear magnetic resonance spectroscopy [52, 56, 57].

3.2 Isolation and Characterization of the Bioactive Compounds and Rosmarinic Acid

Rosmarinic acid is a water-soluble polyphenol. In the isolation of rosmarinic acid, clean rosemary leaves or other species rich in rosmarinic acid can be used. The solvents are very important for a good yield; for instance, methanol, ethanol, *n*-butanol, and water must be used for the extraction of this bioactive compound [43]. Cellulase A and Protamex mixture also can be used in the extraction of rosmarinic acid. Flavourzyme, chamzyme, protamex, papain, and bromelain enzymes can be used in the extraction of rosmarinic acid [58]. Heat-assisted extraction, microwave-assisted extraction, and ultrasound-assisted extraction can be used in the extraction of rosmarinic acid [59]. HPLC, UPLC [58], SephadexLH-20, and silica gel chromatographies were used in the isolation of rosmarinic acid and after isolation, characterization of the isolated rosmarinic acid, ^{1}H NMR, ^{13}C NMR, COSY, HMQC, HMBC 500 MHz, CD3OD spectral methods [43], and liquid chromatography-tandem mass spectrometry was used to analyze rosmarinic acid [60].

3.3 In Vitro Evaluation of the Bioactivity of Phytochemicals Including Rosmarinic Acid on the Oxidative Stress Metabolism

In vitro, inhibitory or activator effects of plant extracts can be evaluated with different techniques and by using various biological sources such as tissues or cells. Different concentrations of plant extracts can be directly added to the enzymatic reaction mixture, and enzyme activities of interest can be measured spectrophotometrically by using pure enzymes, either purchased or isolated [13]. Enzymes can be purified from different biological sources such as the liver, kidney, brain, spleen, lung, and cell lines, or pure enzymes of interest can be purchased from companies. On the other hand, healthy and cancer cell lines are treated with increased concentrations of plant extracts to evaluate their impact on different metabolisms, including oxidative stress, aging, inflammation, glucose, PPP, and energy in both cytosol and mitochondria [14–18].

Anti-aging, anticancer, antiproliferative, anti-inflammatory, health-promoting, antioxidant, and immune-modulatory effects of plant extracts are evaluated by using different cell lines depending on the aim of the study. First, IC50 and cell viability assays must be performed to assess the most effective dosages of plant extracts of interest following isolation and characterization assays are performed. After MTT assays, the most effective dosages are determined and administered to the cell lines for up to 72 h. Depending on the aim of the study, cell lines can be collected every 24 h, as well as 24, 48, and 72 h, to evaluate the time-dependent effects of the plant extracts. Collected cells following treatments are used as the enzyme sources and compared to the untreated control cells to evaluate different impacts of plant extracts such as antiaging, anticancer, antiproliferative, anti-inflammatory, and antioxidant on the cell lines.

3.4 Cell Culture Treatments

Cells are seeded in the flasks or cell culture dishes and supplemented with growth medium (RPMI, DMEM, McCoy's, MEM, modified MEM, IMEM) depending on the cell type supplemented with 10% heat-inactivated fetal bovine serum (FBS) and 1% penicillin-streptomycin. Cells were kept at 37 °C under 5% CO_2 in the incubator. Cells are seeded into the new flasks at 80–90% confluence every 2–3 days to maintain a continuous logarithmic growth phase. Old medium in the flask is aspirated and cells are washed with sterile PBS; the following cells are trypsinized with 0.01% trypsin/EDTA for 2–3 minutes at 37 °C in the CO_2 incubator. After trypsinization, cells are centrifuged at 400 × *g* for 3 min at room temperature, and pellets are collected for further experiments [16].

3.4.1 Evaluation of Bioavailability and IC50 Values of Plant Extracts Via 3-(4,5-Dimethylthiazol-2-yl)-2,5-Diphenyltetrazolium Bromide (MTT)

As described above, cells are trypsinized and collected into complete cell media containing 10% FBS and 1% penicillin-streptomycin. After centrifugation, the cell pellet was resuspended in fresh cell media, and 10 μL cell suspension was mixed with 10 μL of trypan blue to count the live cells. The cell and dye mixture are placed on the hemacytometer, and the cell number is counted under light microscopy. Cells are cultured at 10,000–15,0000 cells/well in 96-well plates depending on the duplication time of the cells in complete cell media. After 24 h incubation, the old medium is aspirated, and the fresh medium is replaced in the 96-well plate. Cells are treated with the increased concentration of plant extract (0.5–400 μg/mL), and untreated cells are used as the control group to be compared with the treated groups.

Cells are incubated for up to 72 h (24, 48, and 72 h), and after the incubation period, old media is aspirated. A total of 160 μL of complete fresh media was added onto the cells, and 40 μL of 3-(4,5-dimethylthiazol-2-yl)-2,5-diphenyltetrazolium bromide (MTT) solution was prepared as 5 mg/mL MTT solution in 1 mL sterile PBS is added into the fresh cell media. The MTT substrate is converted to formazan by viable cells, and quantifying formazan by spectrophotometric methods enables examination of cell proliferation and viability. Cells are incubated with MTT solution for 4 h, and then all media is replaced with 200 μL of the DMSO:EtOH (1:1) solution. 96-well plate is incubated on the shaker at room temperature to dissolve the purple formazan product, indicating the number of viable cells. The formazan amount in each well is measured by its absorbance intensity at 560 nm using a microplate reader. Percent viability is determined by comparing the absorbance average of four biological replicates for each dose, and the absorbance average of untreated control cells is used as 100% viability. Statistical analysis, including cell viability and IC50 value, is performed by one-way ANOVA with Tukey's multiple comparison test of the GraphPad Prism 9.0 software from GraphPad Software, Inc., USA.

3.4.2 Cell Lysate Preparation

Most effective dosages are determined by MTT, and increased concentrations of plant extract are administered into the cells for up to 72 h (24, 48, and 72 h). Cells are seeded into the 25 cm^2 flasks and incubated for 24 h. Following incubation, cells are administered with different concentrations of plant extracts and untreated cells are used as a control group to be compared with treated groups. After incubation, old media is aspirated and flasks are replaced with ice. Buffer containing PBS and protease inhibitors are added onto the cells, and cells are scraped off with a scraper. Cells are sonicated for 4 s with Branson W250 at 40% amplitude and then centrifuged at 4000 × g for 30 min at 4 °C. The supernatant is collected and used as a source of different enzymes found naturally in the cells, such as G6PD, 6PGD, GR, GST, and GPx.

3.4.3 Evaluation of G6PD Activity

The enzymatic assay mixture consists of 10 mM $MgCl_2$, 0.2 mM $NADP^+$, double distilled water (ddH_2O), and 0.6 mM G6P in 100 mM Tris–HCl buffer (pH 8.0). Cell lysate, tissue lysate, or pure G6PD enzyme can be used as an enzyme source, depending on the aim of the study. G6PD activity is measured using a spectrophotometer by following NADPH production monitored at 340 nm for 60 s at 37 °C using the modified method described previously [61]. The reaction is observed for 60 s since it should be linear during reaction time, and 1 unit (U) of activity is defined as the amount of enzyme required to reduce 1 mmol $NADP^+$/min under the assay conditions. Assays are performed in triplicates, and specific activity is defined as the number of units/mg of protein.

3.4.4 Evaluation of 6PGD Activity

The enzymatic assay mixture consists of 10 mM $MgCl_2$, 0.2 mM $NADP^+$, ddH_2O, and 0.6 mM 6PG in 100 mM Tris–HCl buffer (pH 8.0). Cell lysate, tissue lysate, or pure 6PGD enzyme can be used as an enzyme source, depending on the aim of the study. 6PGD activity is measured using a spectrophotometer by following NADPH production monitored at 340 nm for 60 s at 37 °C using the modified method described previously [62]. The reaction is observed for 60 s since it should be linear during reaction time, and 1 unit (U) of activity is defined as the amount of enzyme required to reduce 1 mmol NADP+/min under the assay conditions. Assays are performed in triplicates, and specific activity is defined as the units per mg of protein (units/mg protein) [8].

3.4.5 Evaluation of GR Activity

GR activity is measured by using the modified Staal method reported previously. The reaction mixture contains 100 mM sodium phosphate buffer (pH 7.4), 0.2 mM NADPH, ddH_2O, and 1 mM GSSG. Tissue lysate, cell lysate, or pure enzyme can be used as the enzyme source, depending on the aim of the study. A decrease in the absorbance of NADPH at 340 nm was followed using a spectrophotometer at 37 °C for 60 s. The reaction is observed for 60 s since it should be linear during reaction time, and 1 unit (U) of activity is defined as the amount of enzyme required to reduce 1 mmol $NADP^+$/min under the assay conditions. Assays are performed in triplicates, and specific activity is defined as the number of units/mg of protein [63, 64].

3.4.6 Evaluation of GST Activity

GST enzyme activity is evaluated by measuring the conjugation of GSH with 1-chloro-2,4-dinitrobenzene (CDNB), as described previously [65]. The enzymatic reaction mixture contains 0.2 M sodium phosphate buffer (pH 6.5), 20 mM GSH, 20 mM CDNB, and ddH_2O. Tissue lysate, cell lysate, or pure enzyme can be used as the enzyme source, depending on the aim of the study. A decrease in the absorbance at 340 nm was followed using a spectrophotometer at 37 °C for 30 s. The reaction is observed for 30 s since it should be linear during reaction time. Assays are performed in triplicates, and specific activity is defined as the number of units/mg of protein [28].

3.4.7 Evaluation of GPx Activity

The reaction mixture consists of 200 mmol/L potassium phosphate buffer (pH 7.4), pure GR enzyme (10 U/mL), 200 mmol/L EDTA, 400 mmol/L sodium azide, 2 mmol/L NADPH, 100 mmol/L GSH, and ddH_2O. Tissue lysate, cell lysate, or pure enzyme can be used as the enzyme source, depending on the aim of the study. Samples are incubated for 10 min at 37 °C, and after incubation, 5 μL of 10 mmol/L H_2O_2 is immediately added into the reaction mixture to initiate the reaction. This assay is based on coupled enzyme activity by NADPH and GR. GPx activity is measured spectrophotometrically for 30 s at room temperature [66].

3.4.8 Evaluation of Soluble Protein Concentration in the Cell Lysates

According to the kit instructions, the BCA assay measures soluble protein concentrations in the mitochondria and cytosol lysates by using albumin as standard (Pierce™ BCA Protein Assay Kit).

3.5 Statistical Analysis

Statistical analysis is performed by GraphPad Prism (9.0) by one-way analysis of variance (ANOVA) followed by Tukey's post hoc test for multiple comparisons. The results are represented as the mean ± standard deviation (SD). *P*-value ≤0.05 was considered significant for each data.

4 Notes

4.1 Target Phytochemicals that Could Be Considered for Their Bioactivities by Current Enzymatic Method

4.1.1 Polysaccharides

Polysaccharides are the biggest and most influential group in nature [67]. Polysaccharides must be categorized as biopolymers under a special group that is composed of homopolysaccharides and heteropolysaccharides [68]. This group contains starch, cellulose, and pectin, which have key roles as an energy source that is required in the diet, cellular communication, structural support, antitumor, and energy source of almost all living organisms [69–71]. Polysaccharides can also affect gut microbiota and regulate lipid metabolism in high-fat diets [72]. Polysaccharides have regulatory roles in ferroptosis, fatty liver disease, drug-induced liver injury, and liver cancer [73]; for instance, tea polysaccharides play an antioxidant role in lipid metabolism [74]. Plant carbohydrates starch and cellulose are composed of single sugars glucose polymers that have a great role in the carbohydrate metabolism glycolysis and pentose phosphate pathway and other metabolic pathways of carbohydrate metabolism [75–79]. Increased consumption of plant polysaccharides, such as cellulose and starch can induce activity of glucose transporters and enzyme activies involved in the glucose metabolism [80, 81]. The enzymes regulated by allosteric regulators, hormones, activators, substrates, their catalytic products, and other metabolic products are under the control of their substrates such as polysaccharides [82–84].

4.1.2 Carotenoids

Carotenoids are a group of natural pigments that are found in fruits and vegetables. These molecules have different colors: orange, purple, yellow, and orange [85]. Carotenoids are accepted as antioxidant and prooxidant molecules that are isolated from various plants such as carrots, tomatoes, and turnips. These molecules can reduce the risk of some diseases such as cancer and cardiovascular diseases, which are mediated by reactive oxygen species [86]. Carotenoids have two important properties: the first one is the direct scavenging of radicals and the second one is the increasing activity of enzymes [87]. Paraoxonase 1 (PON1) is one of the antioxidant enzymes. PON1 enzymes have also lipophilic properties and carotenoids increase the activity of PON1 activity. Astaxanthin β-carotene and lycopene can increase the activity of PON1. Carotenoids can affect the gene expression of the PON1 enzyme; for instance, lycopene may increase the PON1 expression by inhibiting PON1 gene methylation [88]. Carotenoids decrease the superoxide dismutase (SOD) and increase the activity of catalase (CAT) enzyme activities [89]. Carotenoids also modulate other antioxidant enzyme activities such as glutathione-S-transferase (GST), acetylcholinesterase (AChE) [90], quinone reductase, glutathione peroxidase [91], 6PGDH, and G6PD [92].

4.1.3 *Phytosterols*

Phytosterols are another important group of phytochemicals found in seeds, fruits, and vegetables. The structure and function of phytosterols are like cholesterol and can be classified as sterols and stanols [93]. Phytosterols have anticancer properties and increase the activity of some antioxidant enzymes such as SOD and GPx. Besides this important anticancer property, they can also decrease the absorption of cholesterol from the intestine, lower the LDL concentration, and protect against cardiovascular diseases [94]. Phytosterols also affect carbohydrate metabolism and increase some antioxidant enzyme activities of CAT, GR, chlorophyllase, and carbonic anhydrase enzyme activities and decrease the hydrolytic enzyme activities [95]. It has been investigated that beta-sitosterol upregulated the PON1 and peroxisome proliferator-activated receptor gamma (PPAR-gamma) gene expression [96]. On the other hand, phytosterols selectively inhibited the cyclooxygenase-1 (COX-1) and/or cyclooxygenase-2 (COX-2) enzyme activities [97]. Consumption of almonds plays a crucial role in health. It has been shown that the consumption of almonds has hypocholesterolemic effects and can reduce the low-density lipoprotein cholesterol (LDL-C). The molecular mechanism may depend on the inhibition of some enzymes via phytosterols such as acyl-CoA:cholesterol acyltransferase (ACAT) and 3-hydroxy-3-methylglutaryl-CoA (HMG-CoA) reductase [98]. Withanolides are derived from the phytosterol pathway [99] and they are important natural molecules. Withanolides can inhibit the acetylcholinesterase and butyrylcholinesterase enzymes. Both enzymes acetylcholinesterase and butyrylcholinesterase are directly correlated with Alzheimer's disease [100]. Alzheimer's disease is one of the important irreversible neurodegenerative disorders that affect aging population [101–104]. Withanolides may have a key role in the treatment of Alzheimer's disease [100, 103].

4.1.4 *Limonoids*

Limonoids are the healthiest group of phytochemicals, and these molecules are classified under the triterpenoids. Limonoids are found in sour fruits. The name limonoid comes from limonin [105]. Citrus fruit contains many active compounds, limonoids. Limonoids upregulate the detoxification enzymes such as GST enzymes' activity [106] Glutathione-S-transferase enzymes are classified under the Phase II detoxification enzymes that protect cells from reactive oxygen species [15, 77, 78, 107, 108]. They have very crucial roles in anticancer therapy such as colon, breast, pancreas, leukemia, and neuroblastoma. Limonoids have antiproliferative and anti-aromatase properties. They can inhibit aromatase and COX-2 enzymes. It has been demonstrated that limonoids can activate caspase-7 enzyme activities, and induction of caspase-7 enzyme activities is directly correlated with the cytotoxicity and antiproliferative effects of limonoids [109]. Limonoids have an inhibitory effect on α-glucosidase enzyme and the activity of these enzymes is needed for the digestion of carbohydrates. Inhibition of α-glucosidase enzyme is very valuable in the treatment of diabetes [110, 111]. The usage of limonin molecules in health may be very useful in inhibiting cancer cell proliferation. These molecules have also positive effects on cardiovascular dysfunction and abnormalities, which are shown in rats and control ductal adenocarcinoma growth and metastasis [112, 113].

4.1.5 Polyphenols

Polyphenols are a well-known group of phytochemicals, and they can be classified according to their number of phenol rings such as phenolic acids, flavonoids, stilbenes, and lignans [114]. They are also classified as flavonoids and non-flavonoid molecules. Flavonols, flavanols, flavones, flavanones, isoflavones, and anthocyanins are known as flavonoids. On the other hand, phenolic acids, hydroxycinnamic acids, lignan, stilbenes, and tannins are grouped under the name of non-flavonoid molecules [115]. Flavonoids have dual activity: in healthy cells, they have antioxidant properties; however, in the cancer cell metabolism, they are pro-oxidants [116]. Flavones have antioxidant and anti-inflammatory properties, and they are accepted as health-promoting phytochemicals [117]. Polyphenols have enormous effects on immune regulations and antioxidant activities on human health. Polyphenolic compounds inhibit phosphatidylinositide-3-kinases, kappa kinase/c-Jun amino-terminal kinases, mammalian target of rapamycin complex 1, and the enzymes that are involved in the eicosanoid's synthesis [118]. It has been investigated that lipoxygenase, maltase, sucrase, isomaltase, and glucoamylase acarbose enzymes are inhibited by epigallocatechin gallate and abietic acid [119–122]. GR enzyme is inhibited by the compounds that are isolated from *Sideritis brevibracteate* [123] and isoorietin, forsythoside B, and verbascoside [124]. On the other hand, the GR enzyme is also inhibited by the colchicine. This phytochemical is classified as an alkaloid [125] and a well-known polyphenol rosmarinic acid [13]. Resveratrol (RES) and curcumin are polyphenolic compounds present in red wine and turmeric, respectively, and both molecules have antioxidant and anti-inflammatory properties [126, 127]. Phytochemicals can exert their effect on enzymes by activating them or inhibiting them [83, 84]; for instance, polyphenols from stevia leaves have antimicrobial and anti-inflammatory properties. Polyphenols from stevia leaves can exert their effects by inhibiting digestive enzymes

[128]. The inhibition of α-glucosidase and α-amylase enzymes are shown using in silico approaches. The inhibition of both enzymes is very valuable; for instance, they can be used in type-II diabetic patients [129]. Cytochrome P450 (P450) enzymes are involved in the drug and xenobiotic metabolism. Dietary polyphenolic compounds can change the expression and the activities of P450 enzymes, which is one of the negative effects of these phytochemicals [130].

4.1.6 Glucosinolates

Glucosinolates have various effects on human health. Myrosinase enzyme produces isothiocyanates after entering the reaction with their substrate glucosinolates. The products of glucosinolates have cancer-preventive effects [131]. Glucosinolates normally protect plants from diseases, and these sulfate-ion- and nitrogen-containing plants are needed for normal plant physiology [132]. The metabolites induce phase II enzymes, which are very important in several cancer types; for instance, isothiocyanates induce apoptosis and inhibit mitosis [133]. The metabolites of glucosinolates inhibit angiotensin-converting enzyme (ACE) and inhibition of this enzyme is very valuable for kidney health and pulmonary artery pressure [134].

4.1.7 Phytoestrogens

Phytoestrogens are plant-derived molecules that can affect the endocrine system; therefore, they are also known as xenoestrogens [135]. Phytoestrogens are also accepted as endocrine-disrupting chemicals that can affect both human and animal endocrine systems [136]. On the other hand, endocrine-disrupting chemicals can affect the activities of metabolic enzymes [17, 18, 40, 61, 137]. Phytoestrogens can inhibit monoamine oxidase enzyme and self-aggregation of amyloid beta protein, which is highly useful for Alzheimer's disease [138]. Soybean seeds are rich in phytoestrogens and one of them is named genistein. Genistein is one of the tyrosine kinase enzyme inhibitors. Inhibition of the tyrosine kinase enzyme has health-beneficial effects such as in the chemoprevention of breast and prostate cancers, cardiovascular disease, and is extremely effective in postmenopause [139]. Phytoestrogens molecules have positive effects on osteoporosis [140]. Phytoestrogens have antioxidant, antiproliferative, antimutagenic, and anticarcinogenic effects on metabolism but the detailed enzyme activation or inhibition is not developed yet [141].

4.1.8 Saponins

Saponins have a characterized aglycone structure that has triterpene or steroid aglycone and an additional one or more sugar chains. The two different structures have different solubilities as the carbohydrate part is water soluble and the sapogenin part is fat soluble [142, 143]. It has been investigated that saponins have inhibitory effects on digestive enzymes such as pancreatic lipase [144], α-amylase, and glucosidase [145]. Saponins are very valuable phytochemicals that can be crucial in the treatment of diabetes [146]. *Panax ginseng* has important roles in various inflammatory diseases and can affect some enzyme expressions such as inducible nitric oxide synthesis (iNOS) and COX-2 [147]. *P. notoginseng* contains panaxatriol saponins and has antioxidant, anti-inflammatory, and antiplatelet properties, and can be used in stroke therapies [148] and can activate the Nrf2 pathway [149].

4.1.9 Anticancer Properties of Plant Extracts

Cancer is the second leading cause of death worldwide, and its incidence is increasing every year. Researchers try to develop molecules to overcome the limitations of conventional chemotherapeutic drugs. Over the past two decades, plant phenolics, including phenolic acids, curcuminoids, flavonoids, tannins, stilbenes, alkaloids, coumarins, quinones, and lignans have gained attention due to their anticancer properties [150]. Colchicine is a proto-alkaloid that acts as a microtubule-disrupting chemical in cancer prevention and can be also used for familial Mediterranean fever (FMF), psoriasis, and Behçet's disease [125, 151]. According to the literature, phenolics also exert immunomodulatory, antioxidant, and anti-inflammatory effects. Antioxidant effects of plant phenolics have been regulated by gene expression, enzyme modulation, epigenetic, and metabolic control that mainly control cell death and proliferation pathways via ROS signaling [150]. On the other hand, flavonoids regulate cell–cell detachment proteins (cadherin, catenin, snail, Zinc finger E-box-binding homeobox 1 (ZEB1), zinc finger protein SNAI2 (SLUG), metastasis associated 1 family member 3 (MTA3), integrin, claudin, Mucin short variant S1 (MUC1), cytokeratin-18, and vimentin), cell migration and tissue barrier proteins (prostate-specific antigen (PSA), matrix metalloproteinases (MMPs), EMMPRIN, urokinase (uPA), the urokinase receptor (uPAR), metalloproteinase inhibitor 1 (TIMP), plasminogen activator inhibitor-1 (PAI), reversion-inducing-cysteine-rich protein (RECK), PNII, Alpha1-AT), cell–matrix adhesion proteins (VASP, integrin, CXCR4, CXCL-12), and angiogenesis (angiopoietin, PECAM-1, VEGF, VEGFR, HIF-1α) to control anti-invasive and antimetastatic properties of cancer cells [152].

4.1.10 Quercetin

Quercetin (3,3′,4′,5,7-pentahydroxyflavone) is a flavonoid found in plants, fruits, and vegetables, including buckwheat, onion, and broccoli. Quercetins are metabolized into glucuronidated, methylated, and sulfated derivatives in the liver, intestine, and kidneys. Quercetin is a powerful antioxidant, regulating glutathione metabolism, ROS production, and antioxidant enzyme activities in vitro. Besides, quercetin exerts anti-inflammatory properties by releasing cytokines, maintaining mast cell stability, and decreasing COX and LOX production [153]. 50, 200, and 500 μg/mL of ethanolic quercetin inhibited cell proliferation in HeLa and MCF-7 cells in vitro. Quercetin induced apoptosis cell cycle arrest and reduced MMP-2/MMP-9 enzyme activity in pancreatic cancer cells PC-3 and leukemia cells in vitro [150]. Quercetin activates caspase-3, −8, −9, Bax, Bad, and cytochrome-c release and inhibits the antiapoptotic proteins, including Bcl-XL, Bcl-2, and Mcl-1 [154]. Moreover, 120–200 mg/kg quercetin administration induced apoptosis and autophagy in prostate, leukemia, and breast cancer in vivo animal models [154]. In MCF-7 cells, quercetin reduced MMP-9 activity and inhibited the protein kinase C (PKC)/ERK/AP-1 signaling cascade involved in cancer metastasis and progression. In epidermal cancer cells, quercetin inhibited EMT markers, MMP-9, and EMT transition, whereas in melanoma cells, MMP-9 activity decreased via the PKC pathway [152]. Intraperitoneal injection of quercetin reduced lung metastasis and breast cancer invasion in animal models [152].

4.1.11 Genistein

Genistein is one of the most widely studied flavonoids in cancer research due to its antimetastatic, antiproliferative, and anti-growth properties. In breast cancer cells, genistein decreased NF-kB, AP-1, and uPA secretion, causing suppression of cell–matrix adhesion and migration. Moreover, MMP-2/MMP-9, CXCR4, CXCL12, MT1-, MT2-, and MT3-MMP were downregulated, whereas alpha 1-AT, TIMP-1, and PNII were downregulated in breast cancer cells upon genistein treatment. 1–37 μM of genistein inhibited invasion and motility of breast cancer cells in vivo of BALB-c animal model [152, 155, 156]. Genistein inhibited TGF-β-mediated phosphorylation of heat shock protein 27 (HSP27) and MAP kinase-activated protein kinase 2 (MAPKAPK2), causing reduced invasion of pancreatic cells. Furthermore, it decreased MMP-2 activity by blocking MEK4 kinase in four pancreatic cancer cell lines, including PC3-M, 1542CPTX, PC3, and 1532CPTX [157]. Genistein inhibited MMP-2/MMP-9 activity, whereas increased TIMP-1 activity was associated with lower invasive and metastatic properties in fibrosarcoma cells. In ovarian cancer cells, genistein decreased expression of CXCR4 and CXCL1, leading to decreased migration. Also, in melanoma cells, genistein reduced adhesion-induced protein tyrosine phosphorylation to inhibit cell–ECM interactions, enabling reduced cell invasion [152, 158].

4.1.12 Apigenin

Apigenin (4′,5,7-trihydroxyflavone) is a dietary flavonoid found in tea, parsley, onions, oranges, celery, thyme, and sweet red pepper. According to the literature, it inhibits cell proliferation, adhesion, motility, invasion, and angiogenesis in cancer cells. 2.5–10 μg/mL apigenin reduces uPA and MMP-9 activities, leading to reduced invasion, migration, and adhesion of breast cancer cells. Also, apigenin inhibits Akt activation and HGF-stimulated integrin β4, causing reduced migration and invasion in breast cancer cells [155, 159]. In prostate cancer cells, apigenin reduced FAK/Src signaling, causing reduced motility of the cells. On the other hand, cell motility and invasion were inhibited in ovarian, cervical, and melanoma cells upon apigenin administration [152].

4.1.13 Anthocyanin

Anthocyanins are flavonoids found in flowers, plants, fruits, and vegetable foods, exerting anti-inflammatory, anticancer, antioxidant, and antiangiogenic properties. Peonidin-3-glucoside and cyanidin-3-glucoside, two anthocyanins, reduced MMP-9, MMP-2, uPA, NF-kB, and AP-1 associated with decreased motility and invasion of hepatocellular carcinoma cells [159]. In non-small lung cancer cells (NSCLCs), anthocyanins reduced MMP-2, uPA, AP-1, and NF-kB activation, whereas upregulated TIMP-2 and PAI associated with reduced invasion and migration. Delphinidin reduced MAPKs, PI3K/Akt/mTOR/p70S6K, Met, NF-kB, and STAT3 and Src signaling in breast cancer cells. Cyanidin-3-glucoside inhibited ErbB2, cSrc, and FAK pathways leading to reduced motility and invasion in breast cancer cells [152]. Anthocyanin reduced MMP-2 and MMP-9 levels while increasing TIMP-1/TIMP-2 levels by regulating MAPK and PKC pathways in prostate and colon cancer cells. Also, anthocyanin downregulated claudin, MMP-2, and MMP-9 by inhibition of p38MAPK and PI3K/Akt pathways in colon cancer cells. Invasion and migration were suppressed by anthocyanin administration in cervical and oral cancers [152, 160].

5 Conclusion

Since antiquity, many studies have proved that phytochemicals have many bioactivities and show significant health-promoting effects such as anticancer, antioxidant, anti-inflammatory, antimicrobial, cardioprotective, anti-allergic, immune-modulatory, antiaging, and hepatoprotective properties. The phytochemicals have crucial roles in detoxifying toxic radicals and pharmaceutical potential. On the other hand, dosing and long-term effects must be determined according to the patient's age, gender, height, weight, and secondary and genetic disease.

References

1. Iyer M, Pal K, Upadhye V (2023) Phytochemicals and cancer. In: Recent frontiers of phytochemicals. Elsevier, pp 295–308
2. Al-Ishaq RK, Overy AJ, Büsselberg D (2020) Phytochemicals and gastrointestinal cancer: cellular mechanisms and effects to change cancer progression. Biomol Ther 10:105. https://doi.org/10.3390/biom10010105
3. Azmir J, Zaidul ISM, Rahman MM, Sharif KM, Mohamed A, Sahena F, Jahurul MHA, Ghafoor K, Norulaini NAN, Omar AKM (2013) Techniques for extraction of bioactive compounds from plant materials: a review. J

Food Eng 117:426–436. https://doi.org/10.1016/j.jfoodeng.2013.01.014

4. Thakur M, Singh K, Khedkar R (2020) Phytochemicals. In: Functional and preservative properties of phytochemicals. Elsevier, pp 341–361
5. Tsao R (2010) Chemistry and biochemistry of dietary polyphenols. Nutrients 2:1231–1246. https://doi.org/10.3390/nu2121231
6. Khatun R, Singh S, Dubey NK, Prasad Das A (2023) A review on marine-based phytochemicals and their application in biomedical research. In: Recent frontiers of phytochemicals. Elsevier, pp 383–395
7. Mohamed HI, El-Beltagi HS, Jain SM, Al-Khayri JM (2021) Date palm (Phoenix dactylifera L.) secondary metabolites: bioactivity and pharmaceutical potential. In: Phytomedicine. Elsevier, pp 483–531
8. Upadhyay S, Dixit M (2015) Role of polyphenols and other phytochemicals on molecular signaling. Oxidative Med Cell Longev 2015:1–15. https://doi.org/10.1155/2015/504253
9. Dave A, Parande F, Park E-J, Pezzuto JM (2020) Phytochemicals and cancer chemoprevention. J Cancer Metastasis Treat 2020. https://doi.org/10.20517/2394-4722.2020.106
10. Kelsey NA, Wilkins HM, Linseman DA (2010) Nutraceutical antioxidants as novel neuroprotective agents. Molecules 15:7792–7814. https://doi.org/10.3390/molecules15117792
11. Cione E, La Torre C, Cannataro R, Caroleo MC, Plastina P, Gallelli L (2019) Quercetin, epigallocatechin gallate, curcumin, and resveratrol: from dietary sources to human microRNA modulation. Molecules 25:63. https://doi.org/10.3390/molecules25010063
12. Vezikov LVSM (2023) Plant alkaloids toxicity. n: StatPearls [Internet] Treasure Island (FL): StatPearls Publishing
13. Tandogan B, Kuruüzüm-Uz A, Sengezer C, Güvenalp Z, Demirezer LÖ, Nuray Ulusu N (2011) *In vitro* effects of rosmarinic acid on glutathione reductase and glucose 6-phosphate dehydrogenase. Pharm Biol 49:587–594. https://doi.org/10.3109/13880209.2010.533187
14. Aydemir D, Ulusu NN (2020) Comment on the: molecular mechanism of CAT and SOD activity change under MPA-CdTe quantum dots induced oxidative stress in the mouse primary hepatocytes. Spectrochim Acta A Mol Biomol Spectrosc 229:117792. https://doi.org/10.1016/j.saa.2019.117792
15. Aydemir D, Sarayloo E, Ulusu NN (2019) Rosiglitazone-induced changes in the oxidative stress metabolism and fatty acid composition in relation with trace element status in the primary adipocytes. J Med Biochem 10. https://doi.org/10.2478/jomb-2019-0041
16. Otahal A, Aydemir D, Tomasich E, Minichsdorfer C (2020) Delineation of cell death mechanisms induced by synergistic effects of statins and erlotinib in non-small cell lung cancer cell (NSCLC) lines. Sci Rep 10:959. https://doi.org/10.1038/s41598-020-57707-2
17. Aydemir D, Oztasci B, Barlas N, Ulusu NN (2020) Influence of the butylparaben administration on the oxidative stress metabolism of liver, kidney and spleen. Turk J Biochem 45:689–694. https://doi.org/10.1515/tjb-2020-0048
18. Aydemir D, Öztaşcı B, Barlas N, Ulusu NN (2019) Effects of butylparaben on antioxidant enzyme activities and histopathological changes in rat tissues. Arch Ind Hyg Toxicol 70:315–324. https://doi.org/10.2478/aiht-2019-70-3342
19. Guan H, Luo W, Bao B, Cao Y, Cheng F, Yu S, Fan Q, Zhang L, Wu Q, Shan M (2022) A comprehensive review of Rosmarinic acid: from phytochemistry to pharmacology and its new insight. Molecules 27:3292. https://doi.org/10.3390/molecules27103292
20. Sik B, Kapcsándi V, Székelyhidi R, Hanczné EL, Ajtony Z (2019) Recent advances in the analysis of rosmarinic acid from herbs in the *Lamiaceae* Family. Nat Prod Commun 14:1934578X1986421. https://doi.org/10.1177/1934578X19864216
21. Swamy MK, Sinniah UR, Ghasemzadeh A (2018) Anticancer potential of rosmarinic acid and its improved production through biotechnological interventions and functional genomics. Appl Microbiol Biotechnol 102:7775–7793. https://doi.org/10.1007/s00253-018-9223-y
22. Petersen M (2003) Rosmarinic acid. Phytochemistry 62:121–125. https://doi.org/10.1016/S0031-9422(02)00513-7
23. Zu G, Zhang R, Yang L, Ma C, Zu Y, Wang W, Zhao C (2012) Ultrasound-assisted extraction of carnosic acid and rosmarinic acid using ionic liquid solution from rosmarinus officinalis. Int J Mol Sci 13:11027–11043. https://doi.org/10.3390/ijms130911027

24. Oh H-A, Park C-S, Ahn H-J, Park YS, Kim H-M (2011) Effect of *Perilla frutescens* var. *acuta* Kudo and rosmarinic acid on allergic inflammatory reactions. Exp Biol Med 236: 99–106. https://doi.org/10.1258/ebm.2010.010252
25. Liu H, Deng R, Zhu C, Han H, Zong G, Ren L, Cheng P, Wei Z, Zhao Y, Yu S, Lu Y (2023) Rosmarinic acid in combination with ginsenoside Rg1 suppresses colon cancer metastasis via co-inhition of COX-2 and PD1/PD-L1 signaling axis. Acta Pharmacol Sin. https://doi.org/10.1038/s41401-023-01158-8
26. Rahbardar MG, Eisvand F, Rameshrad M, Razavi BM, Hosseinzadeh H (2022) In vivo and in vitro protective effects of rosmarinic acid against doxorubicin-induced cardiotoxicity. Nutr Cancer 74:747–760. https://doi.org/10.1080/01635581.2021.1931362
27. Neuberger B, Mello FK, Mallmann MP, da Costa Sobral KG, Fighera MR, Royes LFF, Furian AF, Sampaio TB, Oliveira MS (2023) Beneficial effects of rosmarinic acid in vitro and in vivo models of epileptiform activity induced by pilocarpine. Brain Sci 13:289. https://doi.org/10.3390/brainsci13020289
28. Aydemir D, Hashemkhani M, Acar HY, Ulusu NN (2019) In vitro interaction of glutathione S-transferase-pi enzyme with glutathione-coated silver sulfide quantum dots: a novel method for biodetection of glutathione S-transferase enzyme. Chem Biol Drug Des. https://doi.org/10.1111/cbdd.13614
29. Aydemir D, Ulusu NN (2020) The possible role of the glucose-6-phosphate dehydrogenase enzyme deficiency in the polyneuropathies. J Basic Clin Health Sci. https://doi.org/10.30621/jbachs.2020.1151
30. Aydemir D, Hashemkhani M, Acar HY, Ulusu NN (2020) Evaluation of the biocompatibility of the GSH-coated Ag2S quantum dots in vitro: a perfect example for the non-toxic optical probes. Mol Biol Rep 47. https://doi.org/10.1007/s11033-020-05522-3
31. Forman HJ, Zhang H, Rinna A (2009) Glutathione: overview of its protective roles, measurement, and biosynthesis. Mol Asp Med 30: 1–12. https://doi.org/10.1016/j.mam.2008.08.006
32. Ramsay EE, Dilda PJ (2014) Glutathione S-conjugates as prodrugs to target drug-resistant tumors. Front Pharmacol 5. https://doi.org/10.3389/fphar.2014.00181
33. Fernando PMDJ, Piao MJ, Kang KA, Ryu YS, Hewage SRKM, Chae SW, Hyun JW (2016) Rosmarinic acid attenuates cell damage against UVB radiation-induced oxidative stress via enhancing antioxidant effects in human HaCaT cells. Biomol Ther (Seoul) 24:75–84. https://doi.org/10.4062/biomolther.2015.069
34. Vo QV, Hoa NT, Flavel M, Thong NM, Boulebd H, Nam PC, Quang DT, Mechler A (2023) A comprehensive study of the radical scavenging activity of rosmarinic acid. J Org Chem 88:17237–17248. https://doi.org/10.1021/acs.joc.3c02093
35. Osakabe N, Takano H, Sanbongi C, Yasuda A, Yanagisawa R, Inoue K, Yoshikawa T (2004) Anti-inflammatory and anti-allergic effect of rosmarinic acid (RA); inhibition of seasonal allergic rhinoconjunctivitis (SAR) and its mechanism. Biofactors 21:127–131. https://doi.org/10.1002/biof.552210125
36. Gülçin İ, Scozzafava A, Supuran CT, Koksal Z, Turkan F, Çetinkaya S, Bingöl Z, Huyut Z, Alwasel SH (2016) Rosmarinic acid inhibits some metabolic enzymes including glutathione *S* -transferase, lactoperoxidase, acetylcholinesterase, butyrylcholinesterase and carbonic anhydrase isoenzymes. J Enzyme Inhib Med Chem 31:1698–1702. https://doi.org/10.3109/14756366.2015.1135914
37. Ngo YL, Lau CH, Chua LS (2018) Review on rosmarinic acid extraction, fractionation and its anti-diabetic potential. Food Chem Toxicol 121:687–700. https://doi.org/10.1016/j.fct.2018.09.064
38. McCue PPSK (2004) Inhibitory effects of rosmarinic acid extracts on porcine pancreatic amylase in vitro. Asia Pac J Clin Nutr 13:101–106
39. Duda-Chodak A, Tarko T (2023) Possible side effects of polyphenols and their interactions with medicines. Molecules 28:2536. https://doi.org/10.3390/molecules28062536
40. Aydemir D, Karabulut G, Gok M, Barlas N, Ulusu NN (2019) Data the DEHP induced changes on the trace element and mineral levels in the brain and testis tissues of rats. Data Brief 26:104526. https://doi.org/10.1016/j.dib.2019.104526
41. Tandogan B, Ulusu NN (2010) Purification and kinetics of bovine kidney cortex glutathione reductase. Protein Pept Lett 17:667–674. https://doi.org/10.2174/092986610791112684
42. Sengezer CUN (2007) Three different purification protocols in purification of G6PD from sheep brain cortex. FABAD J Pharm Sci 32: 65–72

43. Kuruüzüm-Uz A, Güvenalp Z, Ströch K, Demirezer LÖ, Zeeck A (2004) Phytochemical and antimicrobial investigation of Echium vulgare growing in Turkey. Biochem Syst Ecol 32:833–836. https://doi.org/10.1016/j.bse.2003.12.001
44. Sasidharan S, Chen Y, Saravanan D, Sundram KM, Latha LY (2011) Extraction, isolation and characterization of bioactive compounds from plants' extracts. Afr J Tradit Complement Altern Med 8:1–10
45. Ekiert HM, Szopa A (2020) Biological activities of natural products. Molecules 25:5769. https://doi.org/10.3390/molecules25235769
46. Sunanta P, Rose Sommano S, Luiten CA, Ghofrani M, Sims IM, Bell TJ, Carnachan SM, Hinkley SFR, Kontogiorgos V (2024) Fractionation and characterisation of pectin-rich extracts from garlic biomass. Food Chem 436:137697. https://doi.org/10.1016/j.foodchem.2023.137697
47. Zainal-Abidin MH, Hayyan M, Hayyan A, Jayakumar NS (2017) New horizons in the extraction of bioactive compounds using deep eutectic solvents: a review. Anal Chim Acta 979:1–23. https://doi.org/10.1016/j.aca.2017.05.012
48. Tobar-Delgado E, Mejía-España D, Osorio-Mora O, Serna-Cock L (2023) Rutin: family farming products' extraction sources, industrial applications and current trends in biological activity protection. Molecules 28:5864. https://doi.org/10.3390/molecules28155864
49. Gil KA, Jokić S, Cikoš A-M, Banožić M, Jakovljević Kovač M, Fais A, Tuberoso CIG (2023) Comparison of different green extraction techniques used for the extraction of targeted flavonoids from edible Feijoa (Acca sellowiana (O.Berg) Burret) flowers. Plan Theory 12:1461. https://doi.org/10.3390/plants12071461
50. Abreu DCA, Figueiredo KC d S (2019) Bromelain separation and purification processes from pineapple extract. Braz J Chem Eng 36:1029–1039. https://doi.org/10.1590/0104-6632.20190362s20180417
51. Santos MB, Sillero L, Gatto DA, Labidi J (2022) Bioactive molecules in wood extractives: methods of extraction and separation, a review. Ind Crop Prod 186:115231. https://doi.org/10.1016/j.indcrop.2022.115231
52. Altemimi A, Lakhssassi N, Baharlouei A, Watson D, Lightfoot D (2017) Phytochemicals: extraction, isolation, and identification of bioactive compounds from plant extracts. Plan Theory 6:42. https://doi.org/10.3390/plants6040042
53. Zhang Q-W, Lin L-G, Ye W-C (2018) Techniques for extraction and isolation of natural products: a comprehensive review. Chin Med 13:20. https://doi.org/10.1186/s13020-018-0177-x
54. Zaitoon A, Lim L-T (2022) Triggered and controlled release of bioactives in food applications. Adv Food Nutr Res 100:49–107
55. Dembitsky VM (2023) Fascinating furanosteroids and their pharmacological profile. Molecules 28:5669. https://doi.org/10.3390/molecules28155669
56. Maestrello V, Solovyev P, Bontempo L, Mannina L, Camin F (2022) Nuclear magnetic resonance spectroscopy in extra virgin olive oil authentication. Compr Rev Food Sci Food Saf 21:4056–4075. https://doi.org/10.1111/1541-4337.13005
57. Lozada-Ramírez JD, Ortega-Regules AE, Hernández LR, Anaya de Parrodi C (2021) Spectroscopic and spectrometric applications for the identification of bioactive compounds from vegetal extracts. Appl Sci 11:3039. https://doi.org/10.3390/app11073039
58. Su C, Pham TTT, Cheng H (2020) Aqueous enzymatic extraction of rosmarinic acid from *Salvia officinalis*: optimisation using response surface methodology. Phytochem Anal 31:575–582. https://doi.org/10.1002/pca.2922
59. Caleja C, Barros L, Prieto MA, Barreiro MF, Oliveira MBPP, Ferreira ICFR (2017) Extraction of rosmarinic acid from Melissa officinalis L. by heat-, microwave- and ultrasound-assisted extraction techniques: a comparative study through response surface analysis. Sep Purif Technol 186:297–308. https://doi.org/10.1016/j.seppur.2017.06.029
60. Su J, Jia F, Lu J, Chen W, Sun H, Liu T, Wu X (2020) Characterization of the metabolites of rosmarinic acid in human liver microsomes using liquid chromatography combined with electrospray ionization tandem mass spectrometry. Biomed Chromatogr 34. https://doi.org/10.1002/bmc.4806
61. Aydemir D, Karabulut G, Şimşek G, Gok M, Barlas N, Ulusu NN (2018) Impact of the Di (2-Ethylhexyl) phthalate administration on trace element and mineral levels in relation of kidney and liver damage in rats. Biol Trace Elem Res 186:474–488. https://doi.org/10.1007/s12011-018-1331-0
62. Pearse BMF, Rosemeyer MA (1975) 6-Phosphogluconate dehydrogenase from

human erythrocytes. Method Enzymol 41: 220–226

63. Aydemir D, Hashemkhani M, Durmusoglu EG, Acar HY, Ulusu NN (2019) A new substrate for glutathione reductase: glutathione coated Ag2S quantum dots. Talanta 194: 501–506. https://doi.org/10.1016/j.talanta.2018.10.049
64. Staal GEJ, Visser J, Veeger C (1969) Purification and properties of glutathione reductase of human erythrocytes. Biochimica et Biophysica Acta (BBA)—Enzymology 185:39–48. https://doi.org/10.1016/0005-2744(69)90280-0
65. Habig WH, Pabst MJ, Jakoby WB (1974) Glutathione S-transferases. The first enzymatic step in mercapturic acid formation. J Biol Chem 249:7130–7139
66. Beutler E (1972) Red cell metabolism. Biochimie 54:759–764. https://doi.org/10.1016/S0300-9084(72)80181-0
67. Mohammed ASA, Naveed M, Jost N (2021) Polysaccharides; classification, chemical properties, and future perspective applications in fields of pharmacology and biological medicine (a review of current applications and upcoming potentialities). J Polym Environ 29:2359–2371. https://doi.org/10.1007/s10924-021-02052-2
68. Song E-H, Shang J, Ratner DM (2012) Polysaccharides. In: Polymer science: a comprehensive reference. Elsevier, pp 137–155
69. Huang G, Huang H (2017) The derivatization and antitumor mechanisms of polysaccharides. Future Med Chem 9:1931–1938. https://doi.org/10.4155/fmc-2017-0132
70. Mishra BP, Mishra J, Paital B, Rath PK, Jena MK, Reddy BVV, Pati PK, Panda SK, Sahoo DK (2023) Properties and physiological effects of dietary fiber-enriched meat products: a review. Front Nutr 10. https://doi.org/10.3389/fnut.2023.1275341
71. Harris PJ, Ferguson LR (1993) Dietary fibre: its composition and role in protection against colorectal cancer. Mut Res/Fund Mol Mech Mutag 290:97–110. https://doi.org/10.1016/0027-5107(93)90037-G
72. Zhao R, Ji Y, Chen X, Ma G, Yao H, Li J, Hu Q, Zhao L (2023) Flammulina velutipes polysaccharides regulate lipid metabolism disorders in HFD-fed mice via bile acids metabolism. Int J Biol Macromol 253:127308. https://doi.org/10.1016/j.ijbiomac.2023.127308
73. Ren Y, Li S, Song Z, Luo Q, Zhang Y, Wang H (2022) The regulatory roles of polysaccharides and ferroptosis-related phytochemicals in liver diseases. Nutrients 14:2303. https://doi.org/10.3390/nu14112303
74. Yang G, Liang X, Hu J, Li C, Hu W, Li K, Chang X, Zhang Y, Zhang X, Shen Y, Meng X (2022) Feeding tea polysaccharides affects lipid metabolism, antioxidant capacity and immunity of common carp (Cyprinus carpio L.). Front Immunol 13. https://doi.org/10.3389/fimmu.2022.1074198
75. Chandel NS (2021) Carbohydrate metabolism. Cold Spring Harb Perspect Biol 13: a040568. https://doi.org/10.1101/cshperspect.a040568
76. Ulusu NN, Gök M, Sayin Şakul AA, Ari N, Stefek M, Karasu Ç (2017) Antioxidant SMe1EC2 modulates pentose phosphate pathway and glutathione-dependent enzyme activities in tissues of aged diabetic rats. Interdiscip Toxicol 10:148–154. https://doi.org/10.1515/intox-2017-0021
77. Gök M, Ulusu NN, Tarhan N, Tufan C, Ozansoy G, Arı N, Karasu Ç (2016) Flaxseed protects against diabetes-induced glucotoxicity by modulating pentose phosphate pathway and glutathione-dependent enzyme activities in rats. J Diet Suppl 13:339–351. https://doi.org/10.3109/19390211.2015.1036188
78. Akbay E, Nuray Ulusu N, Töröner F, Ayvaz G, Taneri F, Aktürk M, Arslan M, Karasu Ç (2004) Effects of rosiglitazone treatment on the pentose phosphate pathway and glutathione-dependent enzymes in liver and kidney of rats fed a high-fat diet. Curr Ther Res 65:79–89. https://doi.org/10.1016/S0011-393X(04)90007-0
79. Ulusu NN, Sahilli M, Avci A, Canbolat O, Ozansoy G, Ari N, Bali M, Stefek M, Stolc S, Gajdosik A, Karasu C (2003) Pentose phosphate pathway, glutathione-dependent enzymes and antioxidant defense during oxidative stress in diabetic rodent brain and peripheral organs: effects of stobadine and vitamin E. Neurochem Res 28:815–823. https://doi.org/10.1023/A:1023202805255
80. Engelking LR (2015) Polysaccharides and carbohydrate derivatives. In: Textbook of veterinary physiological chemistry. Elsevier, pp 124–129
81. Wang B-T, Hu S, Yu X-Y, Jin L, Zhu Y-J, Jin F-J (2020) Studies of cellulose and starch utilization and the regulatory mechanisms of related enzymes in fungi. Polymers (Basel) 12:530. https://doi.org/10.3390/polym12030530

82. Sun XS (2005) Plant materials formation and growth. In: Bio-based polymers and composites. Elsevier, pp 15–32
83. Ulusu NN (2015) Curious cases of the enzymes / Neobiča Istorija Enzima. J Med Biochem 34:271–281. https://doi.org/10.2478/jomb-2014-0045
84. Ulusu NN (2015) Evolution of enzyme kinetic mechanisms. J Mol Evol 80:251–257. https://doi.org/10.1007/s00239-015-9681-0
85. Maoka T (2020) Carotenoids as natural functional pigments. J Nat Med 74:1–16. https://doi.org/10.1007/s11418-019-01364-x
86. Fiedor J, Burda K (2014) Potential role of carotenoids as antioxidants in human health and disease. Nutrients 6:466–488. https://doi.org/10.3390/nu6020466
87. Terao J (2023) Revisiting carotenoids as dietary antioxidants for human health and disease prevention. Food Funct 14:7799–7824. https://doi.org/10.1039/D3FO02330C
88. Otocka-Kmiecik A (2022) Effect of carotenoids on paraoxonase-1 activity and gene expression. Nutrients 14:2842. https://doi.org/10.3390/nu14142842
89. Babin A, Saciat C, Teixeira M, Troussard J-P, Motreuil S, Moreau J, Moret Y (2015) Limiting immunopathology: interaction between carotenoids and enzymatic antioxidant defences. Dev Comp Immunol 49:278–281. https://doi.org/10.1016/j.dci.2014.12.007
90. Rocha F, Yumi Sugahara L, Leimann FV, de Oliveira SM, da Silva BE, Calhelha RC, Barreiro MF, Ferreira ICFR, Porto Ineu R, Gonçalves OH (2018) Nanodispersions of beta-carotene: effects on antioxidant enzymes and cytotoxic properties. Food Funct 9:3698–3706. https://doi.org/10.1039/C8FO00804C
91. Nabi F, Arain MA, Rajput N, Alagawany M, Soomro J, Umer M, Soomro F, Wang Z, Ye R, Liu J (2020) Health benefits of carotenoids and potential application in poultry industry: a review. J Anim Physiol Anim Nutr (Berl) 104:1809–1818. https://doi.org/10.1111/jpn.13375
92. Temel Y, Bengü AŞ, Akkoyun HT, Akkoyun M, Ciftci M (2017) Effect of astaxanthin and aluminum chloride on erythrocyte G6PD and 6PGD enzyme activities *in vivo* and on erythrocyte G6PD *in vitro* in rats. J Biochem Mol Toxicol 31. https://doi.org/10.1002/jbt.21954
93. Wiesner M, Hanschen FS, Maul R, Neugart S, Schreiner M, Baldermann S (2017) Nutritional quality of plants for food and fodder. In: Encyclopedia of applied plant sciences. Elsevier, pp 285–291
94. Woyengo TA, Ramprasath VR, Jones PJH (2009) Anticancer effects of phytosterols. Eur J Clin Nutr 63:813–820. https://doi.org/10.1038/ejcn.2009.29
95. Anwar A, Liu Y, Dong R, Bai L, Yu X, Li Y (2018) The physiological and molecular mechanism of brassinosteroid in response to stress: a review. Biol Res 51:46. https://doi.org/10.1186/s40659-018-0195-2
96. Moustafa EM, Thabet NM (2017) Beta-sitosterol upregulated paraoxonase-1 via peroxisome proliferator-activated receptor-γ in irradiated rats. Can J Physiol Pharmacol 95:661–666. https://doi.org/10.1139/cjpp-2016-0397
97. Akinloye OA, Akinloye DI, Onigbinde SB, Metibemu DS (2020) Phytosterols demonstrate selective inhibition of COX-2: in-vivo and in-silico studies of Nicotiana tabacum. Bioorg Chem 102:104037. https://doi.org/10.1016/j.bioorg.2020.104037
98. Berryman CE, Preston AG, Karmally W, Deckelbaum RJ, Kris-Etherton PM (2011) Effects of almond consumption on the reduction of LDL-cholesterol: a discussion of potential mechanisms and future research directions. Nutr Rev 69:171–185. https://doi.org/10.1111/j.1753-4887.2011.00383.x
99. Singh AK, Kumar SR, Dwivedi V, Rai A, Pal S, Shasany AK, Nagegowda DA (2017) A WRKY transcription factor from *Withania somnifera* regulates triterpenoid withanolide accumulation and biotic stress tolerance through modulation of phytosterol and defense pathways. New Phytol 215:1115–1131. https://doi.org/10.1111/nph.14663
100. Khan SA, Khan SB, Shah Z, Asiri AM (2016) Withanolides: biologically active constituents in the treatment of Alzheimer's disease. Med Chem (Los Angeles) 12:238–256. https://doi.org/10.2174/1573406411666151030112314
101. Ulusu NN (2015) Glucose-6-phosphate dehydrogenase deficiency and Alzheimer's disease: partners in crime? The hypothesis. Med Hypotheses 85:219–223. https://doi.org/10.1016/j.mehy.2015.05.006
102. Ulusu NN, Yilmaz G, Erbayraktar Z, Evlice AT, Aras S, Yener G, Avci A (2015) A Turkish 3-center study evaluation of serum folic acid and vitamin B12 levels in Alzheimer disease. Turk J Med Sci 45:1159–1166. https://doi.org/10.3906/sag-1406-136

103. Erbayraktar Z, Evlice A, Yener G, Ulusu NN (2018) Effects of donepezil on liver and kidney functions for the treatment of Alzheimer's disease. J Integr Neurosci 16:335–346. https://doi.org/10.3233/JIN-170020
104. Erbayraktar Z, Evlice AT, Yilmaz G, Yazici C, Yener G, Ulusu NN (2017) Evaluation of renal function in Alzheimer's disease and geriatric patients: results from a Turkish two-center study. J Med Biochem 36:54–61. https://doi.org/10.1515/jomb-2016-0028
105. Matheyambath AC, Padmanabhan P, Paliyath G (2016) Citrus fruits. In: Encyclopedia of food and health. Elsevier, pp 136–140
106. Shi Y-S, Zhang Y, Li H-T, Wu C-H, El-Seedi HR, Ye W-K, Wang Z-W, Li C-B, Zhang X-F, Kai G-Y (2020) Limonoids from citrus: chemistry, anti-tumor potential, and other bioactivities. J Funct Foods 75:104213. https://doi.org/10.1016/j.jff.2020.104213
107. Ozdemir S, Tandogan B, Ulusu NN, Turan B (2009) Angiotensin II receptor blockage prevents diabetes-induced oxidative damage in rat heart. Folia Biol (Praha) 55(1):11–55
108. Can B, Ulusu NN, Kilinç K, Leyla Acan N, Saran Y, Turan B (2005) Selenium treatment protects diabetes-induced biochemical and ultrastructural alterations in liver tissue. Biol Trace Elem Res 105:135–150. https://doi.org/10.1385/BTER:105:1-3:135
109. Kim J, Jayaprakasha GK, Patil BS (2013) Limonoids and their anti-proliferative and anti-aromatase properties in human breast cancer cells. Food Funct 4:258–265. https://doi.org/10.1039/C2FO30209H
110. Kashtoh H, Baek K-H (2022) Recent updates on phytoconstituent alpha-glucosidase inhibitors: an approach towards the treatment of type two diabetes. Plan Theory 11:2722. https://doi.org/10.3390/plants11202722
111. Peng J-L, Wang J, Kong F-D, Liu Z-Q, Wang P, Gai C-J, Jiang B, Mei W-L, Dai H-F (2016) New phragmalin-type limonoids from Chukrasia tabularis and their α-glucosidase inhibitory activity. Molecules 21:58. https://doi.org/10.3390/molecules21010058
112. Mehmetoglu-Gurbuz T, Lakshmanaswamy R, Perez K, Sandoval M, Jimenez CA, Rocha J, Goldfarb RM, Perry C, Bencomo A, Neela N, Barragan JA, Sanchez R, Swain RM, Subramani R (2023) Nimbolide inhibits SOD2 to control pancreatic ductal adenocarcinoma growth and metastasis. Antioxidants 12:1791. https://doi.org/10.3390/antiox12101791
113. Iampanichakul M, Potue P, Rattanakanokchai S, Maneesai P, Khamseekaew J, Settheetham-Ishida W, Pakdeechote P (2023) Limonin ameliorates cardiovascular dysfunction and remodeling in hypertensive rats. Life Sci 327:121834. https://doi.org/10.1016/j.lfs.2023.121834
114. Pandey KB, Rizvi SI (2009) Plant polyphenols as dietary antioxidants in human health and disease. Oxidative Med Cell Longev 2:270–278. https://doi.org/10.4161/oxim.2.5.9498
115. Di Lorenzo C, Colombo F, Biella S, Stockley C, Restani P (2021) Polyphenols and human health: the role of bioavailability. Nutrients 13:273. https://doi.org/10.3390/nu13010273
116. Kopustinskiene DM, Jakstas V, Savickas A, Bernatoniene J (2020) Flavonoids as anticancer agents. Nutrients 12:457. https://doi.org/10.3390/nu12020457
117. Shahabi M, Idrees R (2023) Flavanols for age-related memory loss: a promising nutritional intervention. Ann Neurosci 30:222–223. https://doi.org/10.1177/09727531231195683
118. Yahfoufi N, Alsadi N, Jambi M, Matar C (2018) The immunomodulatory and anti-inflammatory role of polyphenols. Nutrients 10:1618. https://doi.org/10.3390/nu10111618
119. Ercil D, Ulusu NN, Sakar K (2001) He inhibitory effect on intermedeol on soybean lipoxygenase and its cytotoxicity. Turk J Biochem 26:76–79
120. Ulusu NN, Sakar K, Tezcan EF (2002) (−)-Epicatechin-3-O-gallate and (−)-Epigallocatechin-3-O-gallate are the inhibitors of soybean lipoxygenase. Technol Health Care 10:359–361
121. Kan L, Capuano E, Fogliano V, Verkerk R, Mes JJ, Tomassen MMM, Oliviero T (2021) Inhibition of α-glucosidases by tea polyphenols in rat intestinal extract and Caco-2 cells grown on Transwell. Food Chem 361:130047. https://doi.org/10.1016/j.foodchem.2021.130047
122. Ulusu NN, Ercil D, Sakar MK, Tezcan EF (2002) Abietic acid inhibits lipoxygenase activity. Phytother Res 16:88–90. https://doi.org/10.1002/ptr.983
123. Tandogan B, Güvenç A, Çalış İ, Ulusu NN (2011) In vitro effects of compounds isolated from Sideritis brevibracteata on bovine kidney cortex glutathione reductase. Acta Biochim Pol 58:471–475

124. Tandogan B, Guvenc A, Calis I, Ulusu NN (2013) In vitro effects of isoorientin, forsythoside B, and verbascoside on bovine kidney cortex glutathione reductase. Int J Chem Kinet 45:574–579. https://doi.org/10.1002/kin.20792
125. Tandogan B, Ulusu NN (2010) A comparative study with colchicine on glutathione reductase. Protein J 29:380–385. https://doi.org/10.1007/s10930-010-9263-3
126. Nanavati K, Rutherfurd-Markwick K, Lee SJ, Bishop NC, Ali A (2022) Effect of curcumin supplementation on exercise-induced muscle damage: a narrative review. Eur J Nutr 61:3835–3855. https://doi.org/10.1007/s00394-022-02943-7
127. Bonnefont-Rousselot D (2016) Resveratrol and cardiovascular diseases. Nutrients 8:250. https://doi.org/10.3390/nu8050250
128. Myint KZ, Zhou Z, Shi Q, Chen J, Dong X, Xia Y (2023) Stevia polyphenols, their antimicrobial and anti-inflammatory properties, and inhibitory effect on digestive enzymes. Molecules 28:7572. https://doi.org/10.3390/molecules28227572
129. Riyaphan J, Pham D-C, Leong MK, Weng C-F (2021) In Silico approaches to identify polyphenol compounds as α-glucosidase and α-amylase inhibitors against type-II diabetes. Biomol Ther 11:1877. https://doi.org/10.3390/biom11121877
130. Basheer L, Kerem Z (2015) Interactions between CYP3A4 and dietary polyphenols. Oxidative Med Cell Longev 2015:1–15. https://doi.org/10.1155/2015/854015
131. Orouji N, Asl SK, Taghipour Z, Habtemariam S, Nabavi SM, Rahimi R (2023) Glucosinolates in cancer prevention and treatment: experimental and clinical evidence. Med Oncol 40:344. https://doi.org/10.1007/s12032-023-02211-6
132. Ishida M, Hara M, Fukino N, Kakizaki T, Morimitsu Y (2014) Glucosinolate metabolism, functionality and breeding for the improvement of Brassicaceae vegetables. Breed Sci 64:48–59. https://doi.org/10.1270/jsbbs.64.48
133. IanT J (2002) Glucosinolates: bioavailability and importance to health. Int J Vitam Nutr Res 72:26–31. https://doi.org/10.1024/0300-9831.72.1.26
134. Khaled A, Ahmed E, Mamdouh M, Saad H, Mohamed A, Sobhy M, Piatti D, Sabry M, Saad MA, Sabry OM, Caprioli G (2023) Natural angiotensin converting enzyme inhibitors: a safeguard against hypertension, respiratory distress syndrome, and chronic kidney diseases. Phytother Res 37:5464–5472. https://doi.org/10.1002/ptr.7987
135. Canivenc-Lavier M-C, Bennetau-Pelissero C (2023) Phytoestrogens and health effects. Nutrients 15:317. https://doi.org/10.3390/nu15020317
136. Diamanti-Kandarakis E, Bourguignon J-P, Giudice LC, Hauser R, Prins GS, Soto AM, Zoeller RT, Gore AC (2009) Endocrine-disrupting chemicals: an endocrine society scientific statement. Endocr Rev 30:293–342. https://doi.org/10.1210/er.2009-0002
137. Aydemir D, Aydogan-Ahbab M, Barlas N, Ulusu NN (2023) Effects of the in-utero dicyclohexyl phthalate and di-n-hexyl phthalate administration on the oxidative stress-induced histopathological changes in the rat liver tissue correlated with serum biochemistry and hematological parameters. Front Endocrinol (Lausanne) 14. https://doi.org/10.3389/fendo.2023.1128202
138. Seong SH, Kim B-R, Cho ML, Kim T-S, Im S, Han S, Jeong J-W, Jung HA, Choi JS (2022) Phytoestrogen coumestrol selectively inhibits monoamine oxidase-a and amyloid β self-aggregation. Nutrients 14:3822. https://doi.org/10.3390/nu14183822
139. Dixon R (2002) Genistein. Phytochemistry 60:205–211. https://doi.org/10.1016/S0031-9422(02)00116-4
140. Anderson JJB, Garner SC (1998) 1Phytoestrogens and bone. Bailliere Clin Endocrinol Metab 12:543–557. https://doi.org/10.1016/S0950-351X(98)80003-7
141. Desmawati D, Sulastri D (2019) Phytoestrogens and their health effect. Open Access Maced J Med Sci 7:495–499. https://doi.org/10.3889/oamjms.2019.086
142. Savage GP (2003) Saponins. In: Encyclopedia of food sciences and nutrition. Elsevier, pp 5095–5098
143. Güçlü-Üstündağ Ö, Mazza G (2007) Saponins: properties, applications and processing. Crit Rev Food Sci Nutr 47:231–258. https://doi.org/10.1080/10408390600698197
144. Birari RB, Bhutani KK (2007) Pancreatic lipase inhibitors from natural sources: unexplored potential. Drug Discov Today 12:879–889. https://doi.org/10.1016/j.drudis.2007.07.024
145. Ercan P, El SN (2016) Inhibitory effects of chickpea and Tribulus terrestris on lipase, α-amylase and α-glucosidase. Food Chem 205:163–169. https://doi.org/10.1016/j.foodchem.2016.03.012

146. Hossain U, Das AK, Ghosh S, Sil PC (2020) An overview on the role of bioactive α-glucosidase inhibitors in ameliorating diabetic complications. Food Chem Toxicol 145: 111738. https://doi.org/10.1016/j.fct.2020.111738
147. Im D-S (2020) Pro-resolving effect of Ginsenosides as an anti-inflammatory mechanism of Panax ginseng. Biomol Ther 10:444. https://doi.org/10.3390/biom10030444
148. Li C, Fan C, Zhao J, Di M, Sui C, Han L, Hu L (2020) Panaxatriol saponins promote M2 polarization of BV2 cells to reduce inflammation and apoptosis after glucose/oxygen deprivation by activating STAT3. Inflammation 43:2109–2118. https://doi.org/10.1007/s10753-020-01278-x
149. Huang Y, Yu J, Wan F, Zhang W, Yang H, Wang L, Qi H, Wu C (2014) Panaxatriol saponins attenuated oxygen-glucose deprivation injury in PC12 cells via activation of PI3K/Akt and Nrf2 signaling pathway. Oxidative Med Cell Longev 2014:1–11. https://doi.org/10.1155/2014/978034
150. Maheshwari N, Sharma MC (2023) Anticancer properties of some selected plant phenolic compounds: future leads for therapeutic development. J Herb Med 42:100801. https://doi.org/10.1016/j.hermed.2023.100801
151. Mir SA, Padhiary A, Pati A, Tete SS, Meher RK, Baitharu I, Muhammad A, Nayak B (2023) Potential phytochemicals as microtubule-disrupting agents in cancer prevention. In: Recent frontiers of phytochemicals. Elsevier, pp 225–246
152. Weng C-J, Yen G-C (2012) Flavonoids, a ubiquitous dietary phenolic subclass, exert extensive in vitro anti-invasive and in vivo anti-metastatic activities. Cancer Metastasis Rev 31:323–351. https://doi.org/10.1007/s10555-012-9347-y
153. Dower JI, Geleijnse JM, Gijsbers L, Schalkwijk C, Kromhout D, Hollman PC (2015) Supplementation of the pure flavonoids epicatechin and quercetin affects some biomarkers of endothelial dysfunction and inflammation in (pre)hypertensive adults: a randomized double-blind, placebo-controlled, crossover trial. J Nutr 145:1459–1463. https://doi.org/10.3945/jn.115.211888
154. Primikyri A, Sayyad N, Quilici G, Vrettos EI, Lim K, Chi S, Musco G, Gerothanassis IP, Tzakos AG (2018) Probing the interaction of a quercetin bioconjugate with Bcl-2 in living human cancer cells with in-cell NMR spectroscopy. FEBS Lett 592:3367–3379. https://doi.org/10.1002/1873-3468.13250
155. Farina HG, Pomies M, Alonso DF, Gomez DE (2006) Antitumor and antiangiogenic activity of soy isoflavone genistein in mouse models of melanoma and breast cancer. Oncol Rep 16:885–891
156. Tang S-M, Deng X-T, Zhou J, Li Q-P, Ge X-X, Miao L (2020) Pharmacological basis and new insights of quercetin action in respect to its anti-cancer effects. Biomed Pharmacother 121:109604. https://doi.org/10.1016/j.biopha.2019.109604
157. El Touny LH, Banerjee PP (2007) Genistein induces the metastasis suppressor kangai-1 which mediates its anti-invasive effects in TRAMP cancer cells. Biochem Biophys Res Commun 361:169–175. https://doi.org/10.1016/j.bbrc.2007.07.010
158. Yan C, Han R (1998) Genistein suppresses adhesion-induced protein tyrosine phosphorylation and invasion of B16-BL6 melanoma cells. Cancer Lett 129:117–124. https://doi.org/10.1016/S0304-3835(98)00093-7
159. Lee W-J, Chen W-K, Wang C-J, Lin W-L, Tseng T-H (2008) Apigenin inhibits HGF-promoted invasive growth and metastasis involving blocking PI3K/Akt pathway and β4 integrin function in MDA-MB-231 breast cancer cells. Toxicol Appl Pharmacol 226: 178–191. https://doi.org/10.1016/j.taap.2007.09.013
160. Chen P-N, Kuo W-H, Chiang C-L, Chiou H-L, Hsieh Y-S, Chu S-C (2006) Black rice anthocyanins inhibit cancer cells invasion via repressions of MMPs and u-PA expression. Chem Biol Interact 163:218–229. https://doi.org/10.1016/j.cbi.2006.08.003

Chapter 17

Protocol for Evaluation of Benzofuran from European Butterbur (*Petasites hybridus*) Plant Bioactivity as Anticancer and Immunomodulatory Phytochemical Against Cells' ROS Species

Ali Mokhtar Mahmoud

Abstract

Biochemical researchers have recently been quite interested in benzofuran compounds. Benzofuran molecules have been shown in various studies to possess strong biological effects, such as antitumor, antibacterial, antioxidative, and antiviral activities. The antioxidant activity of benzofuran compounds has been shown in numerous studies. Despite benzofuran biological activities having been the subject of numerous investigations, its activity on metabolism and mitochondria activity have received less attention, considering that most of the free radical and reactive oxygen species production occurs in mitochondria. In this chapter, we will go over the DPPH assay to determine benzofurane antioxidant activity against ROS as well as the techniques for determining how benzofuran affects mitochondria bioenergetics, using extracellular flux analyzer assay including MTT the cytotoxicity assay as the first screening assay, which relies on the activity of NADPH dehydrogenase, a mitochondrial oxidative phosphorylation enzyme.

Key words Benzofuran, Antioxidant, Reactive oxygen species, Mitochondria bioenergetics, MTT

1 Introduction

In traditional medicine, the genus *Petasites* Mill. (Asteraceae) has been known for its multitherapeutic purposes. It is distributed in Europe, Northwest Asia, and North America. Researchers are particularly interested in *Petasites hybridus* species of *Petasites* Mill. genus. This species is enriched with bioactive compounds with promising therapeutic potential, i.e., anti-inflammatory, analgesic, antioxidant, diuretic, antispasmodic, antitumor [1].

Benzofuran derivatives, one of the purified bioactive compounds from *Petasites*, have attracted great interest from researchers in biochemical field [2–6]. Many studies have demonstrated that the majority of benzofuran compounds have potent biological

Mostafa Gouda et al. (eds.), *Plant Chemical Compositions and Bioactivities*, Methods and Protocols in Food Science,
https://doi.org/10.1007/978-1-0716-3938-2_17,

properties, e.g., antitumor, antibacterial, antioxidative, and antiviral properties [3, 7, 8].

Moreover, benzofuran's immunomodulatory properties can be employed to enhance the microenvironment in pathogenesis with protective immune cells that stimulate the immune system's defense against neuropathogenesis, such as Alzheimer's disease (AD). Benzofuran derivatives were able to induce the neuroprotective M2 phenotype of microglia transformation from the pro-inflammatory M1 phenotype [6].

Another health concern is cancer, which remains a national and international health problem regardless of the discovery of several dozens of novel anticancer drugs (natural and synthetic). WHO reports showed constant rates of mortality caused by various types of cancers through the years 1950–2000. There is a wide gap between the disease incidence (0.16%), disease burden (5.6%), and the mortality caused by cancer (12.5%) worldwide. These gaps clearly denote the lack of effective treatment for cancer compared to other mortality causes. The benzofuran family are key compounds in anticancer therapeutic research and exhibit encouraging anticancer properties. Furthermore, numerous synthetic benzofuran compounds have been shown to exhibit strong anticancer properties [9].

It has been shown in 2007 that the source of 974 small molecules drugs for cancer treatment have a main contribution from nature, not only chemotherapeutic agents but also lead compounds provided semisynthesis or totally synthesis of new drugs [10]. On the other hand, the side effect of chemotherapy is very deleterious on the normal cells, so the researchers asked for tackling the cancer cells without hitting the normal cells. Natural products, because of their safety margins and controlled side effects, represent a very good model for treating cancer.

Accordingly, screening of natural products for promising bioactivity is a great initiative field of studies in several countries with diversity of gardenia natural products from plants have been valuable sources for drug discovery [10]. The screening program, which was initiated by Leven et al. (1979), identified many antibacterial, antifungal, antiviral, antiparasitic, and other pharmacologically active substances activities in higher plants [11].

Numerous studies have demonstrated that benzofuran compounds have activity against ROS [12–14]. However, many studies have been done on benzofuran biological activities, but there is unmet gap in studying the benzofuran mechanism of action and its activity on cellular organelles and biomolecular pathways, i.e., mitochondria, glycolysis pathway, and oxidative phosphorylation. Given that free radical and reactive oxygen species are produced mainly by mitochondria, in this chapter we will present the methods for testing the antioxidant activity of benzofuran and its effects on mitochondria bioenergetics.

2 Methods

2.1 Assessing the Antioxidant activity

A widely used colorimetric method for determining a plant or extract's ability to scavenge radicals is the 1,1-diphenyl-2-picrylhydrazyl (DPPH) assay. Antioxidants decrease DPPH (1,1-diphenyl-2-picrylhydrazyl, Sigma-Aldrich, Germany), a free radical. When DPPH reacts with an antioxidant, it loses its ability to function as a free radical and turns yellow instead of violet. Spectroscopy absorbance reader can be used to determine the quantity of reduced DPPH since reduction causes the free radical's color to change [15].

2.1.1 Materials

1,1-Diphenyl-2-picrylhydrazyl (Sigma-Aldrich, Germany).

2.1.2 Protocol of the DPPH Assay

1. Add 20 μL of the sample or standard antioxidant.
2. Add 200 μL of freshly prepared DPPH and mix well.
3. Incubate the mix in the dark for 3–5 min.
4. Read the absorbance at 517 nm.

2.1.3 Data Analysis

Free radical scavenging activity was expressed as inhibition percentage and was calculated using the formula $(A0 - A1)/A0 \times 100$, where A0 was the control absorbance and A1 was the sample absorbance. Antioxidant activity was quantified by a regression analysis of percentage of free radical scavenging (%) versus phenolic compound concentration in the aqueous extract; this was defined as an IC50 value, which is the amount of antioxidant necessary to decrease the initial DPPH radical concentration by 50%.

2.2 The Assessment of Mitochondria Bioenergetics

Mitochondria bioenergetic is measured by High-throughput real-time extracellular flux (XF) analyzer. The XF analyzer measures the activity of the two main cell metabolic pathways; 1) Oxidative phosphorylation by measuring the oxygen consumption rate (OCR) and 2) Glycolysis pathway by extracellular acidification rate (ECAR), it quantifies the proton excretion by optical sensors [16].

3 Protocol of ECAR and OCR Measurements Were Conducted Using the XF96 Extracellular Flux Analyzer

3.1 Materials

Glycolysis (ECAR) was measured by adding glucose, ATP synthase inhibitor oligomycin, and 2-deoxy-D-glucose (2-DG). Mitochondria Oxphos (OCR) was calculated by adding oligomycin, Carbonyl cyanide p-trifluoro-methoxyphenyl hydrazone (FCCP) (uncoupler), and a mixture of antimycin-A and rotenone. These compounds were injected consecutively with a specific time gap and ECAR or OCR values were measured after each injection.

3.1.1 Protocol

1. Cells were seeded at 4 × 10^5/well in XF96 cell culture plates, treated versus control.
2. Prepare the cartridge plate.
3. Incubate cell plates for 24 h in a humidified 37 °C incubator with 5% CO_2.
4. After the 24-h incubation, replace media with appropriate assay medium.
5. Load the cartridge plate with appropriate kit reagents, e.g., metabolic substrates and inhibitors according to kit instruction.
6. Replace the cell plate cover with the cartridge loaded with reagents.

All experiments were performed at 37 °C. Each measurement cycle consisted of a mixing time. ECAR and OCR data points referred to the average rates during the measurement cycles. All compounds were prepared at appropriate concentrations in desired assay medium and adjusted to pH 7.4. In a typical experiment, three baseline measurements were taken prior to the addition of any compound and three response measurements were taken after the addition of each compound.

3.1.2 Data Analysis

The Agilent Seahorse XF Report Generator automatically calculates the Agilent Seahorse XFp OCR and ECAR parameters from Wave data that has been exported to Excel. The report provides a convenient, customizable, and one-page assay summary.

3.2 MTT Assay for Cell Cytotoxicity and NADPH Dehydrogenase Activity

3.2.1 Materials

MTT 0.5 mg/mL, isopropanol, growth media, FBS, benzofuran, and vehicles.

3.2.2 Protocol

1. On the first day, seed cells at appropriate cell number per well in a 96-well plate.
2. After 24 h, treat cells with the drugs.
3. After 24 h, add 50 μL MTT per each well in a 96-well plate on the medium with cells.
4. After 3 h, add 150 μL isopropanol.
5. Incubate with shaking in the dark condition for 30 min at 150 g-force.
6. Measure the absorbance at 620 nm.

References

1. Tzoneva R et al (2021) Anti-cancer effect of Petasites hybridus L. (Butterbur) root extract on breast cancer cell lines. Biotechnol Biotechnol Equip 35:853–861
2. Cabrera-Pardo JR et al (2020) Exploring the multi-target neuroprotective chemical space of benzofuran scaffolds: a new strategy in drug development for Alzheimer's disease. Front Pharmacol 10:492318
3. Miao YH et al (2019) Natural source, bioactivity and synthesis of benzofuran derivatives. RSC Adv 9:27510–27540
4. Napiórkowska M et al (2019) Synthesis of new derivatives of benzofuran as potential anticancer agents. Molecules 24:1529
5. Khodarahmi G, Asadi P, Hassanzadeh F, Khodarahmi E (2015) Benzofuran as a promising scaffold for the synthesis of antimicrobial and antibreast cancer agents: a review. J Res Med Sci 20:1094
6. Montanari S et al (2019) Discovery of novel benzofuran-based compounds with neuroprotective and immunomodulatory properties for Alzheimer's disease treatment. Eur J Med Chem 178:243–258
7. Chand K et al (2017) A review on antioxidant potential of bioactive heterocycle benzofuran: natural and synthetic derivatives. Pharmacol Rep 69:281–295
8. Abbas AA, Dawood KM (2022) Benzofuran as a promising scaffold for the synthesis of novel antimicrobial agents. Expert Opin Drug Discov 17:1357–1376
9. Abbas AA, Dawood KM (2023) Anticancer therapeutic potential of benzofuran scaffolds. RSC Adv 13:11096
10. Schwartsmann G et al (2002) Anticancer drug discovery and development throughout the world. J Clin Oncol 20:47S–59S
11. Jang M et al (1997) Cancer chemopreventive activity of resveratrol, a natural product derived from grapes. Science 275:218–220
12. Cho J et al (2015) Neuroprotective and antioxidant effects of novel benzofuran-2-carboxamide derivatives. Biomol Ther (Seoul) 23:275–282
13. Miceli M et al (2018) Synthesis of benzofuran-2-one derivatives and evaluation of their antioxidant capacity by comparing DPPH assay and cyclic voltammetry. Molecules 23:710
14. Rindhe SS, Rode MA, Karale BK (2010) New benzofuran derivatives as an antioxidant agent. Indian J Pharm Sci 72:231
15. Blois MS (1958) Antioxidant determinations by the use of a stable free radical. Nature 181:4617
16. Mahmoud AM et al (2023) Cannabidiol alters mitochondrial bioenergetics via VDAC1 and triggers cell death in hormone-refractory prostate cancer. Pharmacol Res 189:106683

Chapter 18

Use of RT-qPCR for Measuring the Expression of the Fecal MicroRNA from the Consumption of Pomegranate Fruit Juice

Farid E. Ahmed, Mahmoud A. Mohammad, Nancy C. Ahmed, Laila A. Hussein, and Mostafa Gouda

Abstract

The ongoing process involves the modification and improvement of the stool test that has been validated for the purpose of diagnostic screening of colon cancer (CC). This particular test focuses on the detection of micro(mi)RNAs. The need for this update is primarily driven by the complex nature of fecal density, the variations in stool composition that occur on a daily basis, and the presence of miRNAs from various sources in the stool. These sources include cell-free homogenates, exosomal miRNAs, and stool colonocytes. In a previous study conducted by our team, we performed a microarray miRNA experiment utilizing Affymetrix GeneChip miRNA 2.0 Arrays. The experiment involved the analysis of immunocaptured and enriched stool colonocytes obtained from 15 human subjects. The colon cancer patients were further categorized into different stages, namely TNM stage 0–1, TNM stage 2, TNM stage 3, and TNM stage 4. Each sample was analyzed in triplicates to ensure accuracy and reliability of the results. From this experiment, we successfully identified a panel of 14 mature miRNAs that exhibited preferential expression in association with colon cancer. Out of these miRNAs, 12 were found to be upregulated, specifically miR-19a, miR-20a, miR-21, miR-31, miR-34a, miR-96, miR-106a, miR-133a, miR-135b, miR-206, miR-224, and miR-302. On the other hand, two miRNAs, namely miR-143 and miR-145, were found to be downregulated. In order to further validate our findings, we performed absolute quantitative digital (d) PCR on the 15 stool samples, which spanned the different stages of colon cancer (0–4). This analysis involved the extraction of total small RNA through the process of immunocapture, followed by reverse transcription (RT) and absolute quantification of miRNAs. Additionally, throughout the course of our research, we also conducted mRNA/miRNA sequencing on random samples to ensure that the selected RNAs were indeed of human origin and not of bacterial origin.

Key words Adenocarcinoma, Colon cancer, Colonocyte, Colorectal cancer, TaqMan

1 Introduction

Colorectal cancer (CRC) is observed as the second most prevalent malignant disease in men and the third most common in women, both in developed and developing nations, which includes

Mostafa Gouda et al. (eds.), *Plant Chemical Compositions and Bioactivities*, Methods and Protocols in Food Science,
https://doi.org/10.1007/978-1-0716-3938-2_18,

countries such as the United States and Europe [1, 2]. This significant health issue involves mortality and morbidity associated with a preventable malignant disease that can be effectively managed through regular screening. The early detection of this malignancy would be greatly facilitated if there were accurate, practical, and cost-effective diagnostic biomarkers available. The current screening methods, such as the fecal occult blood test (FOBT), are not sufficiently sensitive [2, 3]. Additionally, they are expensive, can cause side effects, and have low compliance rates. Alternatively, some screening methods, such as colonoscopy, may even result in mortality [4]. Therefore, it would be more advantageous to employ a diagnostic approach that utilizes miRNAs. These molecules are relatively resistant to degradation when extracted from stool using commercially available kits and manipulated thereafter. Such an approach would be more favorable compared to using transcriptomic messenger (m)RNA, mutation DNA, epigenetic, or proteomic-based tests [5].

Stool testing poses a significant challenge because of its composition comprising a multitude of substances and specific inhibitory factors that necessitate their elimination to ensure a successful polymerase chain reaction (PCR). The advantages of stool testing over alternative screening methods, such as blood, are manifold. First, stool testing is truly noninvasive, thereby eliminating the need for any unpleasant cathartic preparation, formal healthcare visits, or disruptions to one's work or routine activities. Additionally, stool testing provides a comprehensive representation of the entire length of the colorectum, enabling the sampling of both the right and left sides of the colon. Unlike blood, which is released intermittently as in guaiac FOBT, it is believed that colonocytes are continuously and abundantly released into the fecal stream. This natural enrichment phenomenon partially negates the necessity for laboratory techniques to enrich for tumorigenic colonocytes, as is the case with blood testing. Moreover, the accessibility to stool screening is essentially unimpeded, as testing can be conveniently performed on mail-in specimens, irrespective of geographical location.

An optimal screening test for colon cancer would be one that possesses the ability to accurately detect advanced adenomas with a notable probability of progressing into a malignant state [6]. When it comes to the clinical management of adenomas, the standard practice involves their removal at the time of detection through the procedure of colonoscopy. However, there exists a significant subset of eligible patients who are reluctant to undergo such a costly and invasive test due to the necessity of bowel preparation, dietary restrictions, the potential for abdominal pain, and the risk of colon perforation [4]. Given the desirability of utilizing a noninvasive test as the initial screening method, molecular approaches [7] were developed as a potential solution. It is postulated that an approach involving miRNA detection in stool could potentially fulfill the criteria for test acceptability, owing to its noninvasive nature, requirement of only a small amount of stool (approximately 1 g), the comparability of its results to those obtained from colon tissue, the absence of the need for consecutive sampling on multiple dates, the possibility of sending samples by mail in cold packs, the ability to differentiate between normal tissue and colon adenomas/carcinomas, the high sensitivity and specificity in detecting advanced polyps, and the potential for automation [5]. The availability of powerful high-throughput approaches for the comprehensive characterization of miRNA at a global level, such as microarrays, along with universally applicable quantification assays for miRNA expression, such as qPCR, suggests that the validation process, which often encounters obstacles [6], would be more efficient for a miRNA assay.

MiRNAs are short, noncoding sequences that consist of 17–27 nucleotides and play a regulatory role in various cellular processes, affecting approximately 30% of mammalian genes. They achieve this regulation by imperfectly binding to the 3′ untranslated region (UTR) of target mRNAs, thereby hindering protein accumulation through mechanisms such as transcription repression or mRNA degradation [8]. It is believed that many miRNAs have the ability to target hundreds of different mRNAs. The availability of powerful high-throughput approaches for the comprehensive characterization of miRNA at a global level, such as microarrays, as well as simpler and universally applicable quantification assays for miRNA expression, such as qPCR, suggests that the validation process, which often encounters obstacles [6], would be more efficient for a miRNA assay. MiRNAs operate within a complex regulatory network, and it is predicted that these molecules collectively regulate thousands of human genes [8]. The process of miRNA synthesis involves the transcription of long primary precursor molecules (pri-miRNA), which are subsequently processed by the nuclear enzyme Drosha and other agents to generate the precursor intermediate miRNA (pre-miRNA). This pre-miRNA is then processed in the cytoplasm by the protein Dicer to generate the mature single-stranded (ss) miRNA [9]. The functions of miRNAs have been shown to include involvement in processes such as development and apoptosis, regulation of cell cycle progression, and thus, specific miRNAs play a critical role in oncogenesis. MiRNAs have also been identified as biomarkers for the classification of both solid [10] and liquid human tumors [11]. Their influence extends to cancer metastasis [12], where they can function either as oncogenes or tumor suppressor genes [13]. It is worth noting that miRNA genes are frequently located in fragile sites and minimal regions associated with loss of heterozygosity, amplification, or common breakpoint regions. This suggests that miRNAs have the potential to accurately classify human cancers of unknown primary origin as

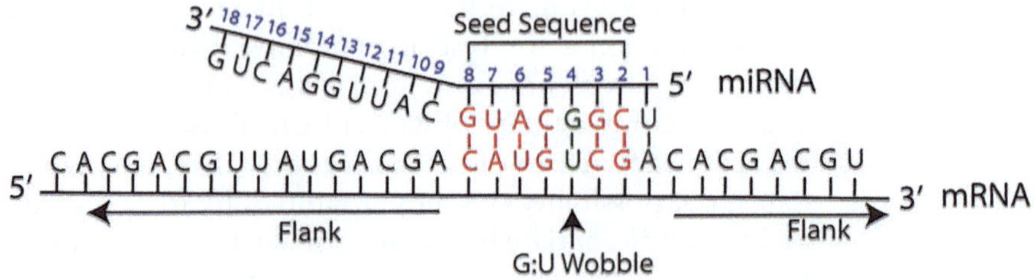

Fig. 1 A schematic of microRNA relationship with its mRNA target as more stable molecular biomarker and their potential interaction. The seed sequence refers to nucleotides in microRNA position numbers 2–8. Flank refers to the mRNA sequence on either side of the region corresponding to the miRNA seed sequence [8]

well as poorly differentiated tumors [14], and they hold promise as biomarkers for cancer diagnosis, prognosis, and response to therapy [11, 15–20] (Fig. 1).

It has been estimated that the daily shedding of colonic epithelial cells from the lower two-thirds of colon crypts is approximately 10^{10} normal adult cells. Each of these cells has a lifespan of 3–4 days. This shedding process provides a compelling and achievable opportunity to develop a screening test for colorectal cancer (CRC) using colonocytes. This is particularly appealing due to the potential use of colonocyte shedding in developing a CRC screening test. In the context of identifying CRC miRNAs in stool colonocytes, the approach discussed in this chapter has the potential to validate novel miRNA/mRNA target pairs within the pathways of interest. By doing so, it may lead to the discovery of cellular functions that are collectively targeted by differentially expressed miRNAs.

2 Materials

2.1 Pomegranate Juice

1. A crucial component of this study was sourced in bulk from the esteemed Obour Public Market located in Cairo, Egypt.
2. The pomegranate fruits, which served as the primary source of the juice, were meticulously peeled to ensure the exclusion of any extraneous matter or impurities.
3. The extraction process itself was conducted using a laboratory pilot press manufactured by Braun, a renowned German company known for its precision and reliability.
4. The extracted juice was then carefully distributed in aliquots, measuring either 100 or 250 g, to facilitate the ease of handling and experimental consistency.
5. These aliquots were specifically housed within airtight, light-proof polyethylene bottles, thereby guaranteeing the

preservation of the juice's chemical properties and preventing any potential degradation or alteration due to external factors.

6. The frozen storage temperature of −20 °C was selected in order to maintain the stability and integrity of the pomegranate polyphenols, which are an essential component of the juice and hold significant scientific interest.
7. This meticulous approach to storage ensures that the pomegranate polyphenols retain their full potency and efficacy throughout the duration of the study, thereby yielding accurate and reliable results.

2.2 Stool Collection Materials

1. Stool is collected in a jacket (Moore Medical, New Britain, CT, USA) to be placed on toilet seat.
2. A clean and sterile plastic spoon or a solid brush is used to collect stool and place it in a small sterile plastic collection jar.
3. RNALater is used as RNA preservation solution.
4. A Stomacher® 400 EVO Laboratory Blender (Seward, UK).

2.3 Target Primers and TaqMan Probes

1. Stem-loop RT primers (Life Technologies/Thermo Fisher Scientific, Waltham, MA, USA) were obtained for each of the 20 selected miRNA genes (miRNAs miRNA-7, miR-9, miR-16a, miRNA-17, miRNA-20a, miRNA-21, miR-29b, miRNA-92a, miR-106a, miRNA-127-5p, miRNA-134, miR-138, miRNA-143, miRNA-146a, miRNA-183, miRNA-196a, miRNA-199a-3p, miRNA-214, miRNA-222, and miR-938).
2. TaqMan with minor grove binder (MGB) probes with non-fluorescence quencher at the 3′ end and the fluorescence dye FAM at the 5′ end (Life Technologies/Thermo Fisher Scientific) were also employed.

2.4 Participants

1. Study subjects were 25 healthy adults, 20–34 years old; exclusion were: absence of metabolic diseases, no use of medication for the last 6 weeks, and no signs of allergy or hypersensitivity to food or ingested material.
2. Compliance with the supplementation in all subjects was satisfactory, as assessed daily, and all subjects continued their habitual diets throughout the study.
3. The research protocol was approved by the institution review board, and all subjects gave written consent prior to their participation in the study.

Table 1
Predictive MicroRNA Index (PMI)

Parameter	PMI sensitivity	PMI specificity
Cancer cases	True positive (TP)	False negative (FN)
Normal subjects	False positive (FP)	True negative (TN)
Equation	100% sensitivity $= \frac{TP}{(TP + FN)} \times 100$	100% specificity $= \frac{TN}{(FP + TN)} \times 100$

3 Method

3.1 Design of the Study

1. A nutrigenomic randomized trial was constructed for the current study.
2. Estimated dietary intake was assessed by three repeated food records, 1 week before they were enrolled in the trial.
3. The average portion size consumed, as well as composition data values from nutrient composition of the food were combined to assess average daily energy and nutrient intakes by the "NutriSurvey" software program.
4. The characteristics of the voluntary subjects who were enrolled in the study, the mean daily energy intake, and selected macronutrients are presented in Table 1. Supplements.
5. MicroRNA expression in stool (B) samples taken from healthy and CC individuals, respectively. The stage of cancer is indicated by the bottom row of the panel. In stool, there were normal healthy individuals and colon cancer (TNM stages 0–4), (stage 0–1), (stage 2), (stage 3), and (stage 4), respectively [21–30].
6. Instances of high expression appear on the right and those with low expression on the left. Expression by stem-loop RT-minor grove-binding qPCR was measured by the comparative cross point (CP) or the E-method on a Roche LightCycler® 480 PCR instrument [31].

3.2 Stool Collection and Storage

1. Stool was obtained from 25 healthy adults, twice at day 0 and 3 weeks after the dietary intervention.
2. All stools were collected with sterile, disposable wood spatulas in cleaned screw top vials (Thermo Fisher Scientific, Inc., Palo Alto, CA, USA).
3. 2 mL of the preservative RNALater (Applied Biosystems/Ambion, Austin, TX, USA) was added to the vials, which prevents the fragmentation of the fragile mRNA molecule [7], and vials were stored at −70 °C until samples were ready for further analysis.

4. Total small RNA, containing miRNAs, was extracted from all frozen samples at once, when ready, and there was no need to separate mRNA containing small miRNAs from total RNA, as small total RNA was suitable to make ss miRNA c-DNA.

3.3 Extraction of Total Small RNA

1. A procedure used for extracting small total RNA from stool was carried out using a guanidinium-based buffer, which comes with the RNeasy isolation Kit® (Qiagen, Valencia, CA, USA) as we have previously detailed [7].
5. DNase digestion was not carried out, as our earlier work demonstrated no difference in RNA yield or effect on RT-PCR after DNase digestion [7].
6. The time to purify aqueous RNA from all of the 25 frozen stool samples was ~3 h.
7. Small RNA concentrations were measured spectrophotometrically at λ 260, 280 and 230 nm, using a Nano-Drop spectrophotometer (Themo-Fischer Scientific).
8. The integrity of total RNA was determined by an Agilent 2100 Bioanalyzer (Agilent Technologies, Inc., Palo Alto, CA, USA) utilizing the RNA 6000 Nano LabChip®. RNA integrity number (RIN) was computed for each sample using instrument's software [7].

3.4 Preparation of ss-cDNA for Molecular Analysis

1. The RT2 miRNA First Strand Kit® from SABiosciences Corporation (Frederick, MD, USA) was employed for making a copy of ss-DNA in a 10.0 μL reverse transcription (RT) reaction.
2. The extracted RNA sample was preserved in a sterile PCR tube, containing 100 ng total RNA, 1.0 μL miRNA RT primer, and ERC mix, 2.0/μL 5X miRNA RT buffer.
3. A volume of 1.0 μL miRNA RT enzyme was mixed with 1.0 μL nucleotide mix, and RNase-free H_2O to a final volume of 10.0 μL.
4. The same amount of total RNA was used for each sample and mixed with the prepared solution in **step 3**. Then, the contents were gently mixed with a pipettor, followed by brief centrifugation. All tubes were then incubated for 2 h at 37 °C.
5. This is followed by heating at 95 °C for 5 min to degrade the RNA and inactivate the RT.
6. All tubes were chilled on ice for 5 min, and 90 μL of RNase-free H_2O was added to each tube. Finished miRNA First Strand cDNA synthesis reactions were then stored overnight at −20 °C [7].

3.5 Use of Cancer RT2 miRNA PCR Array 96-Well Plate to Study miRNAs' Expressions

1. We used a SABiosciences RT2 miRNA qPCR Array Plate System for Human (Qiagen) to analyze miRNA expression using real-time, reverse transcription PCR (RT-qPCR) as a sensitive and reliable quantitative method for miRNA expression analysis.
2. The arrays employ a SYBR Green real-time PCR detection system, which has been optimized to analyze the expression of many mature miRNAs simultaneously.
3. Each 96-well array plate contains a panel of primer sets for 88 relevant miRNA-focused pathways (one universal primer and one gene-specific primer for each miRNA sequence).
4. Four housekeeping genes (Human SNORD 48, 47, and 44 and U6) and two RNA and two PCR quality controls.
5. Duplicate RT controls (RTC) to test the efficiency of the miRNA RT reaction, with a primer set that detects the template synthesized from the built-in miRNA External RNA Control (ERC).
6. There are duplicate RT controls (RTC) to test the efficiency of the miRNA RT process, with a primer set to detect the template synthesized from the kit's built-in miRNA External RNA Control (ERC).
7. There are also duplicate positive PCR controls (PPC) to test the efficiency of the PCR process, using a per-dispensed artificial DNA sequence and the primer set that detects it. The two sets of duplicate control wells (RTC and PPC) also test for interwell, intraplate consistency. The human RT2 miRNA PCR Arrays reflect miRNA sequences annotated by the Sanger miRBase Release 14.
8. Perform real-time quantitative polymerase chain reaction (qPCR).
9. We used RT2 SYBR Green qPCR Master Mix (SBA Biosciences) to obtain accurate results from our qPCR arrays.
10. The following components were mixed in a 15 mL tube for 96-well plate format: 1275 μL of 2X RT2 SYBR Green PCR Master Mix, 100 μL of diluted first strand reaction, 1175 μL of ddH_2O (total volume 2550 μL, of which 2400 was needed for 96 reactions, each well having 25 μL, with 150 μL cocktail remaining.
11. We employed a Roche LightCycler 480® 96-well block PCR Machine (Roche, Mannheim, Germany) to carry out quantitative real-time miRNA expressions. When ready, we removed the needed miRNA qPCR Arrays, each wrapped in aluminum foil, from their sealed bags, added 25 μL of the same cocktail to each well, adjusted the ramp rate to 1 °C/s. We used 45 cycles in the program and employed the Second Derivative Maximum

method, available with the LightCycler 480® software for data analysis (19).

12. We first heated the 96-well plate for 10 min at 95 °C to activate the HotStart DNA polymerase, then used a three-step cycling program (a 15-s heating step at 95 °C to separate the ds DNA, a 30-s annealing step at 60 °C to detect and record SYBR Green fluorescence at each well during each cycle, and a final heating step at 72 °C for 30 s).
13. Each plate was visually inspected after the run for signs of evaporation from the wells. Data were analyzed using the $2^{-\Delta\Delta Ct}$ method (20).
14. Resulting threshold cycle values for all wells were exported to a blank Excel sheet for analysis.
15. We also ran a Dissociation (Melt) Curve Program after the cycling program (21) and generated a first derivative dissociation curve for each well in the plate using the LC (LightCycler's®) software.
16. Statistical and bioinformatics analysis: Gene expressions were standardized by dividing the SNORD48 value while raw melting temperatures were used. Analysis was done using the software R (version 3.1.3), with the package MASS (22).
17. One individual had so many missing values that this case was not used in the analysis so that the number of individuals is 24.
18. For each standardized gene and each melting temperature, a one-way ANOVA was used to obtain a *p*-value.
19. There were four levels of the explanatory variable: Control, Sobya, Pom, and Both. Parallel coordinate plots (parcoord command in *R*) (23) were used to visualize the data for each gene and each melting temperature.
20. Coordinates were ordered using the magnitude of the *p*-value. The two-sample *t*-test was used on gene expression to compare control to Sobya and control to Both (*t*.test command in *R* with var.equal = FALSE).
21. *p*-Values were adjusted to control for false discovery rate. The method is outlined in (24) Benjamini and Yekutieli (*p*.adjust command in *R* with method = 'BY').
22. We have bioinformatically correlated the 2–7 or 2–8 complement nucleotide bases in the mature miRNAs with the untranslated 3′ region of target mRNA (3′ UTR) of a message using a basic algorithm such as Broad's Institute's TargetScan (25): http://www.targetscan.org/archives.html, which provides a precompiled list for their prediction.

4 Statistical Analysis

1. The miRNA expression data is first analyzed with parametric statistics such as Student *t*-test or analysis of variance (ANOVA) test, if the data distribution is random; or with nonparametric Kruskal–Wallis, Mann–Whitney, and Fisher exact tests, if the distribution is not random [30, 32]. If necessary, more complicated models such as multivariate analysis and logistic discrimination [33] can be employed.
2. For the corrected index, cross-validation is used to protect against overfitting [34].
3. The area under the receiver operating characteristic (ROC) curve [in which sensitivity is plotted as a function of (1 − specificity)] can be used to describe the trade-off between sensitivity and specificity [35].
4. Principal component analysis (PCA) method [36], which is a multivariate dimension reduction technique to simplify grouping of genes that show aberrant expression from those not showing expression, or a much-reduced expression, can also be employed for data analysis.
5. If the miRNA gene panel (or a derived Predictive MicroRNA Index, PMI) [37] (Table 1) is better than existing screening methods, all of the data generated can be used to assess the model, so over-fitting is not a concern. The example of absolute quantification for the up- and down-regulated miRNAs in stool samples by RT-PCR system is presented in Fig. 2.

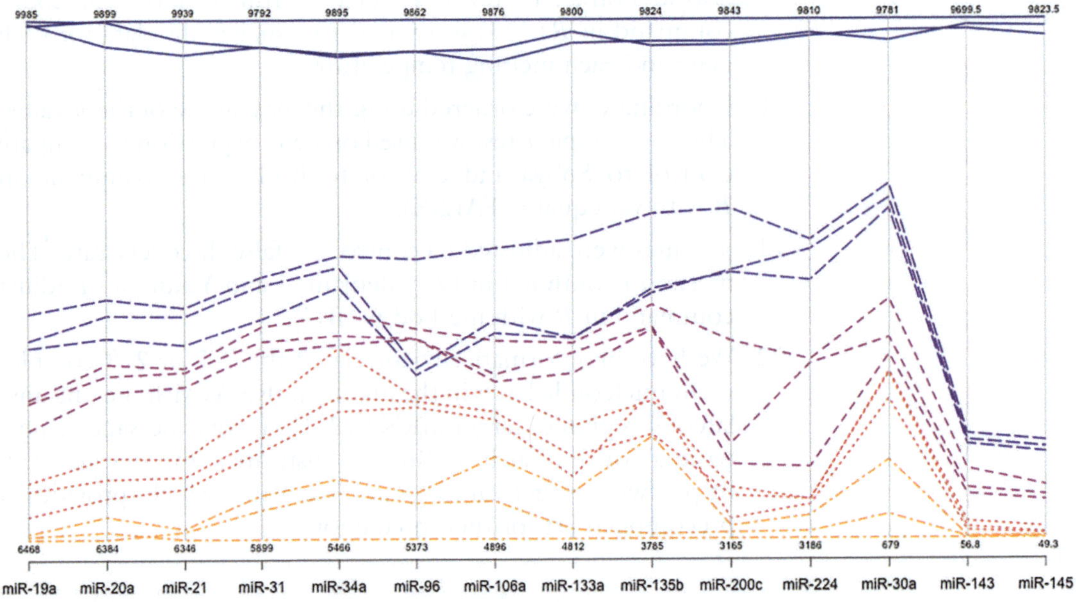

Fig. 2 An example for the absolute quantification results of the up- and down-regulated miRNAs in stool samples by RT-PCR system

5 Notes

1. All solutions utilized in this experiment have been meticulously formulated in a highly controlled environment through the utilization of sterile distilled water, thereby ensuring the utmost purity and precision.
2. All procedures are carried out at room temperature unless otherwise specified.
3. By employing reagents of the highest analytical grade, we can guarantee the accuracy and reliability of our results. It is of utmost importance to note that the addition of sodium azide to the aforementioned reagents has been consciously omitted in order to avoid any potential interference or alteration of the desired outcomes.
4. In order to ensure their long-term viability and integrity, these reagents have been conscientiously stored at a temperature of −20 °C, while the stool samples have been subjected to an even more stringent temperature of −80 °C.
5. This meticulous approach to storage guarantees the preservation of the reagent's chemical composition and the stool's organic integrity.
6. It is essential to highlight that the disposal of waste material adheres strictly to the regulations set forth in waste disposal protocols. The utmost care and attention are given to the disposal process, ensuring its compliance with all established standards and guidelines.
7. Considering that each cell contains ~20 pg total RNA or 0.4 pg mRNA (equivalent to 0.36 pg ss-cDNA), as only a few picograms of cDNA are needed per PCR reaction [18].
8. Isolation of colonocytes from stool, and comparing the Agilent electrophoretic (18S and 28S) patterns to those obtained from total RNA extracted from whole stool, and differential lysis of colonocytes by RT lysis buffer (Quagen), could be constructed as a validation that the electrophoretic pattern observed in stool (18S and 28S) is truly due to the presence of human colonocytes, and not due to stool contamination with *Escherichia coli* (16S and 23S) [38].
9. Employing a two-step RT-PCR format permitting reverse transcription and amplification to be performed separately under optimal conditions is particularly attractive for experiments that require the same RT product to be used for analysis of multiple transcripts [23, 24].
10. Unlike screening for large numbers of messenger (m)RNA, a modest number of miRNAs is used to differentiate cancer from

normal tissue [7, 21], and unlike mRNA [18], miRNAs in stool remain largely intact and stable for detection [31]. Therefore, miRNAs are better molecules to use for developing a reliable noninvasive diagnostic screening for colon cancer.

11. Two main approaches are adopted in commercially available systems: (1) use of a stem-loop RT primer [19] or (2) use an alternative approach (miQPCR) developed by Castoldi and his colleagues [38], which uses T4 ligase activities to extend uniformly miRNAs' 3′ ends by addition of a linkage adapter that is used to anchor and prime cDNA synthesis, followed by qPCR to amplify specifically target amplicons.
12. The stem-loop RT primer approach was developed and commercialized by Applied Biosystems (currently Life Technologies/Thermo) for subsequent TaqMan probe-based qPCR. The 5′-end of the RT primer base-pair with a region several nucleotides from its own 3′-end to create a base-paired stem separated by an unpaired loop. All except the last few nucleotides at the 3′-end of the RT primer are universal, i.e., they contain the same sequence in all miRNA primers. The last few nucleotides that extend 3′ from the stem are complementary to the 3′-end of a target miRNA. Extension of the primer along a miRNA template creates a cDNA that can be amplified with a miRNA-specific forward primer and a universal reverse primer, the latter of which is complementary to the 5′-end of the stem-loop RT primer.
13. All other commercial miRNA qPCR methods, including Sigma's MystiCq® brand, use poly-A tailing to lengthen the miRNA, as described by Shi and Chiang [39]. Poly (A) polymerase (PAP), a template-independent enzyme, catalyzes the transfer of adenosine residues from ATP to the 3′-end of any RNA. The oligo-dT primer includes an adapter sequence at its 5′-end, which enables subsequent qPCR with a forward primer that is complementary to a specific miRNA and a reverse primer that is complementary to the adapter sequence [40, 41].
14. Real-time PCR analysis measures product kinetics by detecting them as they accumulate online at a dynamic range of target molecules at least five orders of magnitude as compared to a traditional PCR format, using Roche's PCR LightCycler's 480™ (LC). The method relies on spectral analysis to distinguish among amplicons, allowing for high-throughput detection of either 96 or 384 samples. Employing a two-step format permits reverse transcription and amplification to be performed separately under optimal conditions, and is used because of the flexibility that this method provides for multiple-parameter analysis [23].

15. Although miRNAs represent a relatively abundant class of transcripts, their expression level could vary among species and tissues [5, 10, 21, 24, 39]. A modified TaqMan-based quantitative real-time PCR assay has been used to quantify the RT product that includes miRNA-specific forward primer, reverse primer, and a dye-labeled hydrolysis TaqMan® probes [21].
16. Better specificity and sensitivity of stem-loop primers compared to conventional linear ones occurs due to base stacking and spatial constraint on the stem-loop structure, which improves the thermal stability and extends the effective footprint of RT primer/RNA duplex that may be required for effective RT from relatively shorter traditional RT primers, and also prevents it from binding double strand (ds) genomic DNA molecules, and therefore prevent amplification of any potential genomic DNA contaminants present in the preparation [19].
17. Some noncoding RNAs that are longer than 100 nucleotides can be detected and quantified by traditional RT-qPCR with the same techniques that are used for mRNA analysis. In contrast, short noncoding RNAs, such as micro and piwi RNAs, are essentially the length of a single PCR primer. As a consequence, a technique that modifies these short RNAs is needed for RT [5, 7, 18, 19, 21, 24, 39].
18. This study emphasized on a commonly accepted dogma of real-time PCR analysis, in which optimal data are achieved when each RT reaction is normalized to contain an equal amount of a starting total RNA [7, 17, 19, 23, 28].
19. The TaqMan minor-groove binding (MGB) probes are employed to increase the Tm of small probes, which also have a 5′ FAM and a 3′ MG [19, 21].
20. Absolute quantification of miRNA is unnecessary because constantly transcribed housekeeping genes effectively serve as internal standards for accurate quantification of miRNA genes of interest; our own experience [7, 18, 19, 23] and others [15, 24, 26, 28] support that conclusion.
21. We used Roche's LightCycler (LC) 480™ for real-time PCR analysis of product as it accumulates (i.e., online) over a large dynamic range of target molecule (i.e., at least five orders of magnitude as compared to traditional end-point PCR) [24, 26], which allows for a higher sample throughput using 96 plates, precise product determination during the log/linear amplification phase, reduced labor, and increased fluorescence.
22. The E method [25] was selected because it produces more accurate results than the $2^{-\Delta\Delta Ct}$ method [26], as it compensates for differences in target and reference gene amplification

efficiency ($E = 10^{-1/\text{slope}}$) either within an experiment or between experiments. Normalization against a housekeeping internal standard or in some cases against several standards is customarily employed in this method.

References

1. Ahmed FE, Gouda MM, Hussein LA, Ahmed NC, Vos PW, Mohammad MA (2017) Role of melt curve analysis in interpretation of nutrigenomics' microRNA expression data. Cancer Genomics Proteome J 14(6):110–124
2. Jemal A, Siegel R, Ward E, Moss SM, Amar SS, Balfour TW, James PD, Mangham CA, Murray T, Xu J, Thun MJ (2007) Cancer statistics, 2007. CA Cancer J Clin 57:43–66
3. Smith A, Young GP, Cole SR, Bampton P (2006) Comparison of a brush-sampling fecal immunochemical test for hemoglobin with a sensitive guaiac-based fecal occult blood test in detection of colorectal neoplasia. Cancer 107:2152–2159
4. Gatto NM, Frucht H, Sundararajan V, Jacobson JS, Grann VR, Neugut AI (2002) Risk of perforation after colonoscopy and sigmoidoscopy: a population based study. J Natl Cancer Inst 95:230–236
5. Ahmed FE (2007) The role of microRNA in carcinogenesis and biomarker selection: a methodological perspective. Exp Rev Mol Diag 7:569–603
6. Davies RJ, Freeman A, Morris LS, Bingham S, Dilworth S, Scott I, Laskey RA, Miller R, Coleman N (2002) Analysis of minichromosome maintenance proteins as a novel method for detection of colorectal cancer in stool. Lancet 359:1917–1919
7. Ahmed FE, Vos P (2004) Molecular markers for human colon cancer in stool and blood identified by RT-PCR. Anticancer Res 24: 4127–4134
8. Lewis BP, Burge CB, Bartel DP (2005) Conserved seed paring, often flanked by adenosines, indicates that thousands of human genes are microRNA targets. Cell 120:15–20
9. Ruby JG, Jan CH, Bartel D (2007) Intronic microRNA precursors that bypass Drosha processing. Nature 448:83–86
10. Cummins JM, Velculescu VE (2006) Implication of microRNA profiling for cancer diagnosis. Oncogene 15:6220–6227
11. Calin GA, Ferracin M, Cimmino A, Dileva G, Shimiz M, Wojcik SE et al (2005) A microRNA signature associated with prognosis and progression in chronic lymphocytic leukemia. N Engl J Med 353:1793–1801
12. Calin GA, Croce CM (2006) MicroRNA signatures in human cancers. Nat Rev Cancer 6: 857–866
13. Chang-Zheng C (2005) MicroRNAs as oncogenes and tumor suppressors. N Engl J Med 353:1768–1771
14. Calin GA, Sevignai C, Dumitru CD, Hyslop T, Noch E, Yendamuri S et al (2004) Human microRNA genes are frequently located at fragile sites and genomic regions involved in cancers. Proc Natl Acad Sci USA 101:2999–3004
15. Schepler T, Reinert JT, Oslenfeld MS, Christensen LL, Silahtaroglu AN, Dyrskjøt L et al (2008) Diagnostic and prognostic microRNAs in stage II colon cancer. Cancer Res 68:6416–6424
16. Lanza G, Ferracin M, Gafa R, Veronese A, Spizzo R, Pichiorri F et al (2007) mRNA/microRNA gene expression profile in microsatellite unstable colorectal cancer. Mol Cancer 6: 54
17. Koga Y, Yasunaga M, Katayose S, Moriya Y, Akasu T, Fujita S, Yamamoto S, Baba H, Matsumura Y (2008) Improved recovery of exfoliated colonocytes from feces using newly developed immunoparamagnetic beads. Gastroenterol Res Practice 2008:1. https://doi.org/10.1155/2008/605273
18. Ahmed FE, Vos PW, James SI, Flake G, Sinar DR, Naziri W, Marcuard SP (2007) Standardization for transcriptomic molecular markers to screen human colon cancer. Cancer Genom Proteom 4:419–432
19. Ahmed FE, James SI, Lysle DL, Dobbs LJ Jr, Johnke RM, Flake G et al (2004) Improved methods for extracting RNA from exfoliated human colonocytes in stool and RT-PCR analysis. Dig Dis Sci 49:1889–1898
20. Matsushita HM, Matsumura Y, Moriya Y, Akasu T, Fujita S, Yamamoto S et al (2005) A new method for isolating colonocytes from naturally evacuated feces and its clinical application to colorectal cancer diagnosis. Gastroenterology 129:1918–1927
21. Chen C, Ridzon DA, Broomer AJ, Zhou Z, Lee DH, Nguyen JT et al (2005) Real-time quantification of microRNAs by stem-loop RT-PCR. Nucleic Acids Res 33:e179

22. LightCycler Software®, Version 3.5 (2001) Roche molecular biochemicals, Mannheim, pp. 64–79
23. Ahmed FE (2005) qRT-PCR: application to carcinogenesis. Cancer Genom Proteom 2: 317–332
24. Schmittgen TD, Lee EJ, Jiang J, Sarkar A, Yang L, Elton TS, Chen C (2008) Real-time PCR quantification of precursor and mature microRNA. Methods 44:31–38
25. Tellman G (2006) The E-method: a highly accurate technique for gene-expression analysis. Nat Methods 3:1–2
26. Livak KJ, Schmittgen TD (2001) Analysis of relative gene expression data using real-time quantitative PCR and the $2^{-\Delta\Delta C_T}$ method. Methods 25:402–408
27. Wegman E (1990) Hyperdimensional data analysis using parallel coordinate. J Am Stat Assoc 85:644–675
28. Peltier HJ, Latham J (2008) Normalization of microRNA expression levels in quantitative RT-PCR assays identification of suitable reference RNA targets in normal and cancerous human solid tissues. RNA 14:844–852
29. Luu-The V, Paquet N, Calvo E, Cumps J (2005) Improved real-time RT-PCR method for high-throughput measurements using second derivative calculation and double correction. BioTechniques 38:287–293
30. Ripley (1997) Classification. In: Encyclopedia of statistical sciences, vol 1. Wiley-Interscience Publication
31. Ahmed FE, Ahmed NC, Vos PW, Bonnerup C, Atkins JN, Casey M et al (2013) Diagnostic microRNA markers to screen for sporadic human colon cancer in stool: I. Proof of principle. Cancer Genom Proteom 10:93–114
32. Moore DS, McCabe GP, Craig B (2009) Introduction to the practice of statistics, 6th edn. W.H. Freeman & Company, St. Louis, MO
33. Nagan CY, Yamamoto H, Seshimo I, Ezumi K, Terayama M, Hemmi H, Takemasa I, Ikeda M, Sekimoto M, Monden M (2007) A multivariate analysis of adhesion molecules expression in assessment of colorectal cancer. J Surg Oncol 95:652–662
34. Efron B, Tibshirani RJ (1993) An introduction to the bootstrap. Chapman and Hall, New York, NY
35. Hanley JA, McNeil BJ (1982) The meaning and use of the area under a receiver operating characteristic (ROC) curve. Radiology 143: 29–36
36. Gabriel KR, Odoroff CL (1990) Biplots in biomedical research. Stat Med 9:469–485
37. DeMuth JP, Jackson CM, Weaver DA, Crawford EL, Durzinsky DS, Durham SJ, Zaher A, Phillips ER, Khuder SA, Willey JC (1998) The gene expression index cmyc x E2F-1/p21 is highly predictive of malignant phenotype in human bronchial epithelial cells. Am J Respir Cell Mol Biol 19:18–29
38. Benes V, Collier P, Kordes C, Stolte J, Rausch T, Muckentaler MU et al (2015) Identification of cytokine-induced mutation of microRNA expression and secretion as measured by a novel microRNA specific qPCR assay. Sci Rep 5:11590
39. Shi R, Chiang VL (2005) Facile means for quantifying microRNA expression by real-time PCR. BioTechniques 29(4):519–524
40. Okugawa Y, Toiyama Y, Goel A (2014) An update on microRNA as colorectal cancer biomarkers: where are we and what's next? Expert Rev Mol Diag 14(8):999–1021
41. Di Cello F, Xie Y, Paul-Satyaseela M, Kim KS (2005) Approaches to bacterial RNA isolation and purification for microarray analysis of *Escherichia coli* K1 interaction with human brain microvascular endothelial cells. J Clin Microbiol 43(8):4197–4199

Correction to: Detecting Apple Valsa Canker (AVC) at an Early Stage Using Micro-SERS Combined with Chemical Imaging Analysis

Shiyan Fang, Yanru Zhao, Yan Wang, Junmeng Li, Fengle Zhu, and Keqiang Yu

Correction to:
Chapter 13 in: Mostafa Gouda et al. (eds.), ***Plant Chemical Compositions and Bioactivities,*** **Methods and Protocols in Food Science,**
https://doi.org/10.1007/978-1-0716-3938-2

Dr. Fengle Zhu affiliation in chapter 13 has been updated as
Zhejiang University of Technology, 288, Liuhe Road, Hangzhou, 310023

The updated version of this chapter can be found at
https://doi.org/10.1007/978-1-0716-3938-2_13

Mostafa Gouda et al. (eds.), *Plant Chemical Compositions and Bioactivities*, Methods and Protocols in Food Science,
https://doi.org/10.1007/978-1-0716-3938-2_19,

Index

A

B

C

E

F

G

H

I

Mostafa Gouda et al. (eds.), *Plant Chemical Compositions and Bioactivities*, Methods and Protocols in Food Science,
https://doi.org/10.1007/978-1-0716-3938-2,